Commerce Raiding

Historical Case Studies, 1755–2009

Bruce A. Elleman
and S. C. M. Paine, Editors

NAVAL WAR COLLEGE PRESS
Newport, Rhode Island

To Willard C. Frank, Jr., scholar and colleague

Naval War College
Newport, Rhode Island
Center for Naval Warfare Studies
Newport Paper Forty
October 2013

President, Naval War College
Rear Adm. Walter E. Carter, Jr., USN

Provost
Amb. Mary Ann Peters

Dean of Naval Warfare Studies
Robert C. Rubel

Naval War College Press
Director: Dr. Carnes Lord
Managing Editor: Pelham G. Boyer

The Newport Papers are extended research projects that the Director, the Dean of Naval Warfare Studies, and the President of the Naval War College consider of particular interest to policy makers, scholars, and analysts.

The views expressed in the Newport Papers are those of the authors and do not necessarily reflect the opinions of the Naval War College or the Department of the Navy.

Contents

Acknowledgments	vii
Foreword, *by John B. Hattendorf*	ix
Introduction, *by Bruce A. Elleman and S. C. M. Paine*	1

CHAPTER ONE   The Breakdown of Borders: Commerce Raiding during the Seven Years' War, 1756–1763   9
by Thomas M. Truxes

CHAPTER TWO   *Guerre de Course* and the First American Naval Strategy   27
by Christopher P. Magra

CHAPTER THREE   French Privateering during the French Wars, 1793–1815   41
by Silvia Marzagalli

CHAPTER FOUR   Waging Protracted Naval War: U.S. Navy Commerce Raiding during the War of 1812   57
by Kevin D. McCranie

CHAPTER FIVE   CSS *Alabama* and Confederate Commerce Raiders during the U.S. Civil War   73
by Spencer C. Tucker

CHAPTER SIX   Two Sides of the Same Coin: German and French Maritime Strategies in the Late Nineteenth Century   89
by David H. Olivier

CHAPTER SEVEN   Missed Opportunities in the First Sino-Japanese War, 1894–1895   105
by S. C. M. Paine

CHAPTER EIGHT	Chinese Neutrality and Russian Commerce Raiding during the Russo-Japanese War, 1904–1905 by Bruce A. Elleman	121
CHAPTER NINE	"*Handelskrieg mit U-Booten*": The German Submarine Offensive in World War I by Paul G. Halpern	135
CHAPTER TEN	The Anglo-American Naval Checkmate of Germany's *Guerre de Course*, 1917–1918 by Kenneth J. Hagan and Michael T. McMaster	151
CHAPTER ELEVEN	Logistic Supply and Commerce War in the Spanish Civil War, 1936–1939 by Willard C. Frank, Jr.	165
CHAPTER TWELVE	The German U-boat Campaign in World War II by Werner Rahn	187
CHAPTER THIRTEEN	The Shipping of Southeast Asian Resources Back to Japan: National Logistics and War Strategy by Ken-ichi Arakawa	209
CHAPTER FOURTEEN	Unrestricted Submarine Victory: The U.S. Submarine Campaign against Japan by Joel Holwitt	225
CHAPTER FIFTEEN	*Guerre de Course* in the Charter Era: The Tanker War, 1980–1988 by George K. Walker	239
CHAPTER SIXTEEN	Twenty-First-Century High-Seas Piracy off Somalia by Martin N. Murphy	253

Conclusion: *Guerre de Course* in the Modern Age   271
by Bruce A. Elleman and S. C. M. Paine

Selected Bibliography   291

About the Contributors   305

Index   311

The Newport Papers   339

Acknowledgments

The editors would like to thank our contributors and the many others who shared their insights and expertise. At the Naval War College, we benefited from the support of Mary Ann Peters, John Garofano, John B. Hattendorf, Robert "Barney" Rubel, John Maurer, and Peter Dombrowski. We owe a considerable debt to Alice Juda and Wayne Rowe for library assistance. We are especially indebted to Andrew Marshall of the Office of Net Assessment for his ongoing support for this project.

On behalf of our contributors, we would like to thank Michael Crawford, John B. Hattendorf, Joshua Smith, Carl Swanson, and Michael A. Palmer for their critical feedback on previous drafts of Chris Magra's essay.

The thoughts and opinions expressed in this publication are those of the authors and are not necessarily those of the U.S. government, the U.S. Navy Department, or the Naval War College.

Foreword

For centuries, attacks on maritime commerce have been consistent features of war at sea. At the same time, a fundamental raison d'être of navies has been the protection of maritime trade against such attacks. From ancient times, piracy has been an issue at sea, and a long tradition of private men-of-war lasted into the mid-nineteenth century.

After 1690, the French navy put into practice a concept of *guerre de course* as an alternative to fleet battle, or *guerre d'escadre*, as a means of dealing with the superior power of Britain's Royal Navy. In the 1870s and 1880s a group of naval thinkers in France, labeled the *Jeune École*, promoted ideas of commerce raiding with high-speed torpedo boats. Other naval theorists—including Alfred Thayer Mahan in the United States, Sir Julian Corbett in Britain, and Raoul Castex in France—concluded from their analyses of history that such commerce warfare was an indecisive method of waging war by relatively weak powers, an approach that was not as effective as one focusing primarily on the victory of one battle fleet over another. During the two world wars of the twentieth century, submarine attacks on maritime trade were extremely effective, leading the great American naval thinker J. C. Wylie to define two different types of strategy: a sequential strategy that leads from one action to another, and a cumulative strategy, such as one involving attrition of merchant shipping in commerce warfare.

Some commentators have argued that in the modern globalized economy, no state would find any advantage in attacking a global interconnected maritime trade that has benefit for all. Yet, as one prescient observer of this subject noted recently, "unlikely threats and outdated practices rear their ugly heads when the situation favors them" (Douglas C. Peifer, "Maritime Commerce Warfare: The Coercive Response of the Weak?," *Naval War College Review* 66, no. 2 [Spring 2013], pp. 83–109, quote at p. 84).

A consideration of the range of historical case studies in this volume provides an opportunity to reflect on the ways in which old and long-forgotten problems might reemerge to challenge future naval planners and strategists.

JOHN B. HATTENDORF, D.PHIL.
Ernest J. King Professor of Maritime History
Chairman, Maritime History Department
Naval War College

Note

The thoughts and opinions expressed in this publication are those of the authors and are not necessarily those of the U.S. government, the U.S. Navy Department, or the Naval War College.

Introduction

In the late nineteenth century, the French *Jeune École,* or "new school," of naval thinking promoted a commerce-raiding strategy for the weaker naval power to defeat the dominant naval power. France provided the vocabulary for the discussion—*Jeune École* and *guerre de course* (war of the chase)—and embodied the geopolitical predicament addressed: France had been a dominant land power, known for its large and proficient army and resentful of British imperial dominance and commercial preeminence. But its navy had rarely matched the Royal Navy in either quantity or quality, and its economy could not support both a preeminent army and navy. So its naval thinkers thought of an economical way out of its predicament. They argued that a *guerre de course* allowed a weaker maritime power, such as France, to impose disproportionate costs on the stronger sea power in order to achieve its objectives. Sadly for France, the strategy did not work as anticipated, and British naval dominance and imperial primacy endured.

The case studies in this book reveal why this was so, and they shed light on the dynamic of rivalries between maritime and continental powers. This issue is an important one in that from the heyday of the British Empire to the present, maritime powers have set the global order, and continental powers have contested it. So the dynamic is still with us, and it is of vital national import to all countries that benefit from the present international order of freedom of navigation, free trade, and the rule of international law.

Commerce raiding, or *guerre de course,* is associated with major wars, such as the U.S. Civil War and the two world wars. Yet in many cases, if not most often, such operations have been conducted with relatively little public awareness. This does not indicate ineffectiveness, however. As a military tactic, commerce raiding has time after time proved itself a most efficient way to exert pressure on an opponent. A few scholars have placed these events in their social, political, and naval contexts, but their studies have been the exception, not the rule. For this reason, this collection should fill a major gap in the academic literature.

This volume will focus on how and why *guerre de course* strategies have been adopted and conducted both in nonwar and in wartime conflicts. Reexamining examples from the eighteenth, nineteenth, and twentieth centuries makes several factors apparent. First, while dominant sea powers have frequently conducted commerce raiding—most notably the American campaign against Japan in World War II—weak naval powers or continental powers have also attempted to cut off opponents' international trade, as the American revolutionaries did in the 1770s, and as Napoleon tried to do from 1803 to 1815, to Britain.

Second, *guerre de course* campaigns are often protracted, especially if the victim, particularly a continental one, opens alternate land lines of communications. In attacks on sea powers, however, speed is essential, as shown by Germany's failure to defeat Great Britain in either world war. The more time a sea power has to create the means to protect its sea trade, the less effective the *guerre de course* strategy will be.

Third, changes in technology have greatly affected commerce raiding—for example, the transition from wood to copper-sheathed ships in the early nineteenth century, the change from coal to oil combustion in the early twentieth century, and the development of airplanes and submarines. Most recently, Somali "pirates" have used small skiffs and handheld GPS devices to capture enormous oil tankers, bringing low-level but highly affordable and dependable technology to the fore.

Commerce raiding has been a traditional mission for all major navies and has played a particularly important role in Western maritime history. Alfred Thayer Mahan highlighted the important if secondary roles of commerce raiders in the American Revolution, the War of 1812, and the U.S. Civil War. Sir Julian Corbett, the unofficial Admiralty historian, emphasized the costs of interfering with international trade, writing in 1907: "The prolonged exercise of belligerent rights" over mercantile shipping, "even of the most undoubted kind, produces an interference with trade that becomes more and more oppressive."* Only seven years after Corbett's book was published, Britain was at war, defending its very existence from a German campaign of unrestricted submarine warfare. Fortunately for London, Washington found German behavior threatening to the rights of neutral powers and eventually declared war on Berlin. In contrast, in World War II it was the U.S. Navy that carried out a thoroughly successful unrestricted submarine warfare campaign, this time against Japan. Postwar, however, the international community tended to band together whenever any country or regional war interfered with international trade, as best shown during the Iran-Iraq Tanker War of the 1980s and the current piracy threat off Somalia.

---

* Sir Julian Corbett, *England in the Seven Years' War* (London: Longmans, Green, 1907 [repr. 1918]), vol. 2, p. 5.

The sixteen case studies in this book reflect the extraordinary diversity of experience of navies attempting to carry out, and also to eliminate, commerce raiding. Because the cases emphasize conflicts in which commerce raiding had major repercussions, they shed light on when, how, and in what manner it is most likely to be effective. The authors have been asked to examine the international context, the belligerents, the distribution of costs and benefits, the logistical requirements, enemy countermeasures, and the operational and strategic effectiveness of these campaigns.

There is a popular—albeit often misguided—image of the commerce raider as the dashing privateer, risking life and limb to bring an enemy prize back to port. The truth is probably much less romantic, and as noted almost all world navies have included commerce interdiction as one of their most basic roles. The chronologically arranged case studies in this volume begin with Thomas M. Truxes's examination of the Seven Years' War, when Great Britain attempted to deprive the French of provisions, supplies, and "warlike stores" from all sources, foreign or domestic. In response, France encouraged the neutral maritime powers—the Netherlands, Denmark, and Spain—both to fill the gap in the French supply train and to test the forbearance of Great Britain. From early in the conflict, there were large-scale interdictions of neutral merchantmen by British warships and privateers, all of which strained Britain's relations with its allies. In addition to problems concerning European neutrals aiding the French, Britain also faced large-scale trading with the enemy by its own subjects in North America and Ireland. The Royal Navy, along with British privateers based in the Bahamas and Jamaica, staged a vigorous but largely unsuccessful campaign to bring an end to this activity.

In the American Revolution, as Christopher Magra persuasively argues, the colonists desperately needed to establish a naval force to prosecute the Revolutionary War. By 30 June 1775 there were twenty-nine British warships stationed off the North American coast, between Florida and Nova Scotia. Without some attempt to develop their own sea power, the united colonies would have lost the war eventually. This chapter, challenging historians who argue that the Americans pursued only a *guerre de course* and privateer-based naval strategy, shows how the Continental Congress leased and paid the expenses of a small number of vessels whose purpose was not strictly to engage enemy merchant shipping. In fact, the naval strategy Americans employed during the Revolutionary War involved deploying publicly controlled small vessels with the intention of eventually engaging larger enemy warships.

French privateering was common during this era. Silvia Marzagalli shows how French shipowners fitted out privateers throughout the eighteenth century and during the wars of 1793–1815. In this latter conflict, in which French shipping and trade were in some instances virtually paralyzed, privateers were fitted out even in ports, such as Nantes or Bordeaux, that had hardly any tradition of such ventures. Important factors affecting the evolution of this activity include French willingness to authorize privateers, even though they competed with the navy for seamen; foreign opportunities to employ ships in other, more profitable trade; and the efficiency of the British navy and privateers in capturing French privateers. The French *guerre de course* was above all an opportunity for profits for merchants and employment for seamen; it was not a decisive element of warfare.

Kevin McCranie shows how during the War of 1812 the American Secretary of the Navy, William Jones, adopted beginning in February 1813 a new oceanic naval strategy that emphasized commerce raiding and de-emphasized ship-on-ship battle, in an effort to preserve America's scarce warships and impose disproportionate expenses on Britain. Jones sought to force the Royal Navy to sustain costly deployments off the U.S. coast, throughout the North Atlantic, and eventually beyond, in other contested waters. The design, execution, and effects of Jones's strategy show how a far weaker power could use a tiny navy effectively against the largest navy in the world.

The American Civil War revealed other uses for commerce raiding. Spencer Tucker focuses on the Confederacy's *guerre de course* strategy to cause serious economic distress in the North so that business interests would clamor for a negotiated end to the war that would in turn bring Southern independence. The Confederacy, without modern shipbuilding facilities and with its ports under blockade, ultimately contracted for eighteen ships abroad. CSS *Alabama,* the most famous Confederate raider, sailed some seventy-five thousand miles and took sixty-four prizes valued at up to six million dollars prior to its destruction in battle with the U.S. Navy screw sloop *Kearsarge* off Cherbourg, France, on 19 June 1864. While the Southern raiders did drive up insurance rates substantially, their major lasting effect was to force a large number of U.S. vessels into permanent foreign registry. More than half of the total U.S. merchant fleet was thus permanently lost to the flag during the Civil War.

David H. Olivier's examination of French and German naval theorists shows how the navies of France and Germany found themselves facing similar strategic situations but arrived at differing rationales for adopting commerce raiding strategies. The French expected their next war to be against the world's most powerful navy, the Royal Navy, so the new French strategy relied on a traditional weapon, the cruiser, in combination with a new threat, the torpedo boat. Together, these warships were to attack British maritime trade, with the primary goal of causing panic in the British business and financial communities. Instead of French military victories, it was to be the desperate pleas of British trade and commercial interests that would force Great Britain to sue for peace. The German navy, by contrast, believed its main foe would be France. Germans too hit on the idea of *guerre de course* but applied it differently. The main purpose behind a German war on French commerce would be to deny France overseas imports, especially weapons. Although both the French and Germans advocated *guerre de course*, then, their methods and their goals differed.

These European discussions of commerce raiding had relevance to the first Sino-Japanese War (1894–95). On the first day of the war, the Japanese navy sank the British-owned-and-operated *Kowshing*, under lease to China to carry Chinese troops to the Korean theater. As S. C. M. Paine shows, this sinking was highly controversial at the time. Large sections of the British public condemned Japan for sinking a British ship, until British courts came down on the side of Japan. Conversely, China's failure to conduct a *guerre de course* against Japan's vulnerable logistical lines arguably cost it the war.

A decade later Japan was at war again, but this time with Russia. As discussed by Bruce A. Elleman, during the 1904–1905 Russo-Japanese War the Russian navy carried out an intense, albeit short-lived, *guerre de course* strategy. Operating simultaneously in the Red Sea and in the Pacific Ocean, Russian commerce raiders attempted to interfere with international trade to Japan. However, Russia's basing in East Asia was inadequate, and China's declaration of neutrality—combined with Japanese insistence that China carry out its full obligations as a neutral power—ensured that Russian commerce raiding would prove ineffective.

In the first of two case studies examining World War I, Paul G. Halpern discusses how a traditional *guerre de course* did not work well for the German navy. A certain number of German cruisers remained at large, auxiliary cruisers managed to put to sea throughout most of the war, and these surface raiders achieved some success—probably greater than allied naval authorities were willing to admit during the war. But that success relative to the volume of allied trade was comparatively small. In contrast, the German U-boats quickly proved their potential as a potent new weapon against allied shipping. However, because Germany's decision to adopt unrestricted submarine warfare affected neutrals, most notably the United States, it widened the war; the entry of the United States into the war ultimately tipped the balance and led to Germany's defeat. In the meantime, and despite Germany's spectacular initial success, allied countermeasures gradually reduced losses to acceptable levels.

In examining Anglo-American relations during World War I, Ken Hagan and Mike McMaster focus on the American and British naval cooperation that began soon after the initiation of U.S. belligerency in April 1917 and lasted through November 1918. In this comparatively brief period the Anglo-American strategy stressed antisubmarine protection extended by convoys to troopships and cargo vessels making the hazardous transatlantic passage from the east coast of the United States to England and continental Europe. They also analyze the importance of the battleships of the Grand Fleet to the containment of U-boats and German surface raiders. At the center of this birth of combined Anglo-American naval operations was Vice Adm. William S. Sims, the commander of U.S. naval forces in Europe. Under his leadership a complex, extensive, and crucially important Anglo-American naval network was constructed almost from scratch, with virtually no prewar planning, to counter the German U-boat threat.

Commerce raiding became important during the Spanish Civil War (1936–39). Willard Frank shows how both the Republicans and Nationalists depended on military supplies imported by sea. Thus, the commerce war became a crucial element in the conduct and outcome of the conflict, in particular since the main suppliers of war materiel were the Germans and Italians for the Nationalists, and the Soviets and Mexicans for the Republicans, while British merchant firms provided most of the imports for the civilian needs of the Republic. A Non-Intervention Committee of European states attempted to contain the conflict by outlawing foreign intervention, and the 1937 British-French Nyon Arrangement countered Italian "piracy" by employing destroyers throughout the Mediterranean to sink on sight any submerged submarines. Yet the Soviets, with costs escalating, eventually halted major aid and effectively abandoned the Republic, dooming it to defeat for lack of weapons. Increased international tension and rearmament combined with the failure of the Non-Intervention system allowed the Nationalists and their allies to evade all controls on their military imports while waging a relentless commerce war.

The German U-boat campaign during World War II met much the same fate as its World War I predecessor. As Werner Rahn explains, the Germans, after achieving early success in sinking the Allied shipping propping up the British economy, conducted from late 1941 onward an increasingly frustrating search for convoys. The U-boat did provide a weapon against enemy shipping up to 1942, but then the general war situation, especially on the Mediterranean and the Eastern fronts, forced the Naval Command to employ its last remaining offensive capability like a "strategic fire brigade." This led to enormous attrition, which undermined the strategic concept of mass concentration in the Atlantic. The U-boat war failed completely in 1943, because the German boats had by then lost their ability to escape from enemy surveillance and increasingly deadly antisubmarine weapons.

In the lead-up to World War II in the Pacific, as Ken-ichi Arakawa shows, the Japanese government was most concerned by the prospect that embargoes by the United States, Great Britain, and the Netherlands would cut off crucial supplies. Merchant shipping was the bottleneck of Japan's wartime economy, and the unexpectedly high shipping losses precluded the transport of sufficient resources from Southeast Asia, resulting in the disintegration of the economy. However, if the war is viewed in its entirety, the picture becomes more nuanced: Japan experienced less-than-predicted shipping losses during the first stage of the war, and not until 1943 did they increase rapidly, owing to increasingly effective U.S. air and submarine attacks. After 1943, the military momentum was on the Allies' side, and the possibility of an ultimate Axis victory decreased. From this time onward, the shipping system transporting southern resources to Japan functioned less and less efficiently.

A major reason for Japan's defeat was the U.S. Navy's unrestricted submarine campaign. Joel Holwitt reveals how the Americans rapidly overcame their cultural aversion to unrestricted submarine warfare. At the beginning of World War II, it was up to the U.S. Navy to inflict maximum damage on the Japanese military and economy. Initially, the submarine force had to overcome timid commanders, inadequate tactics in combat, and serious flaws in the design of its torpedoes. Ultimately, however, the U.S. submarine force seized the initiative and conducted a pitiless commerce campaign that annihilated Japan's merchant marine. The results of this campaign were extraordinary, ranging from the drying up of oil supplies and the almost complete cutoff of imports to the mass starvation of Japanese citizens and soldiers. After the war, Japanese government officials and naval historians assessed that the submarine war was the crucial factor that prevented any hope of a Japanese victory in the Second World War.

The Iran-Iraq Tanker War (1980–88) is one of the most recent examples of commerce raiding. George Walker discusses how the Tanker War eventually involved merchant shipping of many states and the largest wartime deployment of the U.S. and other navies since the Korean War. New legal developments, such as the diversion of shipping for inspection instead of seizure as prizes, and such technological developments as long-range missiles, significantly impacted the conduct of the war. Besides states' traditional interests, this conflict involved intergovernmental organizations—ranging from NATO and the Gulf Cooperation Council, or the Arab League, to the EU/EC—and nongovernmental organizations, including shipping associations, international maritime insurance interests, labor organizations, and human rights and humanitarian law organizations. These groups became important factors leading to the termination of the Tanker War.

The final case study examines the current piracy situation off Somalia. As Martin Murphy highlights, the pirates work from the territory of a single state—albeit a failed state—a fact that distinguishes this type of piracy from strictly private enterprise. In fact, in many ways it resembles a state-sponsored commerce-raiding campaign, in which Somali government officials and clan leaders receive lucrative kickbacks from the pirates even while the pirates gain protection from Somali officials. Until the international community treats these pirate groups as commerce raiders, not pirates, and acts accordingly, there is little likelihood that the situation will improve.

This volume concludes with an analysis of commerce raiding during the past two and a half centuries in terms of the factors of time, space, and force, as well as with respect to positive and negative objectives. The importance of commerce raiding lies not only in the destruction of enemy trade but also in the foreclosure of enemy courses of action. Commerce raiding operations open a potentially efficient way to impose disproportionate costs on the enemy. In a protracted war, its cumulative effects, in combination with those of other military operations, can be decisive. Even in situations short of war, however, attacks on commerce can threaten the orderly growth of global commerce.

Note

The thoughts and opinions expressed in this publication are those of the authors and are not necessarily those of the U.S. government, the U.S. Navy Department, or the Naval War College.

CHAPTER ONE

The Breakdown of Borders
Commerce Raiding during the Seven Years' War, 1756–1763
THOMAS M. TRUXES

War on commerce during the Seven Years' War involved the seizure of trading ships and cargoes in the service of the enemy. Governments on both sides encouraged commerce raiding and provided for the distribution of prizes on generous terms. The most important targets were those that contributed to the enemy's capacity to wage war—cargoes of guns, gunpowder, ammunition, naval stores, shipboard provisions, and the like.

But commerce raiding extended far beyond goods associated with military operations. "In a War between two Nations, each enemy may lawfully take, seize, and possess himself of the Property of his Opponent, wherever it can be found," wrote a British jurist in 1758.[1] The war on commerce encompassed the entirety of the enemy's waterborne trade. In this heavily incentivized activity, the more valuable the cargo, the better.

There are huge gaps in our knowledge of eighteenth-century commerce raiding. Although it is a topic frequently touched on in histories of the period, the accounts that do exist have a regional or local emphasis, and they are typically narrow in scope and heavily anecdotal.[2] What does seem evident is that the gains and losses of one side appear to have been largely offset by the gains and losses of the other.[3] Even if the balance tipped one way or another, as it surely did, it is unlikely that either side derived a significant military advantage from its disruption of the enemy's trade. In the eastern Atlantic, the western Atlantic, and in the Baltic and Mediterranean Seas, both sides exacted heavy tolls. But neither the British nor the French succeeded by commerce raiding in delivering a decisive blow to the enemy's capacity to fight.

The Context for Commerce Raiding during the Seven Years' War

The Seven Years' War is known in the United States as the French and Indian War. It was fought between Great Britain and its allies (notably Prussia) and France and its allies (notably Austria and later Spain). As with the earlier eighteenth-century Anglo-French wars, the Seven Years' War involved dynastic issues, border disputes, and shifts in the balance of power. Even more pressing, however, were territorial concerns on the North American mainland that had been left unresolved at the conclusion of the War of the Austrian Succession, 1744–48.

The Great War for Empire, as the Seven Years' War has been called by one of its best-known historians, was a global conflict in which armies collided in Europe, Africa, the Indian subcontinent, and the Philippines, and there were naval operations with an even longer reach. In the North American and Caribbean theaters, Great Britain and France struggled for control of a vast and rich colonial empire. Fighting erupted in the backcountry of Pennsylvania in 1754, but formal hostilities did not begin until the spring of 1756, when Great Britain declared war on France. Spain's entry into the war in January 1762 on the side of the French was directly associated with the war on commerce.[4]

The mid-eighteenth-century British Empire was a mélange of kingdoms, colonies, and widely dispersed territorial footholds held together by loyalty to a common monarch, broad adherence to a set of legal principles, and participation in a commercial system that encouraged initiative and respected property. The British had been late to establish permanent settlements in the New World, but by the middle decades of the seventeenth century Englishmen, Irishmen, and Scots had found the means to tap the riches of the Americas in the production and marketing of tobacco, sugar, and other semitropical staples.

The rapid accumulation of wealth in British America depended on slave labor, a resilient commercial system, and the capacity of the Royal Navy to protect seaborne trade. By the 1750s, economic expansion was being fueled as well by a rapidly expanding population on the American mainland. Demand for consumer goods in British America had transformed manufacturing in the British Isles and contributed to the emergence of London as Europe's financial capital.[5]

The French were a formidable opponent. The superpower of eighteenth-century western Europe, France had a large population and an abundance of resources.[6] Like the English, the French had a strong economic presence in the New World, one that affected nearly every region in France. Sugar production in the French West Indies exceeded that of the British Caribbean islands, and French territorial holdings on the North American continent—though not well populated—dwarfed those of Great Britain.[7] But not all was well. A large share of France's Atlantic trade was conducted under foreign flags, with the result that the number of sailors available to man the warships of the French navy—necessary for the defense of a far-flung empire—had not kept pace with French commercial expansion.[8]

To doctrinaire mercantilists, trading nations were locked in a perpetual state of undeclared economic war. It was widely believed that there was only so much trade to go around and that colonial commerce belonged exclusively to the mother country. By this reasoning, the gains enjoyed by one colonial empire came at the expense of its rivals. In an age of runaway defense expenditures, governments jealously protected overseas trade, which they valued primarily as a source of revenue. The security of the state was thus intimately linked to the condition of colonial commerce.[9] Trade disruption constituted an immediate and potentially lethal threat to the state, and the undermining of the trade of one's enemies—particularly in time of war—was always justifiable.[10]

The British Campaign against French Wartime Trade

British warships began seizing French trading vessels in July 1755, weeks before news of Gen. Edward Braddock's defeat in western Pennsylvania reached Europe and roughly ten months before Britain's formal entry into the war.[11] As many as three hundred ships, with cargoes valued at approximately thirty million livres, fell into the hands of British cruisers.[12] To critics of the British government's war policy, it was a blatant violation of the "law of nations."[13] From the perspective of the War Ministry, however, the attack on French commerce was defensible as a means of "crippling the enemy's finance at the critical moment of mobilization," according to one contemporary.[14] The French, far behind in their preparations for war, "behaved with studious restraint," says one historian, and "stigmatized the proceedings of the English as simple piracy," writes another.[15]

On 17 May 1756, King George II declared war against the French king, Louis XV. The detained French vessels and cargoes were confiscated, and all French shipping, as well as the shipping of any nation in the service of the French—"the same being taken" by British warships and privateers—"shall be condemned as good and lawful Prize," announced the royal proclamation.[16]

The French returned the favor on 9 June 1756, declaring war on Great Britain and unleashing their own warships and privateers against the British carrying trade.[17] The next month, a parliamentary statute granted the officers and men of British warships and privateers "the sole interest and property of and in all and every ship, vessel, goods and merchandizes, which they shall take [during] the continuance of this war with France (being first adjudged lawful prize in any of his Majesty's courts of admiralty in Great Britain, or in his Majesty's plantations in America)."[18] By statutes and proclamations, the British government streamlined the operation of admiralty and vice-admiralty courts and established an orderly process for the distribution of prizes.[19] Without such reforms, wrote a Boston newspaper, "Privateering may be said to be only a Harvest for Agents and Lawyers."[20]

The Royal Navy was Britain's principal weapon in the war on French commerce. In the eastern Atlantic well over half the prizes taken by the British were seized by the navy.

Among these were such vessels as *Duke of Bourbon,* taken by HMS *Bristol* (fifty guns) in September 1757. The captured French merchantman had been bound from Bordeaux to Saint-Domingue "loaded with Wine, Flour, Oil, Soap, Beef, Pork, and the richest of Bale Goods . . . supposed to be worth £12,000," wrote the *Cork Evening Post*.[21] British naval officers were relentless in their pursuit of prizes. During the first years of the war, for example, the captain of a British warship rowed into the harbor of Brest, where he and his men "boarded a French snow, cut her cables, and brought her clear off though she lay among the men of war."[22]

Private ships of war were "always supplementary" to the warships of the king, writes a British naval historian.[23] Even so, privateers (privately owned fighting ships licensed by the state to seize the property of the enemy) and "letters of marque" (commercial vessels licensed by the state to seize property of the enemy encountered in the normal course of trade) were effective weapons and represented the public face of commerce raiding. In American waters, privateers and "letters of marque" may have had a slight edge over warships of the Royal Navy in the number, if not the value, of seizures.[24] However, competition for prizes between warships and privateers bred distrust on both sides, and naval officers freely expressed contempt for privateersmen, whom they considered seagoing vermin.[25]

In the early months of the war, there was a privateering frenzy on both sides of the Atlantic: "The zeal shown on this occasion produced a fleet of cruisers far exceeding anything attempted in previous wars," writes the historian of Bristol privateering.[26] But he might just as well be speaking about London, Liverpool, Glasgow, Cork, or more than a dozen ports in the British Isles and Channel Islands.[27]

It was no different in British North America, particularly in Newport, Rhode Island, and New York. "The declaration of War having put such spirits in persons here," wrote a New York merchant, "that no Less than 12 Privateers are out and fitting with the greatest dispatch."[28] A North American privateer that caught the public's imagination was the "diminutive" *Herliquin* of New York: "We have had brought into this Port Taken by our Privateers 51 Sail of Prize," a New Yorker told his brother in Liverpool. "But a Little Pilet boat Called the *Herliquin* has been the most Successfull of any having Made 5 good Voyages in 14 months."[29]

British privateers typically operated alone or in pairs. However, the most effective tactics required cooperation and took advantage of the French navy's failure to provide adequate convoy protection. Early in the war it was common for a French convoy to begin its transatlantic crossings with just a single escort, and few escorting vessels provided port-to-port protection. French convoys were frequently abandoned by their escorts or scattered by a strategy of penetration employed by awaiting enemy squadrons.[30] In the early stage of a transatlantic crossing, a westbound French convoy faced harassment by British warships and privateers based in the eastern Atlantic. Those that survived endured a second round of attacks as they entered the West Indian archipelago.[31] The impact on French transatlantic shipping of such experiences was devastating.

In 1756, a Connecticut newspaper wrote, "Our Enemies the French are much straitened for Provisions in their several Settlements on this Continent, as well as in their Islands."[32] The early phase of the war saw a heroic struggle by the French to keep the sea-lanes open. "I hope Heaven shall soon deliver us from this sorrowful Place," wrote a Dutch ship captain from Saint-Domingue in February 1757, "in case now and then a Prize was not brought up, there would really not have been any Bread to eat."[33] By the end of 1757 British warships and privateers had swept the French carrying trade from the sea.[34] After the autumn of 1759, British squadrons based at Port Royal, Jamaica, and English Harbor, Antigua, rarely encountered French warships in Caribbean waters.[35] French merchantmen were loath to put to sea, and when they did—on either side of the Atlantic—they faced likely capture.[36]

London's Policy toward Neutral Shipping

In his declaration of war, George II warned that any vessel carrying "soldiers, arms, powder, ammunition, or other contraband goods" to any territory of the French king "shall be condemned as good and lawful prize."[37] Versailles responded in kind: "Every power at war is naturally attentive to prevent its enemies from carrying on a free trade under the protection of neutral colors," asserted a *Mémoire Instructif* in the summer of 1756: "As the Hollanders are neutral in the present war," it threatened in thinly veiled language, "it is their interest to conform to the regulations of France."[38] The Dutch were in an impossible position. From the British they risked destruction of their commerce at sea; from the French they faced the possibility of invasion through borders that were indefensible against the armies of Louis XV.[39]

By Dutch tradition the seas were free, and compared to that of Great Britain and France, the overseas trade of the Netherlands was unencumbered. Dutch authorities encouraged open markets and the free flow of ships and cargoes through their thriving continental ports, Amsterdam and Rotterdam, as well as strategically located shipping points in the West Indies.[40] The most important of these, the tiny island of Saint Eustatius, is just six miles northwest of Saint Christopher, in the Leeward Islands. For a mile along the crowded shore of Orange Bay, over two hundred warehouses offered an astonishing array of goods to buyers on both sides of the conflict. Saint Eustatius and the Dutch island of Curaçao in the southern Caribbean Sea were busy crossroads of transnational trade, as well as irritants to British, French, and Spanish mercantilists.[41]

Even before the formal declaration of war, a rumor circulated in London, according to a diplomat stationed there, that "France will allow all nations to trade freely with her colonies so that more French sailors may be free to fight, and that French merchants may trade under neutral flags."[42] The British had good reason to be concerned and before long began interdicting Dutch vessels carrying small arms, cannons, gunpowder, shot, and other "warlike stores" mixed with conventional cargoes.[43]

Officials in the Netherlands insisted that the Anglo-Dutch Treaty of 1674 guaranteed the free movement of Dutch ships and goods in time of war. By the middle decades of the eighteenth century, however, the 1674 treaty was seriously out of alignment with the realities of the Atlantic economy. The British—rattled by a string of French victories and faced with the reality that neutral shipping was sustaining the French war effort—asserted that the treaty did not apply to America and unilaterally abrogated the principle of "free ships, free goods" enshrined in the 1674 agreement.[44]

The British had no intention of allowing the unfettered trade of neutrals to threaten the security of Britain. "Considering how widely Commercial Interests are diffused," wrote the author of *The Case of the Dutch Ships Considered* in 1758, "[it] is actually an Impossibility for two great, and Maritime Powers, to engage in a War, but the Intercourse of all the rest must be liable to be disturbed." In such a situation, he added, a neutral state must accept the risks implicit in neutrality: "If the Goods of Enemies may be lawfully seized wherever they are, then it follows, that they certainly may be seized on board the Ships of Neutrals."[45]

In the first two months of the Seven Years' War, no fewer than forty-eight Dutch vessels were taken by the British. "England does not have much regard for Treaty rights when her safety is at stake," said a Spanish diplomat.[46] Britain's assault on the neutral carrying trade, which continued through the war, "renders our Trade very uncertain," wrote a merchant on Saint Eustatius to Bordeaux.[47] From the perspective of London, the shipping and entrepôt services provided by the Dutch and Danes had the potential to turn the war on its head. "What signifies our being masters at sea," commented a London businessman, "if we shall not have liberty to stop ships from serving our enemy?"[48]

Neutral Spain presented an even more difficult problem, as the ministry in London became increasingly obsessed with keeping Spain out of the conflict on the side of France. Spanish merchantmen regularly called at French Atlantic and Mediterranean ports, and in the Caribbean, Spanish vessels were an ordinary sight at Saint-Domingue, Martinique, and Guadeloupe.[49] In spite of clear evidence of cooperation between the Spanish and French, London insisted on respect for Spanish neutrality.[50] From the outset, however, British naval officers and privateer commanders had been taking the law into their own hands, seizing Spanish ships thought to be cooperating with the enemy and—occasionally—entering Spanish territorial waters in pursuit of French merchantmen.[51]

British public opinion demanded that the French be deprived of the protection of neutral flags, even at the risk of enlarging the war.[52] The solution came in the form of "the Rule of 1756," the British assertion that a trade prohibited in peacetime could not be allowed in a time of war.[53] "All the European nations exclude foreigners from their American colonies," wrote Lord Chancellor Hardwicke in September 1756: "The question is whether England shall suffer [the Dutch and the Danes] to trade thither in time of war, without seizure, when the French themselves will not suffer them to trade thither, in time of peace, on that very account."[54] In a stroke, Great Britain set down a sweeping dictum that took on the force of international law.

British interdictions created a diplomatic storm that challenged the forbearance of powerful commercial interests in Amsterdam, Rotterdam, Copenhagen, Cadiz, and other neutral ports.[55] Pushback against the British came in the form of unsanctioned outbursts, each with the potential of creating an international incident. According to the *New-York Gazette,* for example, HMS *Woolwich* (forty-four guns) was cruising off Saint Eustatius late in the war when it "was fired upon from their Batteries, whereby she received considerable Damage, and very narrowly escaped being burnt [having been mistaken for] the Britannia Privateer, Capt. M'Pherson, of Philadelphia, who has intercepted many of their French trading Vessels."[56] About the same time, an Irish newspaper reported that the master of a New York privateer cruising off Curaçao "was taken and carried in there by three Dutch armed vessels" and confined to a dungeon.[57] There were many such incidents, but, in the capitals of northern Europe, cooler heads prevailed.[58]

The French Campaign against British Commerce

On the eve of the Seven Years' War, a British newspaper boasted, "I believe we have nothing to fear for the Trade to the Northward of Dunkirk."[59] The nation paid dearly for such bravado. Two days after George II's declaration of war, Versailles unleashed its corsairs, and five days later the French in America went on an all-out offensive.[60] The government in London, according to one historian, was unprepared for "the spirited manner in which the French commenced the war, and the superiority and activity of their privateers."[61]

The French navy's contribution to the war on commerce was minimal. But when it did strike, it could do so with devastating effect. In 1757, for example, when Adm. Armand de Kersaint's squadron slipped out of Brest and made for the Guinea coast of Africa and the West Indies, Britain's Atlantic trade suffered the consequences.[62] However, the French navy faced severe limits on the availability of ships, crews, and supplies necessary to maintain a credible presence at sea. In 1756 it could provide just fifteen frigates at Brest and Rochefort for coastal protection and convoy duty.[63] Ingrained attitudes about the proper role of fighting ships also kept the navy on the sidelines: "Officers generally sailed under orders which defined battle as an exception," writes a British naval historian.[64] In any case, by November 1759 the Royal Navy had effectively confined the French navy to port.[65]

Despite this, France was able to mount and sustain a vigorous campaign against British commerce. It did so with its large and well equipped fleet of privateers, some comparable in size and sophistication to warships.[66] Among the most dangerous French privateers were those based at Dunkirk, Saint-Malo, and Morlaix, ports that enjoyed close proximity to the English Channel, the waterway through which passed not only the overseas trade of London but a large part of the British coastal trade as well. The privateers of Bayonne and Saint-Jean-de-Luz—farther away but every bit as dangerous—compensated for their geographic disadvantage with their unrelenting pursuit of British merchantmen. Dunkirk and Bayonne, both exempt from the registration of sailors for service in the navy, were close enough to French borders to attract foreigners into their privateer services.[67]

French cruisers operated with impunity in British home waters from the beginning to the end of the war.[68] Early in the fighting, according to one historian, "swift and well armed French privateers found their way into the North passage and the Irish Sea, and kept Liverpool blockaded for many weeks."[69] The *London Evening-Post* reported in 1756 "that it is with great Difficulty the Fisherman of Margate [in East Kent] get any Fish, they being continually chased by French Boats."[70] When Versailles began to take warships out of service in 1760, the number of sailors available to the privateer fleet increased markedly, and so did the boldness of French raids in British waters. "An Order is gone," reported the *Cork Evening Post,* "for two Ships of 40 Guns each, and a Frigate, to sail for the Coast of Ireland, there to cruize for the Protection of our Trade in those Parts, at present much annoy'd by some of the Enemy's Privateers."[71]

It was a similar story in the western Atlantic. Large and heavily manned French privateers, operating in the waters off British North America, were a constant threat.[72] Some of their commanders became quasi celebrities in the American press. Among the most feared was "Monsieur Palanqui," the commander of an eighteen-gun privateer carrying 205 men, who enjoyed a larger-than-life reputation established in the wars of the 1740s, when "he took 69 English Vessels."[73] Even more dreaded was the gallant Capt. Soubier du Chateleau of Saint-Malo, "of late but too well known on our Coast," wrote the *Pennsylvania Gazette* in 1759.[74] Despite the anxiety created by the enemy's presence off New York and other colonial ports, French privateering had little impact on the flow of military supplies into North America, the principal theater of war in the Americas.[75]

Swarms of small and agile privateers dominated French commerce raiding in the Caribbean. Occasionally they coordinated their efforts. In March 1757, for example, an American ship captain who had been captured by a French privateer sloop reported seeing "55 French privateers between Barbuda and Guadeloupe in a chain about a Mile distant from each other."[76] More often they worked alone or in pairs, using the local topography to their advantage. Because of prevailing winds and currents in the eastern Caribbean, British merchantmen were forced to pass within easy striking distance of the French islands, especially Martinique, "whose numerous coves," writes one historian, "sheltered scores of little sloops and schooners ready to dart out when the coast was clear." At least 2,400 British merchant ships were taken in the West Indies during the Seven Years' War by French privateers, most of them based in Martinique.[77]

British countermeasures included convoys, search-and-destroy missions, and raids on enemy privateer bases—all of which put a strain on British naval resources.[78] "I have just received information from Martinique that forty Privateers are ready to put to sea as soon as the Full of the Moon is over," reported the commander of the British squadron at English Harbor, Antigua, in 1759. "Their Lordships may be assured, no Diligence of mine shall be wanting to give proper Protection and Security to the Trade," he wrote, adding, "But the constant Application that is made to me for Convoys [must] take off some of our Cruizers."[79]

There were, as well, dramatic encounters at sea, some involving staggering loss of life.[80] One that caught the attention of the public involved the British privateer *Terrible* (two hundred sailors and twenty-six guns) under the command of Capt. William Death (pronounced "Deeth"). In 1757, *Terrible* lost seventeen men when it took *Grand Alexander*, a richly laden sugar vessel returning from Saint-Domingue. After placing forty of his men as a prize crew on board the French merchantman, Captain Death and *Terrible* encountered a large French privateer, *Vengeance* (360 sailors and thirty-six large cannons), of Saint-Malo. Both commanders were killed in the action that followed, and fewer than ten of the 160 sailors remaining aboard *Terrible* were left uninjured. In all, at least four hundred men were killed or wounded in the engagement.[81]

Perhaps the most dramatic action against French commerce raiding was a June 1758 British incursion at Saint-Malo that destroyed a large number of French privateers and a valuable supply of naval stores with virtually no opposition. According to one account, the "quays and slips were crowded with shipping," and as darkness fell British soldiers (or, quite likely, marines) "stole into the defenseless suburbs, with infantry in support, and began their work. Ship after ship was silently fired; stores, rope walks, and shipyards followed, till the night was red with holocaust." In London, the mission was praised "for the protection it gave our trade by paralyzing one of the most dangerous and active of the French privateer ports."[82] Nonetheless, and despite the Royal

Navy's relentless campaign—and the large number of French privateers captured or destroyed throughout the war—"no absolute success was possible," writes a British naval historian.[83]

The Royal Navy's Response to North American and Irish Trade with the Enemy

With the French carrying trade swept from the sea and the shipping of neutrals under the relentless scrutiny of British cruisers, the French turned to their one remaining option: trading with the enemy.[84] Ports in North America—notably Boston, Newport, New York, and Philadelphia—as well as several in Ireland took on important roles in the French supply chain, by which cargoes of foodstuffs, lumber, and basic supplies were exchanged in the French West Indies for sugar, coffee, and other West Indian produce at fire-sale prices.

This commerce, technically illegal only if there was direct contact between British and French subjects, took four forms. In the first, merchants in British North America and Ireland—and less often in Great Britain—channeled cargoes through the Dutch and Danish West Indies, where local intermediaries steered them into French hands. The business conducted at Dutch Saint Eustatius and Curaçao and the Danish island of Saint Croix grew to huge proportions.

A second venue for trading with the enemy was Spanish Monte Cristi, a sleepy port on the north coast of Hispaniola fewer than twenty miles from the border of French Saint-Domingue. Throughout the war, a fleet of forty-five to sixty Spanish coasters moved vast quantities of goods that had been off-loaded from vessels anchored in Monte Cristi Bay to Fort-Dauphin, Cape François, Port-au-Prince, and other ports in Saint-Domingue. This occurred in full view of British warships patrolling off Monte Cristi. Their commanders were under strict orders to avoid incidents that might upset Britain's delicate diplomatic relationship with neutral Spain.

A third form of trading with the enemy involved merchants in British America doing business under the cover of prisoner-of-war exchanges. Because of a reluctance to shoulder the cost of maintaining prisoners of war, colonial governors granted to ship captains licenses permitting the carriage of French prisoners to Saint-Domingue and elsewhere in exchange for captured British soldiers and sailors. To cover their costs, Americans were allowed to engage in limited trade. From this concession grew a massive flow of goods, mostly from Philadelphia, Newport, and New York, where the cartel trade became a feature of wartime economic life.

In its fourth manifestation, North American vessels trading with the enemy did so without legal cover. In this high-risk game, merchants stood to make enormous profits, having cut out expensive layers of middlemen, or conversely, to lose everything.

In response to this trade, in its various aspects, Parliament passed the Flour Act of 1757, a law that prohibited the carriage of foodstuffs to any port outside the British commercial system. Huge fines were levied against violators, who risked forfeiture of both their vessels and cargoes. The statute applied solely to North American ships and had no impact on British or Irish vessels trading with the enemy. The arbitrary and discriminatory character of the Flour Act left a bitter legacy and was to figure among the colonial grievances in the run-up to the American Revolution.

British privateers occasionally played a supporting role in wartime trade with the French. As French prizes became scarce in the Caribbean and the waters off North America, British naval officers began paying closer attention to American and Irish vessels doing business, directly or indirectly, with the French. The Flour Act made colonial vessels attractive prizes for naval officers, even in cases in which there had been no direct contact with the enemy. To counter this threat, North American merchants hired privateers, or sometimes deployed their own, to capture their own trading vessels returning from the French islands. In these collusive captures, the cooperating privateer commander went through the ritual of capture and placed a prize master, armed with appropriate legal documents, on board the seized vessel, which was then escorted to a safe port. To complicate matters further, court records reveal instances of British privateers carrying goods to the enemy while at the same time cruising against French and neutral shipping. Enemy privateers also figured in this story, by honoring the passports carried by North American and Irish vessels trading within the French islands.

In official London, there was remarkable ambivalence on the subject of trading with the enemy. One need not look far to discover powerful individuals and interest groups that benefited from illicit wartime trade. It was, however, a very different story in the western Atlantic. In 1758, frustrated British naval commanders began to intervene on their own authority. From their perspective, treasonous North American and Irish merchants were provisioning enemy privateers that menaced British shipping and tied up naval resources. Seizure by the Royal Navy—as well as by privateers based in the West Indies, particularly at Nassau, in the Bahamas—meant quick condemnation of ship and cargo in the vice-admiralty courts of Jamaica and other British West Indian islands. But to the chagrin of the captor's officers and the men, who stood to share substantial amounts of prize money, many cases brought before an admiralty appeals court in London were reversed, most often on grounds of insufficient proof of face-to-face contact with the enemy.[85]

Operational and Strategic Effectiveness of the British and French *Guerre de Course*

The goals of commerce raiding varied from one operational theater to another. In the struggle for dominance in North America and the West Indies, both Great Britain and

France placed a high priority on disrupting the enemy's trade in both directions across the Atlantic. The interdiction of westbound commerce deprived the enemy of supplies necessary to sustain the fight, and the seizure of eastbound vessels laden with valuable colonial produce shifted sorely needed tax revenues from the coffers of one state into those of the other. Within the confines of the Caribbean Sea, both sides worked assiduously to strangle the other's interisland commerce and throw its plantation economy into disarray.

In the eastern Atlantic Ocean, Mediterranean Sea, and Baltic Sea, the objectives and tactics of the belligerents differed. The loosely coordinated privateer force of the French sought opportunistically to disrupt British waterborne trade, in whatever way it could, rather than as part of an overarching grand design. Great Britain, which until late 1759 faced the real possibility of a French invasion, used its warships and privateers to keep the enemy's commercial shipping bottled up in port in order to deprive cash-starved France of the benefits of a commercial empire.

Nearly all of the port towns of Great Britain and Ireland felt the sting of French commerce raiding. But none were more than temporarily disturbed by the depredations of enemy privateers. The French were not so fortunate. The *guerre de course* profoundly affected all of the French Atlantic and Mediterranean ports. At La Rochelle, for example, the Seven Years' War threatened the city's economic foundations: "For the Rochelais," the war "precipitated an economic crisis, measured by captured merchantmen, kin and seamen killed or wounded in combat, unpaid debts, soaring costs, and the absolute cessation of the slave trade." It was nearly as bad at Le Havre, Brest, L'Orient, Nantes, Bordeaux, Marseilles, Toulon, and elsewhere on the French Atlantic and Mediterranean coasts, where commercial losses cast a pall over economic life.[86]

In the western Atlantic, neither the British nor the French fully achieved their goals. The Royal Navy possessed adequate resources to halt Dutch, Danish, Spanish, North American, and Irish participation in the French supply chain, but it lacked the political support from London necessary. To the dismay of the Royal Navy, French resourcefulness, the cooperation of neutral states, and the opportunism of merchants in Ireland and North America led to a breakdown of borders that encouraged the free flow of goods in a highly volatile wartime environment.[87] The French, for their part, were disadvantaged by the absence of their navy and their overreliance on a loosely coordinated but aggressive privateer force. French commerce raiding in the western Atlantic created serious disruptions for the British, but it had little impact on the war's outcome.

There were many who benefited from the war on commerce. In fact, commerce raiding acted as a countercyclical stimulant to economic activity in a region negatively affected by war: "The fitting out of commissioned ships involved capital investment, employment of labor, and the consumption of goods produced in the shore-based marine industries," and commerce raiding "generally added to the 'stock' of vessels operating from a port by virtue of the purchase, seizure, or building of ships."[88] Individuals benefited as well, some handsomely. For the officers and men of the Royal Navy, prize money was the greatest attraction of naval service. A lucky captain could "make a fortune overnight," according to one authority on eighteenth-century navies, and "an admiral was almost assured of it."[89] Financial rewards could be substantial for those who manned the privateers, as well as for those who invested in them. There was no payday, however, if a voyage came up empty.[90]

Both sides paid a high price for commerce raiding. It led to a steep rise in transportation charges as the wages of sailors increased, maritime insurance premiums rose, and available cargo space contracted.[91] It also bred uncertainty. An American merchant on the eve of the conflict wrote, "Our produce is falling owing to the apprehensions of a war, . . . [with] people fearing to send out their vessels."[92] But the greatest cost was the loss of life and limb that was a predictable result of armed engagements at sea. If the British and the French stumbled into war over ill-advised forays into commerce raiding, then all of the costs associated with the Seven Years' War must be laid at the feet of what Lord Granville called "vexing your neighbors for a little muck."[93]

It is impossible to discuss this topic without wondering whether mid-eighteenth-century commerce raiding was merely a disguised and legally sanctioned form of piracy. Newspapers were replete with stories that described the pillaging of British, French, and neutral merchantmen, as well as the mistreatment of sailors.[94] Dutch vessels were frequent targets of abuse by British privateers, but the most egregious incidents involved attacks against Spaniards in the Caribbean.[95] Some of these, such as the "piratical" robbery of a Spanish vessel carrying government dispatches by a privateer crew based in Barbados, created international incidents requiring intervention at the highest level. In that instance, the governor of the Leeward Islands reported to London "that this Execution, and the Hanging in Chains [of the bodies of the perpetrators] seem to have struck a general Terror amongst the Privateers; and I flatter myself, that His Catholick Majesty will have no further Cause for Complaint." As for the future, however, "unless some of the Offenders can be prevailed upon to give Evidence for the King . . . under a Promise of Pardon, it will not be possible to convict them."[96]

British parliamentary statutes—notably those of 1708, 1756, and 1759—attempted to establish rules of conduct for the interdiction of trading vessels in the service of the enemy.[97] But it was beyond the capacity of an eighteenth-century state—except where commissioned naval officers were involved—to control the behavior of predator warships in high-stakes confrontations at sea in which lives and fortunes were on the line.

There is no reliable account of the number and value of the prizes taken and condemned by the British and French during the Seven Years' War. However, "there seems no doubt that the French claim to have captured a greater number of vessels than [the British] is justified," writes one authority on eighteenth-century warfare. The value of their captures is less certain. The French, he added, did little harm to British transatlantic convoys, and the enemy's relentless disruption of shipping failed to undermine Great Britain's commercial credit, the weakening of which would constitute "the main strategic value of commerce destruction."[98]

Conclusions

During the Seven Years' War, commerce raiding energized the Atlantic economy, fostered cross border exchanges, and challenged the mercantilist assumptions that governed eighteenth-century commercial policy. It also motivated policy makers in London and Versailles to put their houses in order following the Peace of Paris in 1763.

Under the leadership of the duc de Choiseul, in 1763 the minister of marine and colonies, France initiated an ambitious program to rebuild its navy, strengthen defenses in the West Indies, and restructure French colonial commerce. But in both France and the French Caribbean, officials worked at cross-purposes, and attempts to liberalize the rules governing trade were offset by periods of tightened control and heavy-handed enforcement. French commercial policy grew increasingly out of touch with the realities of the Atlantic world.[99]

Great Britain, likewise, entered a period of adjustment. Victors in the long and costly war, the British initiated a series of postwar reforms intended to discipline colonial commerce and strengthen the revenue of the state. Within weeks of signing the Treaty of Paris, Parliament passed the Customs Enforcement Act of 1763, a statute that called for deputizing the sea officers of the Royal Navy as customs-enforcement agents. A legacy of widespread trading with the enemy during the Seven Years' War, the Customs Enforcement Act helped to create the adversarial relationship between the mother country and the British colonies that culminated twelve years later in the American Revolution.[100]

Notes

The thoughts and opinions expressed in this publication are those of the author and are not necessarily those of the U.S. government, the U.S. Navy Department, or the Naval War College.

1. James Marriot, *The Case of the Dutch Ships Considered* (London: R. and J. Dodsley, 1759), p. 2.

2. For a sample of work touching on commerce raiding during the Seven Years' War, see

Richard Pares, *War and Trade in the West Indies, 1739-1763* (London: Frank Cass, 1963); David J. Starkey, *British Privateering Enterprise in the Eighteenth Century* (Exeter, Devon, U.K.: Univ. of Exeter Press, 1990); James G. Lydon, *Pirates, Privateers, and Profits* (Upper Saddle River, N.J.: Greg, 1970); Gomer Williams, *History of the Liverpool Privateers and Letters of Marque* (London: William Heinemann, 1897); J. W. Damer Powell, *Privateers and Ships of War* (Bristol, U.K.: J. W. Arrowsmith, 1930); Patrick Crowhurst, *The Defence of British Trade, 1689-1815* (Folkestone, Kent, U.K.: Wm. Dawson & Sons, 1975); and Joyce Elizabeth Harman, *Trade and Privateering in Spanish Florida, 1732-1763* (Tuscaloosa: Univ. of Alabama Press, 2004).

3. Stephen Conway, *War, State, and Society in Mid-Eighteenth-Century Britain and Ireland* (Oxford, U.K.: Oxford Univ. Press, 2006), p. 106.

4. Lawrence Henry Gipson, *The British Empire before the American Revolution*, 15 vols. (New York: Knopf, 1939-70). For a critique of Gipson and the term "Great War for Empire," see Francis Jennings, *Empire of Fortune: Crowns, Colonies & Tribes in the Seven Years War in America* (New York: W. W. Norton, 1988), pp. xxi-xxii. A broader perspective can be found in Daniel A. Baugh, *The Global Seven Years' War, 1754-1763: Britain and France in a Great Power Contest* (Harrow, U.K.: Longman, 2011), and Fred Anderson, *Crucible of War: The Seven Years' War and the Fate of Empire in British North America, 1754-1766* (New York: Knopf, 2000).

5. For the emergence and development of the British Empire, see Nicholas Canny, ed., *The Origins of Empire: British Overseas Enterprise to the Close of the Seventeenth Century*, and P. J. Marshall, ed., *The Eighteenth Century*, vols. 1 and 2 of *The Oxford History of the British Empire*, ed. Wm. Roger Louis (Oxford, U.K.: Oxford Univ. Press, 1998).

6. N. A. M. Rodger, *The Command of the Ocean: A Naval History of Britain, 1649-1815* (New York: W. W. Norton, 2004), p. 261.

7. Pierre Henri Boulle, "The French Colonies and the Reform of Their Administration during and following the Seven Years' War" (PhD diss., Univ. of California at Berkeley, 1968), pp. 10-16, 193-206, 215-30; Walter L. Dorn, *Competition for Empire, 1740-1763* (New York: Harper and Row, 1963), pp. 252-54; Paul Butel, *The Atlantic* (London: Routledge, 1999), pp. 149-58.

8. Rodger, *Command of the Ocean*, p. 261.

9. Charles Wilson, *Mercantilism* (London: Historical Association, 1958); Boulle, "French Colonies and the Reform of Their Administration," p. 345.

10. Daniel A. Baugh, *British Naval Administration in the Age of Walpole* (Princeton, N.J.: Princeton Univ. Press, 1965), p. 15.

11. Jonathan R. Dull, *The French Navy and the Seven Years' War* (Lincoln: Univ. of Nebraska Press, 2005), p. 38.

12. The livre *tournois* was a French monetary unit. For its relationship to the British pound sterling, see John J. McCusker, *Money and Exchange in Europe and America, 1600-1775* (Chapel Hill: Univ. of North Carolina Press, 1978), pp. 87-99.

13. Emer de Vattel, *The Law of Nations; or, Principles of the Law of Nature: Applied to the Conduct and Affairs of Nations and Sovereigns* (London: G. G. J. and J. Robinson, 1793).

14. George Bubb Dodington, *The Diary of the Late George Bubb Dodington* (Salisbury, U.K.: E. Easton, 1784), pp. 344-45; Julian S. Corbett, *England in the Seven Years' War: A Study in Combined Strategy* (London: Longmans, Green, 1907 [repr. 1918]), vol. 1, pp. 50-51.

15. I. S. Leadam, *The History of England from the Accession of Anne to the Death of George II* (London: Longmans, Green, 1909), p. 436; William Edward Hartpole Lecky, *A History of England in the Eighteenth Century* (New York: D. Appleton, 1888), vol. 2, p. 487.

16. *London Magazine* (May 1756), p. 237; *Daily Advertiser* (London), 19 May 1756.

17. *Daily Advertiser* (London), 25 June 1756. *Gentleman's Magazine* (June 1756), pp. 296-97; (July 1756), pp. 360-61; (August 1756), p. 411; and (September 1756), p. 452. *New-York Mercury*, 20 September 1767; *Pennsylvania Gazette*, 9 September 1756; Dull, *French Navy and the Seven Years' War*, p. 53.

18. 29 George II, c. 34 (British).

19. Ibid.; *Gentleman's Magazine* (July 1756), p. 44; *New-York Mercury*, 18 October 1756.

20. *Boston Evening Post*, 27 September 1756.

21. Starkey, *British Privateering Enterprise in the Eighteenth Century*, pp. 78-79, 178-80; [John] Dobson, *Chronological Annals of the War, from Its Beginning to the Present Time* (Oxford,

U.K.: Clarendon, 1763), pp. 33–44; *Cork Evening Post*, 24 November 1757.

22. Frederick Hervey, *The Naval, Commercial, and General History of Great Britain* (London: J. Bew, 1786), vol. 5, p. 38.

23. William Laird Clowes, *The Royal Navy: A History from the Earliest Times to the Present* (London: Sampson Low, Marston, 1897–1903), vol. 3, p. 296.

24. Records of the Vice-Admiralty Court for the Province of New York (1685–1775), Record Group 21, U.S. National Archives and Records Administration, Northeast Region (New York City); New York Vice-Admiralty Court Minutes, 1753–1770, MS in Manuscripts and Archives Division, New York Public Library, New York.

25. N. A. M. Rodger, *The Wooden World: An Anatomy of the Georgian Navy* (New York: W. W. Norton, 1996), pp. 185–86.

26. Powell, *Privateers and Ships of War*, p. 184.

27. Starkey, *British Privateering Enterprise in the Eighteenth Century*, p. 165.

28. Cuyler to Vander Heyden, 23 September 1756, "Philip Cuyler Letter Book, 1755–1760," MS in New York Public Library, New York.

29. Beekman to Beekman, 25 November 1757, in *The Beekman Mercantile Papers, 1746–1799*, ed. Philip L. White (New York: New-York Historical Society, 1956), vol. 1, p. 315.

30. John G. Clark, *La Rochelle and the Atlantic Economy during the Eighteenth Century* (Baltimore: Johns Hopkins Univ. Press, 1981), pp. 154–55.

31. Ibid.; log of the Privateer *Duke of Cumberland*, 1758–1760, MS at New-York Historical Society, New York.

32. *(New Haven) Connecticut Gazette*, 12 June 1756.

33. Capt. Louis Ferret to Owners of the ship *America*, 7 February 1757, HCA 45/2 [ship *America*], Public Record Office / The National Archives, Kew Gardens, U.K. [hereafter PRO/TNA].

34. Starkey, *British Privateering Enterprise in the Eighteenth Century*, p. 161

35. Dull, *French Navy and the Seven Years' War*, pp. 115, 141.

36. Clowes, *Royal Navy*, vol. 3, p. 233.

37. *London Magazine* (May 1756), p. 237.

38. *Gentleman's Magazine* (October 1756), p. 460.

39. Alice Clare Carter, *The Dutch Republic in Europe in the Seven Years' War* (London: Macmillan, 1971), pp. 50–68.

40. Charles R. Boxer, *The Dutch Seaborne Empire: 1600–1800* (New York: Knopf, 1965); Wim Klooster, *Illicit Riches: Dutch Trade in the Caribbean, 1648–1795* (Leiden: KITLV, 1998).

41. David Macpherson, *Annals of Commerce, Manufactures, Fisheries, and Navigation* (London: Nichols and Son et al., 1805), vol. 3, p. 161. See Thomas M. Truxes, *Defying Empire: Trading with the Enemy in Colonial New York* (New Haven, Conn.: Yale Univ. Press, 2008), pp. 57–60.

42. Jean O. McLachlan, "The Uneasy Neutrality: A Study of Anglo-Spanish Disputes over Spanish Ships Prized 1756–1759," *Cambridge Historical Journal* 6, no. 1 (1938), p. 56.

43. Robert Beatson, *Naval and Military Memoirs of Great Britain from 1727 to 1783* (London: Longman, Hurst, Rees, and Orme, 1804), vol. 2, p. 45.

44. Richard Pares, *Colonial Blockade and Neutral Rights, 1739–1763* (Oxford, U.K.: Oxford Univ. Press, 1938), pp. 187–90; Carter, *Dutch Republic in Europe in the Seven Years' War*, pp. 103–12.

45. Marriot, *Case of the Dutch Ships Considered*, p. 2.

46. Abreu to Wall, 24 August 1756, quoted in McLachlan, "Uneasy Neutrality," p. 57; *Boston Evening Post*, 11 October 1756.

47. Morgan to Thouron, Jr., 21 May 1758, HCA 45/2 [ship *Novum Aratum*], PRO/TNA.

48. Nicolas Magens, *An Essay on Insurances* (London: J. Haberkorn, 1755), vol. 1, p. 435.

49. Pares, *War and Trade in the West Indies*, pp. 556–62.

50. Pitt to Governors, 16 September 1757, in *Correspondence of William Pitt*, ed. Gertrude Selwyn Kimball (New York: Macmillan, 1906), vol. 1, pp. 105–106; McLachlan, "Uneasy Neutrality," pp. 58–60.

51. E. S. Roscoe, ed., *Reports of Prize Cases Determined in the High Court of Admiralty . . . from 1745 to 1859* (London: Stevens and Sons, 1905), vol. 1, pp. 7–8; McLachlan, "Uneasy Neutrality," pp. 62, 70.

52. "Our very good friends the Dutch are, according to their wonted custom, contriving every scheme and practicing every method

to engross the trade and supply our enemies." *London Evening-Post*, 24 July 1756.

53. Pares, *Colonial Blockade and Neutral Rights*, pp. 180–204; anon. to Blakes, 18 December 1758, ADM 1/235, PRO/TNA; Marriot, *Case of the Dutch Ships Considered*, pp. 7–8.

54. Pares, *Colonial Blockade and Neutral Rights*, p. 197.

55. *Boston Evening Post*, 27 September 1756.

56. *New-York Gazette, or Weekly Post-Boy*, 22 April 1762; Clowes, *Royal Navy*, vol. 3, p. 242 note.

57. *Cork Evening Post*, 14 July 1760.

58. ADM 1/307, f. 358, PRO/TNA; Pares, *Colonial Blockade and Neutral Rights*, pp. 255–309.

59. *New-York Mercury*, 9 February 1756.

60. Boulle, "French Colonies and the Reform of Their Administration," p. 94 note.

61. Williams, *History of the Liverpool Privateers*, p. 87.

62. Richard Middleton, "British Naval Strategy, 1755–62," *Mariner's Mirror* 75 (1989), p. 355.

63. *New-York Mercury*, 9 August 1756; Dull, *French Navy and the Seven Years' War*, p. 60.

64. Rodger, *Command of the Ocean*, p. 273; Clowes, *Royal Navy*, vol. 3, p. 296.

65. Conway, *War, State, and Society in Mid-Eighteenth-Century Britain and Ireland*, p. 88.

66. Clowes, *Royal Navy*, vol. 3, p. 296; Dobson, *Chronological Annals of the War*, pp. 18, 25.

67. Dull, *French Navy and the Seven Years' War*, p. 61; Crowhurst, *Defence of British Trade*, pp. 454–64. See Truxes, *Defying Empire*, p. 55.

68. Conway, *War, State, and Society in Mid-Eighteenth-Century Britain and Ireland*, p. 88.

69. Williams, *History of the Liverpool Privateers*, p. 86.

70. *London Evening-Post*, 16 September 1756.

71. *Cork Evening Post*, 9 June 1760. For French privateer commander François Thurot's daring raid at Carrickfergus on the northeast coast of Ireland in February 1760, see Gipson, *British Empire before the American Revolution*, vol. 8, pp. 24–27.

72. *New-York Mercury*, 15 November 1756.

73. *Cork Evening Post*, 24 November 1757.

74. *Pennsylvania Gazette*, 1 March 1759.

75. See Truxes, *Defying Empire*, pp. 54–55.

76. *New-York Mercury*, 2 May 1757.

77. Rodger, *Command of the Ocean*, p. 277.

78. Ibid., pp. 288–89; Corbett, *England in the Seven Years' War*, vol. 1, pp. 392–95; *Berrow's Worcester Journal*, 4 August 1757.

79. Moore to Clevland [*sic*], 3 October 1759, SP 42/41, ff. 459–60, PRO/TNA.

80. *New-York Mercury*, 3 January 1756; Dobson, *Chronological Annals of the War*, pp. 14, 21–22.

81. Hervey, *Naval, Commercial, and General History of Great Britain*, vol. 5, pp. 81–82.

82. Rodger, *Command of the Ocean*, p. 270; Corbett, *England in the Seven Years' War*, vol. 1, pp. 276–77.

83. Rodger, *Command of the Ocean*, p. 277.

84. For trading with the enemy during the Seven Years' War, see Truxes, *Defying Empire*.

85. William Burrell, *Reports of Cases Determined by the High Court of Admiralty and Upon Appeal Therefrom, . . . Together with Extracts from the Books and Records of the High Court of Admiralty and the Court of the Judges Delegates, 1584–1839*, ed. Reginald G. Marsden (London: William Clowes and Sons, 1885).

86. Clark, *La Rochelle and the Atlantic Economy*, pp. 151, 155–56.

87. Adm. Charles Holmes, "Memorial Respecting Monto Christi," December 1760, Adm. 1/236, ff. 156–63, PRO/TNA.

88. David J. Starkey, "The Economic and Military Significance of British Privateering, 1702–83," *Journal of Transport History* 9, no. 1 (March 1988), pp. 52–55.

89. Baugh, *British Naval Administration in the Age of Walpole*, pp. 112–18.

90. Crowhurst, *Defence of British Trade*, pp. 463–64. See Thomas M. Truxes, ed., *Letterbook of Greg & Cunningham, 1756–57: Merchants of New York and Belfast* (Oxford, U.K.: Oxford Univ. Press, 2001), pp. 48–49, 58, 74–75.

91. James Pritchard, *Louis XV's Navy: A Study of Organization and Administration* (Montreal: McGill-Queen's Univ. Press, 2009), p. 80.

92. William S. Sachs, "The Business Outlook in the Northern Colonies, 1750–1775" (PhD diss., Columbia Univ., New York, 1957), p. 66.

93. Corbett, *England in the Seven Years' War*, vol. 1, pp. 60–61.

94. *New-York Mercury,* 16 May 1757; *(New Haven) Connecticut Gazette,* 5 February 1757. Such accounts, especially if they involved the abuse of the Dutch, were often dismissed as fabrications; see *Dublin Journal,* 14 September 1756. There was nothing exceptional about the experience of the ship *Amsterdam.* On a voyage from Cork, Ireland, to Saint Eustatius, the vessel "met with no less than 14 different Privateers [which] pilfered and plundered her of many of her Cabin stores." Examination of John Govan, the Master, 17 November 1757, HCA 45/1 [ship *Amsterdam*], PRO/TNA.

95. "Account of Goods Taken from Dutch Ships by Privateers [1758]," and "Memorial . . . with Respect to Several Seizures [1759]," T 1/392, ff. 13–15, PRO/TNA; *New-York Mercury,* 21 March 1757; Berthold Fernow, ed., *New York (Colony) Council: Calendar of Council Minutes, 1668–1783* (Harrison, N.Y.: Harbor Hill Books, 1987), p. 443; deposition of Juan D'Miranda, 11 March 1757, ADM 1/234, f. 635, PRO/TNA.

96. Thomas to Holdernesse, 7 November 1757, and Thomas to Pitt, 13 January 1758, CO 152/46, f. 158, PRO/TNA.

97. 6 Anne, c. 65 (British); 29 George II, c. 34 (British); 32 George II, c. 25 (British).

98. Corbett, *England in the Seven Years' War,* vol. 2, p. 375.

99. Boulle, "French Colonies and the Reform of Their Administration," pp. 564–673; Dorothy Burne Goebel, "The 'New England Trade' and the French West Indies, 1763–1774: A Study in Trade Policies," *William and Mary Quarterly* 20, no. 3 (July 1963), pp. 331–72.

100. 3 George III, c. 22 (British); Neil R. Stout, *The Royal Navy in America, 1760–1775* (Annapolis, Md.: Naval Institute Press, 1973). See Truxes, *Defying Empire,* pp. 173, 176, 179, 186.

Guerre de Course and the First American Naval Strategy
CHRISTOPHER P. MAGRA

In 1755, a twenty-year-old John Adams wrote a letter to a friend stating that since the North American colonies had "all the naval stores of the nation in our hands, it will be easy to obtain the mastery of the seas, and then the united force of all Europe will not be able to subdue us."[1] Perhaps the substance of this letter reflects nothing more than the chest-thumping posturing of an ambitious youth who had dreams of elevating America to prominence through sea power. After all, the thirteen North American colonies that eventually became part of the United States of America were still firmly embedded in the British Empire in 1755, the colonists relied on the British navy for protection, and any separation of the colonies from the mother country was not going to be easy. Yet Adams never stopped believing resource-rich Americans could attain "mastery of the seas." Twenty years later, the Massachusetts delegate to the Continental Congress helped convince skeptical Founding Fathers that sea power was possible and even necessary to attain during the American Revolution. Adams was so heavily involved at the start of the conflict in formulating the first American naval strategy that he has been lauded as "the real father of the American navy."[2]

There is a general consensus among Revolutionary War historians that the colonists primarily pursued a maritime strategy of privateering during the Revolution.[3] It is true that Americans did not construct a large squadron of floating fortresses, nor did they engage the enemy in traditional line-of-battle tactics, in which a fleet of sizable warships in column delivered broadsides against an enemy fleet in the same formation. But Adams's definition of a navy was never that expansive: "I don't Mean 100 ships of the Line[;] ... [instead] this Term might be applied to any naval force consisting of several Vessels, tho[ugh] the Number, the Weight of Metal, or the Quantity of Tonnage may be small."[4] To achieve this goal, it is generally understood, the colonial government relied largely on the private sector. Civilian vessel owners secured letters of marque and converted small commercial vessels into warships, whose efforts they focused on harassing

British shipping, with the intention of taking cargo-rich prizes for sale in the marketplace. Crews shared in the prize money as their reward.[5]

While these generalizations may be valid for the conflict as a whole, however, they do not apply to the early months of the Revolutionary War in 1775, when the American government had not yet deferred to privateers and so did not adhere to a plan of action strictly based on *guerre de course*. In these months smaller, privately owned vessels were indeed converted to warships. But these vessels were leased to, and their expenses were reimbursed by, the Continental Congress, an arrangement that made them at least the temporary property of the American government. Very few naval historians have bothered to differentiate between *guerre de course* involving privateers and that waged by government-controlled warships.[6] Moreover, these vessels were not solely intended to engage enemy merchant shipping for commerce raiding. At the very beginning of the Revolutionary War at least, the naval strategy Americans developed involved deploying public—that is, government-leased—vessels with the broad intention of engaging the enemy's commercial shipping and its warships as well. This was the very first American naval strategy. It was bold, and it was daring, just like Adams's 1755 letter.

Early American Naval Strategies

Scholars have debated whether John Adams was correct to insist that the colonists needed some measure of sea power to prevent the American war for independence from being defeated.[7] By 30 June 1775, there were twenty-nine British warships stationed off the coast of North America between Florida and Nova Scotia. These warships carried a total of 584 guns and 3,915 men.[8] Vice Adm. Samuel Graves, with eight of these warships, patrolled the New England coastline.[9] In October, thirty-five British naval vessels, including twelve ships of the line, patrolled the coast of North America. Fifteen of these thirty-five, and no fewer than seven of the twelve of the line, were positioned in New England waters near Adams's Massachusetts home.[10] At that time the British military, however, pursued in general a strategy of pacification as opposed to conquest, and its goal was not to destroy the revolutionaries.[11]

The Admiralty ordered its commanders to support the army in North America by engaging colonial positions on land, transporting and evacuating British troops, blockading the coastline, and cutting off American access to transatlantic military supplies. By these means the navy was to help the army suppress colonial dissent. Full naval force was, arguably, never brought to bear on the colonies, however, because of a perpetual fear of French invasion.[12] Yet the British vessels stationed off the coast of North America provided major leverage over the colonists. Without some attempt to develop their own sea power, the United Colonies might have eventually lost their bid for independence. During the first three years of revolution, the American navy sent only four frigates to sea, and none of these were operational at the very start of the military conflict, making the leasing of privately owned ships essential if a naval force was to be assembled.[13]

Most naval historians of the Revolutionary War equate privateering with commerce raiding, or cruiser warfare.[14] In this strategy, small and maneuverable vessels target an enemy's merchant shipping with hit-and-run tactics. The purpose of this strategy is to wreak economic havoc on the enemy through the loss of valuable cargoes and vessels and through increased maritime insurance rates. In theory, economic distress then motivates the enemy's business interests to pressure the government to end the conflict.[15] This cumulative strategy is "a type of warfare in which the entire pattern is made up of a collection of lesser actions, but these lesser or individual actions are not sequentially interdependent[;] . . . the thing that counts is the cumulative effect."[16]

Cutting seaborne supply, carried in many instances throughout the eighteenth century by unarmed, unescorted private contractors on which militaries relied for transoceanic logistical support, could also adversely impact the combat effectiveness of an adversary's ground forces.[17] According to Alfred Thayer Mahan, overreliance on privateers and *guerre de course* limited the size and scope of the Continental Navy, both through loss of manpower to the private sector and through reduced naval construction, and it left British ships of the line in command of the Atlantic Ocean throughout the Revolution. Mahan considered this strategy indecisive and inconclusive, even calling it a "secondary" naval operation, as it did not in his opinion eliminate the central naval threat.[18]

American leaders discussed naval strategies over the course of 1775, and America's very first naval strategy developed out of these discussions. Instead of beginning construction of a new navy, and rather than immediately moving to issuance of letters of marque, representatives at the Continental Congress in Philadelphia debated whether or not to fund a navy. On the one hand, there were those who supported the formation of an American navy, including Rhode Island legislators. The Rhode Island General Assembly was a revolutionary political body, residing in a colony exposed to seaborne assaults. Its members were, by and large, prominent merchants and shipowners who were very concerned that the Royal Navy could and would attack their property at any moment. On 26 August 1775 they sent instructions to their delegates at the Continental Congress, Samuel Ward and Stephen Hopkins, to make the first formal motion in Congress for "the building and equipping of an American fleet, as soon as possible . . . at the Continental expenses." In particular, Rhode Islanders wanted "a fleet of sufficient force, for the protection of these colonies, and for employing them in such manner and places as will most effectively annoy our enemies, and contribute to the common defense of these colonies."[19]

On the other hand, there were leaders in Congress who felt that the costs of a naval force outweighed the benefits, while others still fervently hoped for reconciliation with the mother country.[20] Samuel Chase, a representative from Maryland, famously stated, "It is the maddest idea in the world to think of building an American fleet[;] . . . we should mortgage the whole continent."[21] Such concerns proved so forceful that they retarded American naval construction at the start of the conflict, ensuring that the colonists would not pursue traditional line-of-battle naval tactics in the Revolutionary War.

But even naysayers like Chase were willing to compromise, since they did want some measure of sea power. They simply had their own ideas as to the necessary level and the best way to achieve it. John Adams kept notes on key debates within the Continental Congress. According to his notes, Chase stressed the importance of military intelligence—he wanted to know what the enemy was planning and executing, and he wanted this knowledge before the American leadership took any extensive action. That is why the delegate from Maryland proposed allocating funds for "two swift sailing vessels" for the purposes of "gaining Intelligence" instead of setting aside money for the construction of an American navy.[22] In fact, such ships would be necessary for a navy as well.

Creating an American Navy

Christopher Gadsden, a former purser in the Royal Navy and a member of the Continental Congress from South Carolina, also supported forming an American navy. He is best known for opposing the building from scratch of a completely new fleet of large warships. The "extensiveness" of building, arming, manning, and maintaining an American navy concerned Gadsden. Yet Gadsden insisted it was "absolutely necessary" that "some Plan of Defense by Sea should be adopted."[23] Initially, he emphasized the role of maritime military forces in coastal defense.

Following a private meeting with Adams, then acting as the representative of Massachusetts, Gadsden shifted from a strategy predicated on defense to a bolder, more offensive-minded plan of attack. The details of this meeting are unknown, but afterward—as Adams reported—Gadsden was "confident that We may get a Fleet of our own, at a cheap Rate." Perhaps owing to his talks with Adams, Gadsden concluded that smaller commercial vessels could be inexpensively converted into warships and that they could effectively offer some measure of sea power. According to Adams, Gadsden held that such a "cheap" navy could "easily take their [i.e., Great Britain's] Sloops, schooners and Cutters [smaller vessels], on board of whom are all their best Seamen, and with these We can easily take their large Ships, on board of whom are all their impressed [that is, forcibly conscripted] and discontented Men."[24] Adams later stated his own belief that "two or three Vessels of 36 and twenty Guns, well armed and manned might attack and carry a 64 or a 70 or a 50 Gun Ship."[25]

Gadsden's strategy was based on the large number of impressed colonists in the British crews. He maintained that such men would not put up much of a fight, especially against fellow colonists. Gadsden's prior experience in the Royal Navy may have contributed to the development of this particular naval strategy. Gadsden would have known directly or indirectly that when an enemy vessel and crew were captured, the size of the prize crew assigned to it and the disposition of prisoners were precisely calculated to minimize the chances that prisoners would attempt to retake the prize. These ratios may have been on Gadsden's mind as he weighed the odds that a small vessel of Americans could capture a larger vessel that was manned, at least in part, by impressed American seamen.

Some British naval officers during the Revolutionary War echoed Gadsden's insights. For example, on 20 November 1775 Vice Admiral Graves informed the Admiralty Board that he had authorized only with great reluctance the impressment of colonial seamen, as "they will seize every Opportunity of making their Escape, or of assisting their Countrymen in Rebellion."[26]

Adams later transmitted the plans that he and Gadsden had worked out to Elbridge Gerry, a fish merchant from Marblehead, Massachusetts, who was a member of the Massachusetts Provincial Congress, which organized and administered much of the military resistance in and around Boston throughout 1775. Gerry almost certainly informed the members of the Provincial Congress of Adams's correspondence, as this body was at that very moment debating whether or not Massachusetts should use public funds to arm its own vessels for war. On 20 June 1775, the Provincial Congress resolved

> that a number of armed Vessels, not less than six, to mount from eight to fourteen carriage guns, and a proportionable number of swivels [swivel guns], &c. &c. be with all possible dispatch provided, fixed, and properly manned, to cruise as the Committee of Safety, or any other person or persons who shall be appointed by this Congress for that purpose, shall from time to time order and direct, for the protection of our trade and sea-coasts against the depredations and piracies of our enemies, and for their annoyance, capture, or destruction.

The actual military conversion of commercial vessels was "ordered to subside for the present," but this program did get under way in Massachusetts later in August.[27] It is likely that the pecuniary-minded members of the Provincial Congress wanted to wait to see whether the Continental Congress would formally approve the Rhode Island proposal. After all, if the Continental Congress had agreed to create a national navy at public expense, that would have saved the Massachusetts Provincial Congress from paying for its own.

The first American naval strategy, then, was worked out among delegates to the Continental Congress in June and July 1775. This strategy was communicated to Massachusetts, then under siege by the British forces. The plan at this time involved arming and manning smaller commercial vessels that could be fitted out quickly and at low cost. The goal was to capture successively larger enemy warships, free their impressed sailors, and thereby disrupt enemy supply lines. Since it focused on freeing colonists held against their will, such a strategy was not simply commerce raiding. The American strategy, to be sure, involved hit-and-run tactics against enemy merchant vessels carrying military supplies and trade goods. But like commerce raiding, this American strategy was intended to weaken British sea power. To accomplish this task, the Continental Congress turned to armed schooners.

The Colonial Use of Armed Schooners

As a result of these early maritime debates, the Continental Congress deployed armed schooners against the British over the course of 1775. These armed schooners have erroneously been called privateers. But Congress did not officially authorize privateering until the very end of 1775. Additionally, while privateers always remained under the command of captains who worked for civilian merchants, the American government established lease agreements with other merchants for use of their vessels. The government thereby assumed a significant degree of control over those vessels. A number of fishing vessels, for example, were leased directly to the Continental Congress, making them the temporary property of the United Colonies.

On 18 July 1775, the Continental Congress officially sanctioned the conversion of commercial shipping into armed vessels. America's national leaders resolved "that each colony, at their own expense, make such provision by armed vessels or otherwise, as their respective assemblies, conventions, or committees of safety shall judge expedient and suitable to their circumstances and situation for the protection of their harbors and navigation on their sea coasts, against all unlawful invasions, attacks, and depredations, from cutters and ships of war."[28] Orders were sent to the Massachusetts Provincial Congress, and this body assigned John Glover the task of finding vessels to arm.[29]

In August 1775, Glover succeeded in assembling five of the six armed vessels the Provincial Congress had authorized in June. The Committee of Safety, which was affiliated with the Provincial Congress, furthermore ordered "that Colonel John Glover" use his authority in Marblehead "for the prevention of Intelligence" leakage to the British patrol vessels in the harbor.[30] The leased vessels were all fishing schooners, they all belonged to fish merchants in Marblehead, and they were all converted into warships in the harbor of the nearby town of Beverly. The five schooners were *Hannah, Franklin, Hancock, Lee,* and *Warren*.[31]

Hannah, of "78 tons" burden and ten years old, was the property of Glover himself, leased to the Continental Congress on 24 August.[32] Glover had purchased it in 1769. In typical fashion, *Hannah* and its crew had transported fish and lumber to Barbados in the winter months between 1770 and June 1775.[33] The ship returned from these voyages bearing muscovado sugar and West Indian rum in its hold.[34] Glover now leased the fishing vessel to "the United Colonies of America," or, in other words, the Continental Congress. It is important to reiterate that the Marblehead fish merchant did not lease the schooner to the Massachusetts Provincial Congress, nor did he lease it directly to General George Washington. Such a lease underscores *Hannah*'s role as the first "American," as opposed to state, naval vessel.[35]

Silas Deane, a Connecticut representative at the Continental Congress, one of the early members of its Marine Committee, and a man remembered for having referred to the creation of an American fleet as "a Favorite object of mine," believed that getting colonial vessels "into Continental pay" (i.e., leased) was one of the first steps toward the realization of this "Favorite object."[36] What is more, in his orders General Washington explicitly reminded Nicholson Broughton, *Hannah*'s captain, that it was Congress that had paid his salary and that as "a Captain in the Army of the United Colonies of North America," Broughton personally fell under the commander in chief's authority. Moreover, as "the schooner *Hannah*" had been "fitted out & equipped with Arms, Ammunition and Provisions at the Continental Expense," Broughton was doubly beholden to Washington.[37]

On 2 September 1775 Washington's first set of fighting instructions for Broughton were issued, and they parted in significant details from a purely *guerre de course* strategy. These orders were arguably the first naval instructions in American history. Washington did not simply and solely instruct Broughton to target the enemy's shipping and supplies, with the ultimate goal of putting economic pressure on Great Britain. Certainly, Broughton was told to seize "arms, ammunition or provisions" bound to or from the British forces at Boston. He was "to avoid any engagement with any armed vessel of the enemy," and Washington clearly stated that "the design of this enterprise" was "to intercept the supplies of the enemy." However, *Hannah* was clearly meant to weaken Britain's fighting capacity as well, since it was also ordered to seize enemy "soldiers." In this regard, Washington gave Broughton specific instructions on the care of "prisoners you may take." This was not an indirect form of warfare. Moreover, Broughton was to "be very particular and diligent in your search after all letters and other papers tending to discover the designs of the enemy, or of any other kind, and to forward all such to me as soon as possible." Washington clearly envisioned *Hannah* as an intelligence-gathering tool, not simply a blunt economic weapon.[38]

On 4 October 1775, Washington assigned Stephen Moylan, the Muster Master General, to assist Glover in arming the leased vessels for war. Both men were to report either to Col. Joseph Reed, Washington's military secretary, or to the commander in chief directly.[39] The two reported on 9 October 1775 that the terms of the contracts they had negotiated with vessel-owning merchants included a requirement that the merchants "shall put their vessels in the same good order & Condition which they would be obliged to do, were they hired to take in a Cargo for the West Indies or elsewhere." For their part, Glover and Moylan agreed "that what extra expense may accrue from the nature of their present employment must be a public Charge." The vessel owners wanted extra sails, over and above the "three sails, Mainsail, foresail, & jib . . . sufficient for the Voyages they usually Make," to be made "a public Charge."[40]

In light of these instructions, Washington's strategic objectives for the armed schooners can best be summarized as cutting military supply lines, gathering military intelligence, and reducing the enemy's fighting capacity. Only the first objective is fully consistent with *guerre de course*. The terms of the ship contracts confirm that all conversion expenses were paid with public funds, which distinguished them from simple privateers.

Operating Methods of the Armed Schooners

The armed schooners funded by the Continental Congress operated in a manner that was not strictly privateering. William Falconer, the author of a maritime dictionary in 1769, defined a privateer as a privately owned vessel, fitted out and armed in wartime "to cruise against and among the enemy, taking, sinking or burning their shipping" in exchange for shares of any captured prizes.[41] Robert Gardiner focused more on potential profits, defining privateers as "free-enterprise warships, armed, crewed and paid for by merchants who gambled on the dividend of a valuable capture."[42]

To be sure, there is ample evidence that contemporaries regarded the fleet of armed schooners fitted out at Beverly as a collection of privateers. For example, "Manly, A Favorite New Song in the American Fleet," composed in Salem, Massachusetts, in March 1776, referred to the captain of the armed schooner *Lee*, John Manley, as a "Privateer."[43] Out of exasperation, Washington once even went so far as to refer to the men on the schooners as "our rascally privateersmen" in a letter to his secretary Col. Joseph Reed.[44] Such evidence, combined with the facts that the fishing schooners remained privately owned and the crews at least earned some prize shares, has led several naval historians to consider the vessels armed at Beverly to be privateers. Gardiner, for example, describes the "handful of Marblehead fishing schooners, armed with four or six tiny 4 pdrs and 2 pdrs," not as representing "the beginnings of a national navy" but as being privateers "conceived with a specific raiding purpose in mind."[45] Following this line of reasoning, the refitted ships would have been profit-driven business ventures and little more.

Nevertheless, there are compelling reasons to argue that the fishing schooners armed for war in late 1775 were not privateers. First and foremost, most of the prize money from the sale of prizes they took went not to the vessel owners, as would normally have been done with privateers, but to the American government to recoup outfitting costs. On 25 November 1775, the Continental Congress established formal rules regarding prize shares for privateers, colony (state) naval vessels, and Continental Navy vessels. The owners of privateers were to get all prize money associated with their captures, military or commercial. For colony/state vessels, the government entity funding the ship was to get two-thirds of the prize money and the crew the remainder. This same distribution applied to Continental Navy vessels, with the Continental Congress getting two-thirds of the prize shares. If, on the other hand, "the Capture be a Vessel of War," then whether operating under a colony, a state, or the Congress, the captors received one-half.[46]

Second, in a sharp break with tradition, the crews of the armed schooners were given wages in addition to prize shares, and these wages were paid directly by the Continental Congress. The standard practice for privateers in the late eighteenth century, by contrast, was to give crews food but not wages.[47] The commander in chief of the American forces specifically referred to those same vessels, in a letter to the Continental Congress at the end of 1775, as "the Continental armed vessels."[48]

All of this evidence indicates that the small collection of fishing vessels armed at Beverly, Massachusetts, were not just "rascally privateersmen."[49] In fact, they represented the first American naval warships. Considering the times, this fact should not be overly surprising. There was an established naval tradition in the early-modern Atlantic world of arming fishing vessels for war.[50] Moreover, during the Revolution most of the vessels engaged in combat at sea with the British were small in size.[51] Although small, they represented the origins of what would one day become the world's largest and most powerful navy.

Conclusions

It is true that during the Revolutionary War Americans unleashed the private sector and the profit motive on their enemy's commercial shipping and military transports. The Continental Congress printed and issued over two thousand letters of marque licensing entrepreneurial American shipowners to engage enemy-flagged vessels between 25 November 1775, when Congress first authorized the practice of privateering, and the peace that ended the war in 1783. On 7 September 1776, Beverly merchants made public the following handbill: "Now fitting for a Privateer, In the harbor of Beverly, the Brigantine *Washington*. . . . Any Seaman or Landsman that has an inclination to make their Fortunes in a few months may have an opportunity by applying to John Dyson."[52] Boston merchants printed similar advertisements in the local newspaper as late as 13 November 1780, under the title "An Invitation to all brave Seamen and Marines, who have an inclination to serve their country and make their Fortunes." The Boston merchants shrewdly added that those who signed on for a cruise with the privateer would receive "that excellent Liquor called Grog, which is allowed by all true seamen to be the Liquor of Life."[53]

In exchange for these letters of marque, shipowners could sell legally appropriated prizes, including vessels and nonmilitary cargoes. Prize courts adjudicated whether the vessel had been an enemy vessel and whether or not the cargo was contraband. During the course of the war, American privateers captured and sold over six hundred prizes.[54] Nearly all of these prizes were British merchant vessels either carrying trade goods across the Atlantic for sale in overseas marketplaces or contracted to transport military stores across the ocean for the army and navy. These statistics are in keeping with the traditional naval history of the American Revolution, which focuses on the widespread use of privateers.

However, it is important to emphasize that privateering did not play a large role in the initial American naval strategy. War broke out on 19 April 1775, but Congress did not authorize privateering until the end of the year. As a result, the vast bulk of the letters of marque Congress sold during the war were issued after the first year of the conflict. With regard to the American turn to the use of a *guerre de course* strategy, Mahan would later write, "The colonists could make no head against the fleets of Great Britain, and were consequently forced to abandon the sea to them, resorting only to a cruising warfare, mainly by privateers."[55] But the widely held belief that Americans relied only on privateering throughout the entire Revolutionary War is inaccurate. Americans did not initially intend to wage a large-scale privateering war against the British. In fact, at the start of the Revolutionary War publicly funded, government-controlled warships were used. Therefore, the naval strategy that the Americans initially adopted was not strictly one of *guerre de course*. In truth, the first American naval strategy was to support smaller, cheaper oceangoing vessels to target the enemy's commercial vessels, military supplies, soldiers, sailors, and warships.

Notes

The thoughts and opinions expressed in this publication are those of the author and are not necessarily those of the U.S. government, the U.S. Navy Department, or the Naval War College.

1. John Adams to Nathan Webb, 12 October 1755, in *Old Family Letters,* ed. Alexander Biddle (Philadelphia: J. B. Lippincott, 1892), pp. 5–6.

2. Nathan Miller, *Sea of Glory: A Naval History of the American Revolution* (Charleston, S.C.: Nautical and Aviation, 1974), p. 39. For more along these lines, see Frederick H. Hayes, "John Adams and American Sea Power," *American Neptune* 25, no. 1 (January 1965), pp. 35–45; and Carlos G. Calkins, "The American Navy and the Opinions of One of Its Founders, John Adams, 1735–1826," U.S. Naval Institute *Proceedings* 37 (1911), pp. 453–83.

3. Alfred Thayer Mahan, *The Influence of Sea Power upon History, 1660–1783* (1890; repr. New York: Dover, 1987), pp. 344–47, 539–40; Mahan, *The Major Operations of the Navies in*

the *War of American Independence* (London: Sampson Low, Marston, 1913); Kenneth J. Hagan, "The Birth of American Naval Strategy," in *Strategy in the American War of Independence,* ed. Donald Stoker, Kenneth J. Hagan, and Michael T. McMaster (New York: Routledge, 2010), pp. 37, 39; Hagan, *This People's Navy: The Making of American Sea Power* (New York: Free Press, 1991), pp. 16–17; Raymond G. O'Connor, *Origins of the American Navy: Sea Power in the Colonies and the New Nation* (Lanham, Md.: University Press of America, 1994), pp. 22–26; William M. Fowler, Jr., *Rebels under Sail: The American Navy during the Revolution* (New York: Scribner's, 1976), pp. 244–46; Miller, *Sea of Glory,* chap. 16; William Bell Clark, *Ben Franklin's Privateers: A Naval Epic of the American Revolution* (Baton Rouge: Louisiana State Univ. Press, 1956); Gardner W. Allen, *A Naval History of the American Revolution* (Boston: Houghton Mifflin, 1913), vol. 1, pp. 42–51; William J. Morgan, "American Privateering in America's War for Independence, 1775–1783," *American Neptune* 36, no. 2 (April 1976); Don Higginbotham, *The War of American Independence: Military Attitudes, Policies, and Practice, 1763–1789* (Boston: Northeastern Univ. Press, 1971 [repr. 1983]), pp. 331–51, esp. pp. 332, 338; Robert H. Patton, *Patriot Pirates: The Privateer War for Freedom and Fortune in the American Revolution* (New York: Pantheon, 2008). For a dissenting voice, see James C. Bradford, "John Paul Jones and Guerre de Razzia," *Northern Mariner / Le marin du nord* 13, no. 4 (October 2003), pp. 1–15.

4. John Adams to James Warren, Philadelphia, 19 October 1775, in *Naval Documents of the American Revolution* [hereafter *NDAR*], vol. 2, p. 528. His broad definition of a navy will be used throughout this chapter.

5. Carl E. Swanson, *Predators and Prizes: American Privateering and Imperial Warfare, 1739–1748* (Columbia: Univ. of South Carolina Press, 1991).

6. Hagan, "Birth of American Naval Strategy," pp. 36–37, 39; Allen, *Naval History of the American Revolution,* vol. 1, chaps. 2, 3.

7. J. R. Dull, "Was the Continental Navy a Mistake?," *American Neptune* 44 (1984), pp. 167–70; W. S. Dudley and M. A. Palmer, "No Mistake about It: A Response to Jonathan R. Dull," *American Neptune* 45 (1985), pp. 244–48. I am grateful to Michael Palmer for these references.

8. Vice Adm. Samuel Graves to Gen. Thomas Gage, Boston, 30 June 1775, in *NDAR,* vol. 1, p. 785.

9. Vice Adm. Samuel Graves to Gen. Thomas Gage, Boston, 5 July 1775, in *NDAR,* vol. 1, pp. 819–20.

10. CO 5/122/35, Public Record Office / The National Archives, Kew Gardens, U.K.

11. Jeremy Black, "British Military Strategy," in *Strategy in the American War of Independence,* ed. Stoker, Hagan, and McMaster, pp. 58–72.

12. John Reeve, "British Naval Strategy: War on a Global Scale," in *Strategy in the American War of Independence,* ed. Stoker, Hagan, and McMaster, pp. 73–99. For more on British naval strategy during the American Revolution, see N. A. M. Rodger, *The Command of the Ocean: A Naval History of Britain, 1649–1815* (New York: W. W. Norton, 2004), pp. 327–57; Richard Buel, Jr., *In Irons: Britain's Naval Supremacy and the American Revolutionary Economy* (New Haven, Conn.: Yale Univ. Press, 1998); David Syrett, *The Royal Navy in American Waters, 1775–1783* (Aldershot, U.K.: Scolar, 1989); Daniel A. Baugh, "Why Did Britain Lose Command of the Sea during the War for America?," in *The British Navy and the Use of Naval Power in the Eighteenth Century,* ed. J. Black and P. Woodfine (Leicester, U.K.: Leicester Univ. Press, 1988), pp. 149–69.

13. Richard B. Morris, introduction to *The American Navies of the Revolutionary War: Paintings,* by Nowland Van Powell (New York: G. P. Putnam, 1974), p. 17. For more on late-eighteenth-century frigates, see Fowler, *Rebels under Sail,* pp. 190–223; and Robert Greenhalgh Albion and Jennie Barnes Pope, *Sea Lanes in Wartime: The American Experience, 1775–1942* (New York: W. W. Norton, 1942), p. 22.

14. Mahan, *Influence of Sea Power upon History,* pp. 344, 539–40; Hagan, *This People's Navy,* pp. 1–6, 16–17; O'Connor, *Origins of the American Navy,* p. 26; Fowler, *Rebels under Sail,* pp. 22–23; Frank C. Mevers, "Naval Policy of the Continental Congress," in *Maritime Dimensions of the American Revolution* (Washington, D.C.: Naval History Division, U.S. Navy Dept., 1977), p. 5; Miller, *Sea of Glory,* chap. 16; Robert Gardiner, ed., *Navies and the American Revolution, 1775–1783* (London: Chatham, in association with the National Maritime Museum, 1996), pp. 66–69; Albion and Pope, *Sea Lanes in Wartime,* pp. 25–26; and Allen, *Naval History of the American Revolution,* vol. 1, p. 50. Miller, *Sea*

of Glory, even defines privateers as "privately owned commerce raiders" (p. 8).

15. Bernard Brodie, *A Guide to Naval Strategy* (Princeton, N.J.: Princeton Univ. Press, 1944), p. 137. For more on *guerre de course,* see Geoffrey Symcox, *The Crisis of French Sea Power, 1688–1697: From the Guerre d'Escadre to the Guerre de Course* (The Hague: Martinus Nijhoff, 1974); Michael P. Gerace, *Military Power, Conflict and Trade* (London: Frank Cass, 2004), pp. 26–27; Gardiner, ed., *Navies and the American Revolution,* pp. 66–69; Fowler, *Rebels under Sail,* pp. 22–23; and Albion and Pope, *Sea Lanes in Wartime,* pp. 25–26.

16. J. C. Wylie, *Military Strategy: A General Theory of Power Control,* 3rd ed. (Annapolis, Md.: Naval Institute Press, 1989), p. 23. My thanks to John B. Hattendorf for this reference.

17. David Syrett, *Shipping and Military Power in the Seven Years War: The Sails of Victory* (London: Univ. of Exeter Press, 2008); Syrett, *Shipping and the American War, 1775–1783* (London: Athlone, 1970); R. Arthur Bowler, *Logistics and the Failure of the British Army in America, 1775–1783* (Princeton, N.J.: Princeton Univ. Press, 1975); Albion and Pope, *Sea Lanes in Wartime.*

18. Mahan, *Influence of Sea Power upon History,* p. 539.

19. "Journal of the Rhode Island General Assembly," Providence, 26 August 1775, in *NDAR,* vol. 1, p. 1236.

20. O'Connor, *Origins of the American Navy,* pp. 16–17.

21. John Adams, *Diary and Autobiography of John Adams,* ed. L. H. Butterfield, Adams Papers (Cambridge, Mass.: Belknap of Harvard Univ. Press, 1961), vol. 2, p. 198.

22. Ibid.

23. Ibid.

24. John Adams to Elbridge Gerry, Marblehead, Philadelphia, 7 June 1775, in *NDAR,* vol. 1, pp. 628–29.

25. Adams to Warren, p. 528.

26. Vice Adm. Samuel Graves to Philip Stephens, in HMS *Preston,* Boston, 20 November 1775, in *NDAR,* vol. 2, p. 1083.

27. Journal of the Provincial Congress of Massachusetts, Watertown, 20 June 1775, in *NDAR,* vol. 1, p. 724. O'Connor contends that "this proposal was never implemented," but he did not consider the conversion of fishing vessels. See O'Connor, *Origins of the American Navy,* p. 14.

28. Worthington C., Ford, ed., *Journals of the Continental Congress, 1774–1789* (Washington, D.C.: U.S. Government Printing Office, 1904–37, vol 2, p. 189.

29. Journal of the Provincial Congress of Massachusetts, p. 724. Glover was a fish merchant from Marblehead involved in the fishing industry and the colonel of the port town's regiment. For more on Glover, see George Athan Billias, *General John Glover and His Marblehead Mariners* (New York: Henry Holt, 1960); Ashley Bowen, *The Journals of Ashley Bowen (1728–1813) of Marblehead,* ed. Philip C. F. Smith (Boston: Colonial Society of Massachusetts, 1973), vol. 2, p. 657.

30. Minutes of the Massachusetts Committee of Safety, Cambridge, 27 April 1775, in *NDAR,* vol. 1, p. 229.

31. Appraisal of the *Speedwell* [renamed *Hancock*], Beverly, 10 October 1775; Appraisal of the *Eliza* [renamed *Franklin*], Beverly, 10 October 1775; Appraisal of the *Two Brothers* [renamed *Lee*], Beverly, 12 October 1775; Appraisal of the *Hawk* [renamed *Warren*], Beverly, 12 October 1775; all in *NDAR,* vol. 2, pp. 387, 412–13.

32. *John Glover's Colony Ledger,* item 729½, Marblehead Museum & Historical Society, Marblehead, Mass.

33. There has been disagreement about the schooner's size. Fowler, *Rebels under Sail,* describes *Hannah* as "a typical New England fishing schooner of about seventy tons" (p. 29). Hearn, Billias, and Clark follow *Glover's Colony Ledger* in listing it at "seventy-eight tons": Chester G. Hearn, *George Washington's Schooners: The First American Navy* (Annapolis, Md.: Naval Institute Press, 1995), p. 10; Billias, *General John Glover and His Marblehead Mariners,* p. 74; and William Bell Clark, *George Washington's Navy: Being an Account of His Excellency's Fleet in New England Waters* (Baton Rouge: Louisiana State Univ., 1960), p. 3. A smaller figure of forty-five tons is supported by Philip C. F. Smith and Russell W. Knight, "In Troubled Waters: The Elusive Schooner *Hannah,*" *American Neptune* 30, no. 2 (April 1970), pp. 15, 22, and app. 2, p. 41.

34. Smith and Knight, "In Troubled Waters," app. 2, pp. 41–43.

35. *John Glover's Colony Ledger.* While the amount and the form of payment varied from vessel

to vessel and colony to colony, the rate "per ton per month" was standard. See Minutes of the Connecticut Council of Safety, Lebanon, 3 August 1775, in *NDAR,* vol. 1, p. 1054; and Stephen Moylan and Col. John Glover to George Washington, Salem, 9 October 1775, in *NDAR,* vol. 2, p. 368.

36. Silas Deane to Thomas Mumford, Philadelphia, 15 October 1775, *NDAR,* vol. 2, p. 464.

37. George Washington's Instructions to Capt. Nicholson Broughton, 2 September 1775, in *NDAR,* vol. 1, p. 1287.

38. "Instructions to Captain Nicholson Broughton to proceed on a cruise, in the Schooner Hannah, against Vessels in the service of the Ministerial Army, 2 September 1775," series 4, vol. 3, p. 633, *American Archives* [hereafter AA]; and George Washington Papers, 1741–1799, series 4, General Correspondence, 1697–1799, image 1071, Library of Congress, Washington, D.C., memory.loc.gov/ammem/gwhtml/gwhome.html.

39. Col. Joseph Reed to Col. John Glover, Head Quarters, Cambridge, 4 October 1775, in *NDAR,* vol. 2, pp. 289–90; and Col. Joseph Reed to Col. John Glover and Stephen Moylan, Camp at Cambridge, 4 October 1775, in ibid., p. 290.

40. Moylan and Glover to Washington, p. 368.

41. William Falconer, *A New Universal Dictionary of the Marine* (London: T. Cadell and W. Davies, 1769 [repr. 1815]), p. 353.

42. Gardiner, ed., *Navies and the American Revolution,* p. 66. According to Albion and Pope, *Sea Lanes in Wartime,* "profits were the *raison d'être* of privateers" (pp. 23–24).

43. *Manly. A favorite new song, in the American fleet. Most humbly addressed to all the jolly tars who are fighting for the rights and liberties of America. By a sailor* (Salem, Mass.: printed and sold by E. Russell, upper end of Main-Street, 1776), Early American Imprints, 1st series, Evans 43057.

44. George Washington to Col. Joseph Reed, Camp at Cambridge, 20 November 1775, in *NDAR,* vol. 2, p. 1082.

45. Gardiner, ed., *Navies and the American Revolution,* pp. 66–67. Syrett similarly refers to Washington's schooner fleet as "the American cruiser offensive"; Syrett, "Defeat at Sea," p. 16. Octavius T. Howe, *Beverly Privateers in the American Revolution* (Cambridge, Mass.: J. Wilson and Son, [1922?]).

46. Journal of the Continental Congress, Philadelphia, 25 November 1775, in *NDAR,* vol. 2, p. 1133.

47. Albion and Pope, *Sea Lanes in Wartime,* p. 23.

48. Letter from General Washington to the President of Congress, Cambridge, 4 December 1775, p. 180, series 4, vol. 4, AA.

49. Ibid.

50. The Spanish Armada, which remains one of the most famous flotillas in all of recorded history, included Basque, Portuguese, and Spanish fishing vessels. Michael Barkham, "Spanish Ships and Shipping," in *Armada, 1588–1988,* ed. M. J. Rodríguez-Salgedo (London: Penguin Books, in association with the National Maritime Museum, 1988), pp. 151–63. England's Rump Parliament relied on "shallops and ketches," vessels primarily used to catch fish, in addition to ships of the line, to defend its newfound sovereignty from royalists at home and abroad. Bernard Capp, *Cromwell's Navy: The Fleet and the English Revolution, 1648–1660* (Oxford, U.K.: Clarendon, 1989), p. 4.

51. Hearn, *George Washington's Schooners;* Fowler, *Rebels under Sail.*

52. Cited in Howe, *Beverly Privateers in the American Revolution,* p. 338 note 2.

53. *Boston Gazette,* 13 November 1780.

54. Hagan, "Birth of American Naval Strategy," pp. 36–37.

55. Mahan, *Influence of Sea Power upon History,* p. 344. Academic naval historians who follow Mahan on this point include Fowler, *Rebels under Sail,* pp. 262–63, and Miller, *Sea of Glory,* p. 282. Dissenters on this point include Hagan, *This People's Navy,* pp. 17–20; Syrett, *Shipping and the American War;* Clark, *George Washington's Navy* and *Ben Franklin's Privateers;* and Dudley W. Knox, *The Naval Genius of George Washington* (Boston: Riverside, 1932). Also see Morgan, "American Privateering in America's War for Independence." None of these dissenters disagrees with Mahan's original case that Americans primarily pursued a *guerre de course* strategy during the Revolution. But the dissenters argue that it was more successful than Mahan described, for various reasons.

French Privateering during the French Wars, 1793–1815

SILVIA MARZAGALLI

In Old Regime France, war was a recurrent experience. Between 1689 and 1815 France was at peace no more than one year out of two. As a major continental power the French crown invested heavily in its army, but the navy too played an important role in the strategy for warfare, especially after France became a colonial power in the seventeenth century and fought to defend its overseas territories and economic interests. The French navy was in charge of securing merchant shipping, destroying the enemy's navy, and eventually organizing raids and the occupation of enemy territories. These tasks, however, were only one aspect of war at sea. The capture of the enemy's merchant ships represented the other relevant feature.[1] But if the navy took part in the capture of the enemy's private merchant ships, this was not its top priority.[2] These activities were mostly the preserve of privateers, who were granted letters of marque authorizing them to raid enemy property at sea.

Significant efforts were made in Europe beginning in the late seventeenth century to reduce plundering and violence by soldiers on civil populations;[3] however, the legitimacy of the maritime equivalent—the capture of private property at sea—was not seriously questioned before the early 1790s. By declaring private property a natural right (article 2), however, the Declaration of the Rights of Man and of the Citizen in 1789 implicitly opposed any kind of predatory activity, including at sea. Moreover, article 12 contested the use of public force for the benefit of private interest, a principle that could be applied against privateering. In 1792, French revolutionaries intensively debated the legitimacy and the utility of privateering, but once the war against maritime powers broke out in 1793 France once again authorized shipowners to fit out vessels and raid enemy shipping. Not until the Paris international convention in 1856 did France officially agree to ban privateering.[4]

The French Wars (1793–1815), the last conflict pitting France against Great Britain in a century-long struggle for the control over colonies and trade, also marked the last concrete experience of French privateering before its abolition in 1856.[5] This chapter seeks to understand privateering both as an element within a long tradition and as the product of specific circumstances created by the French Revolution and Napoleon. Only ships that were fitted out by private individuals to raid enemy merchant ships will be considered, thus excluding the captures made by the French navy and by merchant ships fitted out *"en guerre et marchandises"* (for war and trade).[6] In particular, it will show that from the delivery of the letter of marque granting the right to outfit a privateer to the final court procedure establishing whether the prize was valid or not, government authorities after 1793 controlled all phases of privateering. They also occasionally either encouraged or restricted it, according to the prevailing national interests. In addition, it will discuss the practical organization of a campaign, as well as the consequences of privateering both in France and abroad.

French Privateering: A Multifaceted Activity

The first historians of privateering celebrated the brave actions of certain outstanding privateers.[7] Later research has quantified the relevance of this activity for French ports;[8] other work has stressed its effects on international relations.[9] More recently, attention has been paid to its broader economic implications.[10] Privateering did not concern the French government only as an element of warfare; for captains, sailors, and shipowners and their families, it represented an essential alternative to peacetime shipping and trade. The activity was a source of potential profit in times of war, not only for those who invested in it but for all the privateers' crew members. Prizes contributed to provisioning local markets where their cargoes were sold, and prize ships themselves fed a lively secondhand market.

Ultimately these various economic interests influenced the extent of privateering.[11] The calculation of costs and expectations of benefits determined the willingness of shipowners to fit out privateers, although occasionally the French government intervened by subsidizing shipowners—for instance, providing naval munitions or other forms of assistance to lower their costs.[12] The absence of decent alternatives to earn livings at sea made fishermen and sailors, and more generally unemployed young men, particularly willing to participate, despite the high risk of being captured and taken to England as a prisoner.[13]

The impact of French privateering during the French Wars went beyond the economic interests of shareholders and crews. At times, the numbers of privateers and their degree of success affected international shipping and trade and influenced merchants' strategies, among both belligerents and neutrals. Their activities provoked public responses outside France. The effectiveness of the British measures adopted to control trade routes and to protect shipping against raiding affected in turn the success of French privateering. Raiding at sea did not concern belligerents only; privateering generated international tensions with neutral powers, and the number of prizes depended also on the capacity of neutral countries to enforce their rights and defend their interests.

Although privateering was essentially the result of private initiative, the French government had a close interest in supporting it. This was the case not only because capturing ships and goods at sea was considered an important naval tactic, adversely affecting the enemy's trade and interests while weakening the manning of its navy, but also because the government was concerned to avoid indiscriminate plundering of foreign property. Finally, in cases of seizure, the state received a portion of the prizes. For all these reasons, the French government closely regulated privateering.

The Short 1793 Privateering Experience

Although revolutionary France was involved in war from 1792, privateering resumed only after the declaration on 1 February 1793 of war against Great Britain and the Netherlands, followed a few days later by the declaration of war against Spain. At first, privateers could raid enemy merchant ships only, but in May 1793 France declared that goods belonging to enemy subjects on neutral ships could be seized as well—a factor that considerably extended the potential profitability of privateering, given the importance of neutral shipping in wartime. In making this decision, France adopted a much-contested 1756 British measure. Whereas prior to 1756 it was commonly admitted that the flag protected the cargo—implying that enemy goods, with the exception of contraband of war, could be safely shipped by neutrals—at the beginning of the Seven Years' War the British navy and British privateers started to capture and condemn enemy goods on neutral vessels.

Once war was declared in 1793, the French Convention encouraged its citizens to fit out privateers, and it offered some incentives; for instance, it temporarily gave up its own right to a share in the prize money.[14] Given the disruption of the French navy in the early 1790s, due to the emigration of many of its aristocratic officers, privateering seemed potentially the most effective means to affect enemy shipping and trade.[15]

France enjoyed a long tradition of privateering. Many sailors had served on French privateers during the American war of independence and had acquired solid experience. Although the golden age of French privateering had occurred under Louis XIV and the glorious times of Jean Bart, eighteenth-century conflicts maintained both the memory and the practice of privateering across generations.[16] As the war began, the Convention could therefore easily tap existing representations of brave privateers: "Merchant navy! Under the reign of despotism . . . you gave birth to Jean Bart, Duquesne and Duguay-Trouin; what will you not be able to do under the realm of Equality!"[17]

French shipowners were very responsive, in particular because prizes were welcome in a time of severe food shortages. Over five months Dunkirk fitted out forty-nine privateers, Marseille twenty-seven, Saint-Malo twenty-three, Bayonne fifteen, Boulogne thirteen, Bordeaux twelve, and Nantes ten; many other metropolitan ports sent out a few additional privateers.[18] Furthermore, privateers under French colors were fitted out in the French West Indies, as well as in neutral ports, both in Europe and in the United States. This factor occasionally provoked diplomatic tensions, as in the case of French plenipotentiary minister Edmond-Charles Genêt at Charleston, South Carolina, who was a bit too eager in his support of privateers.[19]

On 22 June 1793, a governmental decision to stop all privateering rapidly chilled the sudden French enthusiasm for commerce raiding. This decision was taken because the French navy badly needed experienced seamen; both activities tapped the same workforce of trained sailors, and the navy's needs were even more dramatic than before, because civil war in France had made it impossible to recruit sailors in some parts of the country.[20] This government embargo on privateering lasted until 15 August 1795.

French Privateering against Neutral Ships

French privateering revived again under the Directory (1795–99), with 1797 and 1798 representing the peak years. In January 1798 (on 19 Nivôse, Year VI, in the revolutionary calendar), France adopted legislation affecting neutral trade that increased opportunities to take prizes. Besides enemy ships, French privateers were now authorized to raid neutral ships carrying enemy goods. The novelty was that not only the cargo but the neutral ship itself was now a legally valid prize.[21] An essential factor determining the willingness of investors to fit out privateers was the extent to which ships and goods could legally be seized. Whenever legislation extended the boundaries of potential prey, the number of privateers increased.

This legislation had profound consequences on neutral shipping. Between July 1796 and June 1797 French privateers captured 316 American vessels, mostly in the West Indies.[22] This situation worsened considerably in 1798, and the Quasi-War against the United States represented a golden opportunity for French privateers to raid the booming American shipping industry.[23] The reestablishment of diplomatic relations between the two countries after the Mortefontaine Treaty in 1800 did not stop predations on neutral ships, and French courts condemned American vessels as good prizes throughout the French Wars. Ulane Bonnel lists 1,434 American prizes from 1797 to 1813, but this number is almost certainly an underestimate.[24] Other neutral merchant ships experienced a similar fate. The list of cases judged by the Prize Council in Paris contains an impressive variety of flags, some of them of entities—like the independent small states of Papenburg, Kniphausen, and Oldenburg—that had never possessed important merchant fleets before the conflict, a fact that reflected the massive increase of neutral shipping in time of war.[25]

Judgment of the Prizes

The French government, besides regulating privateering by issuing letters of marque, judged the legitimacy of prizes taken. Since the end of the seventeenth century and as recently as the war of American independence, this judgment had been made by the Admiralty or by an extraordinary commission called the Conseil des Prises, or Prize Council, established at the beginning of each conflict. The Revolution suppressed the Admiralty, and the last Conseil des Prises, created in 1778, had been dissolved in 1788. In 1793 and again in 1795, after some fluctuation during the Terror, authority to judge maritime prizes was conferred on the Tribunals of Commerce, which existed in all major French cities.[26] These courts were composed of local merchants and shipowners, who were likely to possess shares in privateers. When a prize arrived in a foreign port, the decision as to its legitimacy was taken by the French consul or, between 1796 and 1800, by the consulate chancery. In both cases, many abuses were reported. Departmental courts in France handled appeals.

A series of peace treaties in Europe and with the United States followed Bonaparte's coup d'état in November 1799. The immediate result was a decline in the number of potential prizes. Napoleon also introduced a major change in jurisdiction: by a decree of 27 March 1800 the decision on the legitimacy of neutral prizes was given to a Conseil de Prises in Paris, whereas local commissions were in charge only of enemy ships, when the prize was not contested by the captured captain. Tribunals of Commerce oversaw locally the accounts of privateers' campaigns and the distribution of profits and losses according to the law. They also ruled on conflicts between managing owners and crews.[27]

Managing Privateering

Privateers were fitted out by managing owners, who handled all the organizational, administrative, and financial issues. First, they acquired the ships, which could be newly built French vessels, former prizes, or secondhand French or foreign ships; after 1803, privateers had to be French-built by law, and the use of foreign-built ships required special permission. Patrick Crowhurst has clearly shown the wide variety of types and sizes of privateers, by exploiting the records of hundreds of French privateers captured by the British and condemned by the High Court of Admiralty. Whereas in the eastern half of the English Channel, privateers were mostly below twenty tons, on the high seas they were generally over eighty tons and could be as big as five or six hundred tons.[28] The average was less than a hundred tons, however, which was the case for the 327 privateers fitted out in Saint-Malo.[29] The choice of vessel depended on the kind of campaign and the intended targets.

After acquiring ships, managing owners equipped them with cannons, small arms, and all the victuals that were required for their cruises. They completed the necessary administrative steps to obtain letters of marque, which authorized the captains to raid for six months; eventually letters were issued for up to twenty-four months. Managing owners also hired captains and crews, which consisted mostly of men aged between eighteen and twenty-eight—conscripted seamen, novices, foreign sailors, and untrained volunteers. An extremely high crew ratio of one man per ton, comparable only to vessels bound for Newfoundland's on-shore fisheries, was common on privateers.

Many neutral seamen carried into French ports as part of prize crews or arriving there at the end of merchant voyages were tempted by potentially high profits to embark on French privateers. Numerous American sailors, for instance, served on French privateers during the Quasi-War and often ended up raiding American ships.[30] As a result, most "French privateers were manned by a polyglot crew drawn from throughout the French coastal ports as well as Northern Europe, the Mediterranean and even North America."[31] Crew members might receive a fixed salary. They were also granted a part of the net profit of the cruise, but this system could oblige them to help cover losses as well. Crew members who became disabled during a cruise were entitled to half pay for the rest of their lives, as were seamen in the navy.[32] The French government received 5 percent of the proceeds of the privateers' campaigns for this purpose.

Fitting out privateers was normally financed by shares, as in other risky maritime activities, such as insurance or the slave trade. Shareholders were liable only for the amounts of their shares, whereas the responsibility of the managing owner was unlimited. The value of the shares was generally between a thousand and five thousand francs (F 5 = U.S.$1 at that time), but it was eventually possible to acquire a half or a quarter share, so that even small merchants and people who did not belong to the commercial world could invest in privateering. Shares were eventually sold outside the city in which a privateer was fitted out, notably in Paris.

The managing owner generally acquired a controlling percentage of the shares. Daniel Lacombe, for instance, owned half the shares of *Venus,* fitted out in Bordeaux in 1803.[33] The master was often a shareholder, a factor that likely increased his zeal: Captain Laveille owned fifty thousand francs in *Psyché,* which he commanded, just as much as did "Balguerie junior," a merchant of Bordeaux, whereas Louis Chaurand, the managing owner in Nantes, where the ship was actually fitted out, held only five thousand francs on this privateer.[34] Some managing owners fitted out only one privateer or two in their entire careers, while others did so systematically. The Bastarrèche brothers of Bayonne, for instance, fitted out twenty privateers from 1793 to 1813, and Jacques Conte of Bordeaux fitted out thirty-two privateering campaigns from 1796 to 1815, representing 15 percent of all the privateering based in Bordeaux.[35]

The managing owner kept accounts of expenses and receipts, which were validated by the Tribunal of Commerce and produced to shareholders. Receipts included both sales of prizes and ransoms obtained for captains of boarded ships. It was legal, and sometimes considered convenient by the captains of both the privateer and the boarded ship, to conclude a mutual agreement by which the prize continued its voyage and paid a fixed sum upon arrival. Hostages would be taken on board the privateer in the meantime, and the legitimacy of the prize and the ransom would be later judged by the court.[36]

After deducting the costs of fitting out the privateer and other expenses—such as repairs, legal costs, other operating outlays, the part due to the French government, and the commission for the managing owner (generally 2 percent)—the profits were shared between shareholders and the crew, who received (as groups) two-thirds and one-third, respectively. Whereas shareholder profit was divided proportionally to the number of shares owned, crew profits were allocated among crew members according to a predetermined scale. For example, the captain might receive twelve parts, a sublieutenant four, regular seamen one and a half, a cabin boy one, etc.[37]

The cost of fitting out a privateer varied considerably, according to the size of the ship and the crew. Both depended on the areas in which they meant to operate and the kinds of ships they intended to capture. French privateers raided in the West Indies, in European waters, in the Mediterranean, and in the Indian Ocean.[38] While some campaigns at sea lasted a number of months, in other instances privateers made campaigns of only two or three days before sailing back to their home ports.

Was Privateering Profitable?

The impact of privateering was multifaceted. Economically, the activity affected the home ports and home countries of privateers, as well as enemy and neutral shipping and trade interests. Socially, it offered opportunities for rapid upward mobility.[39] Strategically, it forced the enemy to adopt measures to counter the threat to its merchant ships. Diplomatically, it could profoundly affect international relations.

To assess the economic impact of privateering in this period it would be necessary to know, among other things, the total number of French privateers fitted out throughout the French Wars, the number of ships they took, and the profitability of their cruises. Unfortunately, the relevant data are scattered and unsatisfactory.[40] Patrick Villiers has computed that 1,542 letters of marque were issued in France from 1803 to 1814, but as he does not discuss sources, it is unclear whether privateers fitted out outside French metropolitan ports are included in this total. According to Villiers, French ships took 5,600 prizes from 1793 to 1802 and 5,500 during the Napoleonic Wars, but it is again unclear whether prizes carried and judged in colonial and foreign ports are included in these figures.[41] Villiers also does not take into account prizes that were recaptured at sea, a phenomenon that became more frequent over time and obviously heavily influenced the profitability of privateering.[42] According to a Parisian merchant-house report from 1807, three prize ships out of four were freed before reaching the ports to which privateers intended them to sail.[43] Contemporaries also believed that some of the prize seizures were faked, organized jointly by French and British merchants to circumvent the prohibition on importation of British goods into France.

Available data show that the geography of ports fitting out privateers during the French Wars changed from earlier wars. Privateering was no longer confined to ports with strong traditions of privateering. Certainly, these ports were still active, although Boulogne supplanted Dunkirk in the Napoleonic Wars.[44] But cities like Bordeaux, which had fitted out few privateers in the past conflicts, organized 209 campaigns during the French Wars;[45] Cherbourg, which also had no strong tradition, fitted out a total of eighty-seven privateers.[46] The latter benefited from the tight British blockade of major ports in the Channel, whereas in ports like Bordeaux merchants had been used to investing in the colonial and slave trades, which they could no longer pursue, and were looking for profitable alternatives to replace them. Merchants in French colonies were extremely active as well. On Mauritius (a French colony until 1810) local merchants fitted out a total of 122 privateers;[47] Guadeloupe's privateers infested the West Indies, with 114 privateers fitted out in 1797–98 alone. This island had replaced Martinique as the core of colonial privateering after Martinique fell into British hands in 1794.[48]

Assessing the profitability of French privateering is an extremely difficult task. It is moreover doubtful whether an overall national average would be significant, as variability was extremely high. Shareholders generally invested in different privateers to reduce risk, but the results were unpredictable. They ranged from the capture of privateers immediately after departure to successful six-month cruises producing a number of extremely valuable prizes. Many expeditions produced losses, and others ended with hardly any profit despite taking some prizes, but shareholders occasionally had extremely high returns: Bordeaux shipowner Daniel Lacombe lost 81 percent on *Gascon,* which was fitted out in Lorient and captured during its first cruise in 1809, but he received a 759.5 percent return on *Rôdeur,* which was fitted out at Passages in 1808.[49]

Privateering could be an important factor in social mobility. The former slaver captain Robert Surcouf of Saint-Malo made a fortune as a privateer in the Indian Ocean in the 1790s. His privateering campaign in 1800 on board *Confiance,* with a 130-man crew, led to the capture of nine British ships, including the 1,200-ton East Indiaman *Kent*, with 440 men.[50] Captain Delattre in Dunkirk took sixty-five prizes between 1793 and 1805, which produced seven million francs.[51] Jacques Conte in Bordeaux earned approximately 1.2 million francs in commissions on the privateers he fitted out, and this amount does not include his profit as a shareholder.[52] But these were exceptional exploits, and many privateers were captured by the British before taking any prizes at all. According to Patrick Villiers, this happened to about half of the privateers, with an average of seven of ten Bordeaux privateers taken.[53] Under these circumstances, people either invested in or joined privateers as crew members because fabulous profits could change their lives, if they were among the lucky ones.

British Countermeasures against French Privateering

As time passed, it became increasingly difficult for French privateers to avoid being taken by the British navy. The main factor affecting profitability, however, was less the overall effectiveness of the enemy's counterstrategies than the ability of privateers to take one or more valuable prizes before being captured. For this reason, it was in Britain's long-term interest to retain all captured privateers as prisoners, so that they could neither transmit their knowledge to a new generation of privateers nor embark anew themselves.

If privateering was a lottery in terms of profits for shareholders and crew, counterstrategies had a major impact on the fate of thousands of sailors. From 1793 to 1800 the British took 743 French privateers and over 26,500 seamen; 130 privateers and 7,094 men were taken in 1797 alone.[54] The British captured more than forty thousand seamen from French ships (see the figure). The common sailors were kept for many years in prisons, or moored ships used as prisons, where health conditions were extremely poor. Officers, who were usually of higher rank socially, were mostly entitled to parole.[55] Apart

Seamen Captured on French Privateers, 1793–1813

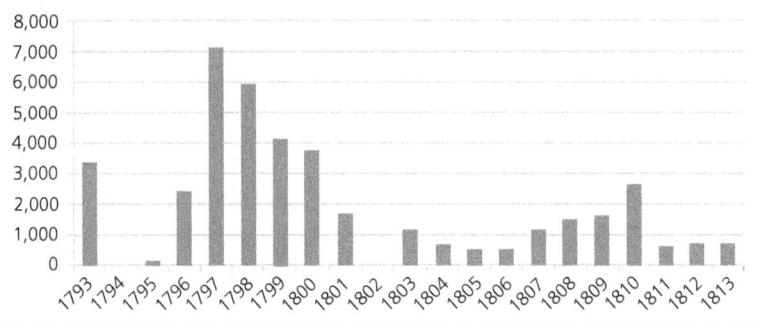

Source: data provided from Crowhurst, *French War on Trade*, app., pp. 207–209.

from humanitarian consideration for the fates of these prisoners and the distress of their families, the temporary loss of thousands of seamen affected the French navy. This strategy had a cost for Britain too—the estimated expense of holding French prisoners between 1803 and 1815 was six million pounds.[56]

Although the effectiveness of British defense at sea increased over time, the rapid rise in the number of French prisoners in Britain roughly reflects the growth trends of French privateering itself, which prospered in the late 1790s and subsequently declined. It is difficult to ascertain, however, whether this decline depended on the decreasing profits of the activity—as some contemporary sources suggest—or on the existence of lucrative alternatives for investors, notably through recourse to neutral trade before 1807, and the revival of maritime trade under the system of licenses after 1810.[57]

Whatever the reasons, French privateering did not significantly affect British trade over the long run. Thousands of British ships were taken as prizes, but predation amounted to only 2 to 2.5 percent at most, a figure that on the average doubled the normal losses at sea to other factors.[58] From 1796 to 1814 British imports doubled, exports grew by half, and reexports increased threefold.[59] Nevertheless, locally and temporarily, privateers could affect costs: Lloyd's insurance rates for shipping to the Lesser Antilles increased by from 10 percent to 33 percent from 1794 to 1797 because of the effectiveness of privateers from Guadeloupe.[60]

To protect its shipping, Britain adopted a set of countermeasures. On the defensive side, the traditional wartime policy of convoying was reestablished in 1793, became compulsory for most British trade in 1798, and was resumed in 1803 after the termination of the Peace of Amiens. West Indian as well as Baltic convoys could be as large as eight hundred or a thousand ships, many of which did not sail under the British flag. Other effective means of protecting British ships were patrolling the Atlantic coasts and blockading French ports, notably Brest, thus preventing French privateers (and the French navy) from getting out and prizes from being sent into port.[61]

British patrols were indeed effective, and it became increasingly difficult over time to cross the Atlantic without running into British privateers or frigates. *Benjamin Morgan*, a 291-ton merchant ship fitted out in Bordeaux under American colors, left France on 5 August 1803. The ship was captured the next day by the British frigate *Diamond*. On its way as a prize to Plymouth, however, it was freed by the French privateer *Adventure*, and its captain proceeded to Philadelphia, arriving on 30 September. Before its safe arrival, however, the ship was visited by three different British privateers and was eventually stopped by the French privateer *Alert*, which discharged two American sailors whom it had on board.[62]

Defensive and offensive policies to counter French privateering did incur costs to the British government. Importantly, N. A. M. Rodger calls these added costs "the principal achievement of the French war on trade."[63] Still, while these costs possibly contributed to the difficulties occasionally experienced by the British economy during the French Wars, they did not result in its collapse.[64]

Meanwhile, French privateers negatively affected neutral shipping as well, eventually more than British shipping. Only 27 percent of prizes taken into Guadeloupe in 1799–1801 were British; in fact, 58 percent sailed under the U.S. flag.[65] Neutral ships were often visited by belligerents at sea and were occasionally plundered;[66] if captured, ships (with their cargoes) were eventually ransomed or taken as prizes. However strong the levy on neutral shipping was, however, it did not prevent neutral powers from gaining enormously from wartime profiteering.[67] Even during the Quasi-War, when the French massively attacked American shipping, privateering did not even slow the spectacular growth rate of American trade.[68]

Neutral ships could protect themselves against privateering by, for instance, joining a British convoy, which liberally protected them.[69] While this might defend them from attacks during one voyage, however, it made them subject thereafter to French seizure. Another common means of protection consisted in double papers or forged papers and in false itineraries. Whereas direct trade between two enemy ports on a neutral ship made it vulnerable to privateering, even a short call in a neutral port increased the chances of a neutral ship boarded at sea being allowed to continue its voyage unharmed.

The existence of countermeasures and the absence of significant impact on global trade aggregates should not imply that privateering had no effect at all on shipowners, merchants, and seamen. French privateering was of major concern to British shipowners. Moreover, privateering was a constant source of diplomatic tension, in particular with the United States. The countermeasures adopted by British merchants and their massive recourse to neutral shipping ultimately led Napoleon to adopt strict measures to halt neutral trade.[70]

Conclusions

This brief analysis of French privateering during the French Wars has shown the complex nature of the phenomenon and the wide variety of approaches adopted by belligerent and neutral countries to halt it. Although France's privateers ultimately proved ineffective in depriving the Royal Navy of seamen, causing major financial losses, or undermining the British national economy, French privateering was far from a marginal activity.

During this two-decade period, French legal predation at sea affected individuals, cities, national economies, and international relations. Hundreds of shareholders invested in privateering, and thousands of sailors and their families depended on it for their livelihoods. For all the crews of neutral and belligerent merchant ships, as well as for the crews of privateers, these years of conflict were marked by the enormous risk on every voyage of being stopped, searched, and perhaps captured or, if lucky, ransomed. For almost a quarter of a century, therefore, privateering represented a major aspect of maritime life.

Although public criticism regarding the legitimacy and the utility of privateering arose as early as 1792, there is no evidence that privateering was seriously challenged or contested throughout the French Wars. The fact that French shipowners applied (unsuccessfully) for letters of marque even after 1815 seems to prove that the eventual abolition of privateering in 1856 came about as a result of political and juridical discourse rather than from economic factors. In fact, investors who backed privateering ships widely considered predation at sea a thoroughly acceptable way to make potentially enormous profits.

Notes

The thoughts and opinions expressed in this publication are those of the author and are not necessarily those of the U.S. government, the U.S. Navy Department, or the Naval War College.

1. On the French navy, see Martine Acerra and Jean Meyer, *Histoire de la Marine française des origines à nos jours* (Rennes, Fr.: Editions Ouest-France, 1994); for a bibliographical essay, Hervé Couteau-Bégarie, *L'histoire maritime en France* (Paris: Economica, 1995).

2. In contrast to previous conflicts, the French navy took a consistent number of prizes during the war of independence and represented a serious competitor for private enterprises: Patrick Villiers, *Marine royale, corsaires et trafic dans l'Atlantique de Louis XIV à Louis XVI* (Dunkirk, Fr.: Société dunkerquoise d'histoire et d'archéologie, 1991), vol. 2, pp. 671–73, 753–54.

3. John A. Lynn, *Giant of the Grand Siècle: The French Army, 1610–1715* (Cambridge, U.K.: Cambridge Univ. Press, 1997).

4. On the contradictions between the principles claimed by the French Revolution and privateering and on the evolution of maritime rights leading to its abolition in 1856, see Florence Le Guellaff, *Armements en course et droit de prises maritimes (1792–1856)* (Nancy, Fr.: Presses Universitaires de Nancy, 1999), pp. 37, 71–195.

5. France did not issue letters of marque in any of the conflicts in which the country was involved after 1815, despite the shipowners' requests; ibid., pp. 131–55. Historians have suggested that France and Great Britain fought a second "Hundred Years' War" between 1689 and 1815: Jean Meyer and John S. Bromley, "La seconde guerre de Cent Ans (1689–1815)," in *Dix siècles d'histoire franco-britannique. De Guillaume le conquérant au marché commun*, ed. Douglas Johnson, François Bédarida, and François Crouzet (Paris: Albin Michel, 1979), pp. 153–90; François Crouzet, "The Second Hundred Years War: Some Reflections," *French History* 10, no. 4 (1996), pp. 432–50.

6. Le Guellaff, *Armements en course et droit de prises maritimes*, pp. 203–205.

7. Napoléon Gallois, *Les corsaires français sous la république et l'Empire*, 2 vols. (Le Mans, Fr.: Julien, Lanier, 1854); Henry Ribadieu, *Histoire maritime de Bordeaux. Aventures des corsaires*

et des grands navigateurs bordelais (Bordeaux, Fr.: Dupuy, 1854). Jean de Maupassant, "Les corsaires à l'exposition de Bordeaux," *Revue Philomatique de Bordeaux* (1907), pp. 358–80; Henry Malo, *Les derniers corsaires. Dunquerque 1715–1815* (Paris: Emile-Paul Frères, 1925). Also Édouard Ducéré, "Les corsaires basques et bayonnais sous la République et l'Empire," *Bulletin de la Société des sciences et arts de Bayonne* (1895), pp. 161–208, 317–63, 489–540; (1896), pp. 1–48; (1897), pp. 25–156, 161–208, 309–44; (1898), pp. 39–79, 85–160 (published in book form [Bayonne, Fr.: A. Lamaignère, 1898], as listed in the bibliography). In other papers, however, de Maupassant examines privateering also from juridical and economic points of view: de Maupassant, "Le corsaire la *Dorade* et l'affaire de la *Juliana*, 1796–1798," *Revue Philomatique de Bordeaux* (1913), pp. 275–88, 340–47, and (1914), pp. 176–86; "Balguerie junior contre Duntzfeld. Affaire de l'*Antoinette*, 1799–1800," *Revue Philomatique de Bordeaux* (1917), pp. 25–36, 65–77, 124–43, 190–99, 227–40, 269–76. Attention to economic aspects and to the importance of the phenomenon is also manifest in Abbé F. Robidou, *Les derniers corsaires malouins. La course sous la Révolution et l'Empire, 1793–1815* (Paris: Oberthür, 1919), and Léon Vignols, "La course maritime: Ses conséquences économiques, sociales et internationales," *Revue d'Histoire Économique et Sociale* (1927), pp. 196–230.

8. See, for instance, Paul Butel, "L'armement en course à Bordeaux sous la Révolution et l'Empire," *Revue Historique de Bordeaux* (1966), pp. 24–64; Nicole Charbonnel, *Commerce et course sous la Révolution et le Consulat à La Rochelle: Autour de deux armateurs—les frères Thomas et Pierre-Antoine Chegaray* (Paris: PUF, 1977); Joseph-Edouard Even, "Les corsaires du bassin de l'Adour sur les côtes septentrionales de l'Espagne, sous la Révolution et l'Empire," special issue, *Revue de Pau et du Béarn* (1987), pp. 177–85.

9. Ulane Bonnel, *La France, les États-Unis et la guerre de Course (1797–1815)* (Paris: Nouvelles Éditions Latines, 1961).

10. The most comprehensive study of French privateering 1793 to 1815 is still Patrick Crowhurst, *The French War on Trade: Privateering, 1793–1815* (London: Scolar, 1989).

11. For a recent bibliographical essay on the state of the art on French privateering, see Sylviane Llinares and Philippe Hrodej, "La mer et la guerre à l'époque moderne," *Revue d'histoire maritime*, nos. 10–11 (2010), pp. 317–39, esp. 329–31.

12. For a detailed analysis of the assistance provided by the French state to privateering, see John S. Bromley, "The Loan of French Naval Vessels to Privateering Enterprises (1688–1713)," in *Les Marines de Guerre européennes, XVIIe–XVIIIe siècles*, ed. Martine Acerra, José Merino, and Jean Meyer (Paris: Presses Universitaires de la Sorbonne, 1985), pp. 65–90. France did not lend any vessels for privateering during the French Wars except in September 1797, and that decision was revoked in December of the same year. See Le Guellaff, *Armements en course et droit de prises maritimes*, pp. 210–11.

13. The French Newfoundland fisheries and slave trade were particularly affected by warfare, as ships were extremely vulnerable to attack. Both Newfoundland ships and slavers could be easily converted into privateers, and their crews were eager to find an alternative occupation.

14. Crowhurst, *French War on Trade*, p. 6.

15. On the French navy during the Revolution and the Empire, see Henri Legohérel, *Histoire de la Marine française* (Paris: PUF, 1999), pp. 67–70; and Martine Acerra and Jean Meyer, *Marines et Révolution* (Paris: Editions Ouest-France, 1988), pp. 149–64.

16. French privateering decreased in intensity throughout the eighteenth century. Saint-Malo fitted out 424 privateers in the Nine Years' War (1688–97) and 425 during the War of the Spanish Succession (1702–13) but only sixty-nine during the Seven Years' War (1756–63) and fifty-two during the war of independence (1778–83); Villiers, *Marine royale*, vol. 1, pp. 131, 145, 316, 351, and vol. 2, p. 661. This was a national trend, although the decline of privateering in Saint-Malo was particularly severe. The only significant exception was Dunkirk, which fitted out 198 privateers during the war of independence, compared to 145 during the Seven Years' War; see ibid. See also Patrick Villiers, *Les corsaires du Littoral de Philippe II à Louis XIV, Boulogne, Calais et Dunkerque 1560–1715* (Villeneuve d'Ascq, Fr.: Presses Universitaires du Septentrion, 1999).

17. Address of the National Convention, 23 February 1793, cited by Ducéré, "Les corsaires basques et bayonnais sous la République et l'Empire" (1895), p. 170.

18. Crowhurst, *French War on Trade*, p. 7. Figures for Bordeaux in Michel Casse, "La course à Bordeaux en 1793. Un état de la question," in *Bordeaux et la Marine de Guerre,* ed. Silvia Marzagalli (Bordeaux, Fr.: Presses Universitaires de Bordeaux, 2002), pp. 125–51. I have taken Casse's figures also for Bayonne, which are higher than those in Crowhurst.

19. In foreign ports, French consuls and official representatives delivered letters of marque to privateers. The excessive zeal of Genêt, who arrived at Charleston in April 1793, became a source of political tension. Genêt organized a tribunal to judge the legitimacy of prizes of French privateers carried to ports in the United States. He was extremely active in recruiting seamen for French and American privateers, to which he delivered letters of marque. His activity was perceived as an attack on American sovereignty, and American shipping was hindered by these privateers, who suspected everybody of carrying British property. President Washington asked for and obtained Genêt's recall. Melvin H. Jackson, *Privateers in Charleston, 1793–1796: An Account of a French Palatinate in South Carolina* (Washington, D.C.: Smithsonian Institution, 1969), pp. 1–7.

20. The recruitment of sailors for the French navy relied on the temporary conscription of seamen. Privateers were allowed to recruit conscripted seamen in a proportion that varied over time; see Le Guellaff, *Armements en course et droit de prises maritimes,* pp. 224–31. In August 1793 conscription for the army hit all men aged between eighteen and fifty, a decision that reduced even more the number of available seamen; see ibid., p. 239.

21. Bonnel, *La France, les États-Unis et la guerre de Course,* pp. 35–41, 91.

22. Anne Clauder, *American Commerce as Affected by the Wars of the French Revolution and Napoleon, 1793–1812* (Clifton, N.J.: A. M. Kelley, 1932 [repr. 1972]), p. 44.

23. On the growth of American shipping, see ibid.; Silvia Marzagalli, "Establishing Transatlantic Trade Networks in Time of War: Bordeaux and the United States, 1793–1815," *Business History Review* 79, no. 4 (2005), pp. 811–44; and Marzagalli, "American Shipping into the Mediterranean during the French Wars: A First Approach," *Research in Maritime History* 44 (2010), pp. 43–62.

24. More than half (approximately 70 percent) of my own list of American ships or crews carried to Bordeaux during the same period is not mentioned in Bonnel's work. See Bonnel, *La France, les États-Unis et la guerre de Course,* pp. 318–407, who has exploited the archives of the Conseil de Prises in Paris. Silvia Marzagalli, *Bordeaux et les États-Unis, 1776–1815: Politique et stratégies négociantes dans la genèse d'un réseau commercial* (Geneva: Droz, forthcoming), draws on a variety of American and French sources: protests in Bordeaux, 1795–1799, Record Group [hereafter RG] 84 (vol. 144), vols. 1–3, National Archives and Records Administration, Washington, D.C. [hereafter NARA]; Register of consular acts, RG 84, n° 36, Bordeaux Consulate; F^{12}1657B, Archives Nationales, Paris [hereafter AN]; and notary records, 3E 13091, 3E 13093, 3E 31369, 3E 35923–35924, Archives Départementales de la Gironde [hereafter AdG]. Greg H. Williams, *The French Assault on American Shipping, 1793–1813: A History and Comprehensive Record of Merchant Marine Loss* (Jefferson, N.C.: McFarland, 2009), provides additional material but lacks scientific rigor and contains many mistakes; it should therefore be used with extreme caution. Even eliminating all ships mentioned by Williams only, almost 40 percent of the prizes contained in my book are absent in Bonnel's work.

25. The list is in Le Guellaff, *Armements en course et droit de prises maritimes,* annex I, III–LVI.

26. From June 1794 to October 1795, the decision was taken by the executive power (notably the Committee of Public Safety), which received the case from local judges of the peace in the ports; see ibid., pp. 408–12. In 1800 the competences of the judges of peace were transferred to the Navy Administration; see ibid., p. 535.

27. Crowhurst, *French War on Trade,* p. 105; Le Guellaff, *Armements en course et droit de prises maritimes,* pp. 440–532. The possibility of appealing against the decision of the Conseil des Prises varied over time.

28. Crowhurst, *French War on Trade,* pp. 46–83.

29. Le Guellaff, *Armements en course et droit de prises maritimes,* p. 208; Robidou, *Les derniers corsaires malouins, passim*.

30. John Troop, master of the bark *Favourite* of New York, arriving at Bordeaux on 19 September 1797, lamented the desertion of three of his men (out of ten) who signed on to the privateers *Bonaparte* and *Furet*. Protest, 5 October 1797, pp. 126–27, RG 84 (vol. 144), vol. 2, NARA.

31. Crowhurst, *French War on Trade*, p. 175.
32. Le Guellaff, *Armements en course et droit de prises maritimes*, pp. 207–43.
33. Fonds Delpit, p. 145, Archives Municipales de Bordeaux [hereafter AMB]. Other examples are in Le Guellaff, *Armements en course et droit de prises maritimes*, p. 276.
34. Fonds Delpit, p. 144, AMB. Most shareholders were from Bordeaux, although the biggest among them was a well-known manager-owner from Bayonne, Bastarrèche.
35. Crowhurst, *French War on Trade*, pp. 162–64, who misspells the name as "Bastérreche"; Michel Casse, "Un armateur en course bordelaise sous la Révolution et l'Empire: Jacques Conte, 1753–1836," in *Bordeaux, porte océane, carrefour européen*, Actes du 50ᵉ congrès de la Fédération Historique du Sud-Ouest (Bordeaux, Fr.: FHSO, 1999), pp. 411–26.
36. On the precise legal frame of ransom, see Le Guellaff, *Armements en course et droit de prises maritimes*, pp. 312–17. In some instances, a prize was neither taken nor ransomed. If the privateer could not take the prize into port or the captured ship was in too bad shape to ransom, the ship was sunk after transferring crew and valuable objects to the privateer.
37. Example provided by Crowhurst, *French War on Trade*, p. 149. In Guadeloupe, the captain received four parts, each seaman one, and each cabin boy a half part; Frédéric Régent, *Esclavage, métissage, liberté. La Révolution française en Guadeloupe, 1789–1802* (Paris: Grasset, 2004), p. 307.
38. According to Patrick Villiers, 16 percent of French privateers were fitted out in the Mediterranean. See Patrick Villiers, *Les corsaires: des origines au traité de Paris du 16 avril 1856* (Paris: J. P. Gisserot, 2007), p. 51.
39. Besides the sudden fortune it offered successful managing owners and captains of privateers, privateering represented a major turning point in the formation of urban societies in the French West Indian colonies and allowed for an unprecedented upward mobility for former slaves and people of color. See Régent, *Esclavage, métissage, liberté*, pp. 309–12.
40. Data on privateers fitted out in seven major French ports from 1793 to 1801 in Acerra and Meyer, *Marines et Révolution*, pp. 265–66, are attributed to Crowhurst, but the latter does not provide all of them, and his data do not match with those provided by Acerra and Meyer. Data for 1793, for instance, are different for five out of seven ports. See Crowhurst, *French War on Trade*, p. 7.
41. Villiers, *Les corsaires: des origines au traité de Paris*, pp. 51, 123. In Guadeloupe alone, 1,239 prizes were judged from 1794 to 1798 and from December 1799 to 1801. Data between October 1798 and November 1799 are missing: Anne Pérotin-Dumon, "Economie corsaire et droit de neutralité. Les ports de la Guadeloupe pendant les guerres révolutionnaires," in *L'Espace Caraïbe, théâtre et enjeu des luttes impériales du XVIIᵉ au XIXᵉ siècle*, ed. Paul Butel and Bernard Lavallé (Bordeaux, Fr.: Maison des Pays Ibériques, 1996), pp. 251–52.
42. Villiers, *Les corsaires: des origines au traité de Paris*, p. 52.
43. Letter of Lefebure & Cie. to the Minister of the Interior, 28 July 1807, F12507, folder 8, AN.
44. Patrick Villiers, "La Guerre de Course," in *Napoléon et la Mer, un rêve d'Empire*, ed. Jean-Marcel Humbert and Bruno Ponsonnet (Paris: Seuil, 2004), p. 135; Villiers, *Les corsaires: des origines au traité de Paris*, p. 51.
45. Butel, "L'armement en course à Bordeaux," pp. 24–64. A list of Bordeaux privateers is in the annex of the otherwise unexploitable narrative of Daniel Binaud, *Les corsaires de Bordeaux et de l'estuaire. 120 ans de guerres sur mer* (Biarritz, Fr.: Atlantica, 1999).
46. Villiers, *Les corsaires: des origines au traité de Paris*, pp. 70–71.
47. Villiers, "La Guerre de Course," p. 134.
48. Pérotin-Dumon, "Economie corsaire et droit de neutralité," p. 252. See also Michel Rodigneaux, *La guerre de course en Guadeloupe, XVIIIᵉ–XIXᵉ siècles, ou Alger sous les Tropiques* (Paris: L'Harmattan, 2006); and Régent, *Esclavage, métissage, liberté*, pp. 306–12.
49. Silvia Marzagalli, *"Les boulevards de la fraude": Le négoce maritime et le Blocus continental, 1806–1813—Bordeaux, Hambourg, Livourne* (Villeneuve d'Ascq, Fr.: Presses Universitaires du Septentrion, 1999), pp. 109–17.
50. On the career of Robert Surcouf and his brother Nicolas, see Auguste Toussaint, *Les frères Surcouf* (Paris: Flammarion, 1979). In 1801, Surcouf settled in Saint-Malo, where he fitted out fourteen privateers during the Napoleonic Wars; he commanded himself a last expedition, to the Indian Ocean, in 1807. On the campaign of the *Confiance*, see Jeanne

Kaeppelin, ed., *Surcouf dans l'océan Indien: Journal de bord de la "Confiance"* (Saint-Malo, Fr.: Cristel, 2007).

51. Villiers, "La Guerre de Course," p. 133.

52. Casse, "Un armateur en course bordelaise sous la Révolution et l'Empire," p. 416.

53. Ibid.; Villiers, "La Guerre de Course," p. 132.

54. Alfred T. Mahan, *The Influence of Sea Power upon History 1660–1783* (London: Sampson Low, 1890), p. 208, citing statements of contemporary British authorities; Crowhurst, *French War on Trade*, pp. 207–208.

55. Details about prisoners in Crowhurst, *French War on Trade*, pp. 173–98. See also Le Guellaff, *Armements en course et droit de prises maritimes*, pp. 576–80; and Patricia K. Crimmin, "Prisoners of War and British Port Communities, 1793–1815," *Northern Mariner / Le marin du nord* 6, no. 4 (October 1996), pp. 17–27.

56. Crimmin, "Prisoners of War and British Port Communities," p. 18.

57. For decreasing profits, "Les expéditions de ce genre, pendant la guerre actuelle, n'ont pas joui d'un succès aussi brillant et le port de Bordeaux a eu à regretter la perte de plusieurs corsaires" [Expeditions of this kind have not enjoyed such a brilliant success during the present war, and Bordeaux regrets the loss of a number of privateers]; Report on the port activities in Bordeaux, first semester 1807, s.d., 8 M 182, AdG.

Privateering could produce high returns even in the late phase of the Napoleonic era. Some examples of high profits are given in Marzagalli, *"Les boulevards de la fraude,"* pp. 109–17; Karine Audran, "Les négoces portuaires bretons sous la révolution et l'Empire. Bilan et Stratégies. Saint-Malo, Morlaix, Brest, Lorient et Nantes, 1789–1815" (unpublished PhD diss., Univ. of Lorient, 2007), vol. 1, p. 254. On the overall context of shipping and on licenses, see François Crouzet, *L'économie britannique et le blocus continental, 1806–1813*, 2nd ed. (Paris: Economica, 1987); Frank Edgar Melvin, *Napoleon's Navigation System: A Study of Trade Control during the Continental Blockade* (New York: D. Appleton, 1919 [repr. 1970]).

58. Mahan, *Influence of Sea Power upon History*, p. 209; the figure has been taken over by Paul M. Kennedy, *The Rise and Fall of British Naval Mastery* (New York: Scribner's, 1976), and cited by Crowhurst, *French War on Trade*, p. 31, who suggests that it is an overestimation. Nicholas Rodger, however, validates this figure from contemporary estimation: N. A. M. Rodger, *The Command of the Ocean: A Naval History of Britain, 1649–1815* (London: Penguin Books, 2005), pp. 559–60.

59. Crowhurst, *French War on Trade*, p. 31.

60. Pérotin-Dumon, "Economie corsaire et droit de neutralité," p. 254.

61. Crowhurst, *French War on Trade*, pp. 69–70.

62. Carl C. Cutler, *Greyhounds of the Sea* (Annapolis, Md.: Naval Institute Press, 1984), pp. 19–20.

63. Rodger, *Command of the Ocean*, p. 560.

64. For a detailed analysis of the British economy in the Napoleonic era, see Crouzet, *L'économie britannique et le blocus continental*.

65. Anne Pérotin-Dumon, *La ville aux îles, la ville dans l'île, Basse-Terre et Pointe-à-Pitre, Guadeloupe, 1650–1815* (Paris: Karthala, 2000), pp. 244–57.

66. Marine protests offer a good source for this aspect of privateering. See, for instance, the protest of the first mate of *Molly* of Georgetown, Captain Stangler, which left Bordeaux for Mauritius in July 1796 and was plundered by the Bordeaux privateer *Adventure* on the way to Hamburg; Consular Register-book for Protests, Declarations, Passports etc., Bordeaux, 21 April 1796–30 December 1797, p. 126, 18 July 1797, RG 76, vol. 2, NARA.

67. On the increase and profits of Swedish shipping in warfare, see Leos Müller, *Consuls, Corsairs, and Commerce: The Swedish Consular Service and Long-Distance Shipping, 1720–1815* (Uppsala, Swed.: Uppsala universitet, 2004).

68. Douglass C. North, "The United States Balance of Payments, 1790–1860," in *Trends in American Economy in the Nineteenth Century* (Princeton, N.J.: Princeton Univ. Press, 1960), pp. 573–627.

69. Rodger, *Command of the Ocean*, p. 559.

70. On French legislation toward neutral trade, see Marzagalli, *"Les boulevards de la fraude,"* pp. 106–109, 204.

Waging Protracted Naval War
U.S. Navy Commerce Raiding during the War of 1812
KEVIN D. MCCRANIE

During the first six months of the War of 1812, the thirty-eight-gun British frigates *Guerriere, Macedonian,* and *Java* struck their colors to frigates of the U.S. Navy. Over the remainder of the war, however, the Americans failed to duplicate such results.[1] A partial explanation stems from the Royal Navy's ability to adapt. Leading the world's largest navy, the Admiralty had significant flexibility that allowed it to dispatch reinforcements to the North American Station. The one ship of the line and five frigates present in June 1812 became by the middle of the next year ten ships of the line and sixteen frigates, as well as one modified ship of the line known as a "razee." By late 1814, the strength on the station had increased to twelve ships of the line, two razees, and twenty-nine frigates.[2] Additional roving squadrons patrolled critical sea lines of communication. Convoys received stronger escorts, and in July 1813 the Admiralty directed its frigate captains to avoid single combat with the largest American frigates, such as *Constitution*.[3]

Though British material strength, as well as choices about deployments and rules of engagement, minimized Royal Navy losses, a leadership change on the American side also contributed. The appointment in the United States of William Jones as Secretary of the Navy in January 1813 led to the development of a new oceanic naval strategy. Jones realized that the United States entirely lacked ships of the line, the battleships of the day, for fleet-on-fleet engagements with the British. Smaller warships were also scarce commodities, so ship-on-ship battles or squadron-sized encounters were also discouraged, in an effort to preserve America's scarce warships while imposing significant costs on the British navy and protracting the naval war.[4]

By stressing single-ship cruises targeting British commerce rather than attempting to meet the Royal Navy in battle, Secretary Jones sought to husband the strength of the U.S. Navy, even while forcing the British navy to sustain costly deployments off the American coast, throughout the North Atlantic, and eventually beyond—a geographically expansive contested zone that proved impossible for the British to reduce as long as the Americans had operational warships. The design, execution, and effects of Jones's strategy, as well as its long-term results, illustrated the U.S. Navy's opportunities and challenges at war with the largest navy in the world.

The American Decision to Adopt Commerce Raiding

Strategic direction for the U.S. Navy emanated from the office of the Secretary of the Navy. Benjamin Stoddert, appointed to this position in 1798, set the precedent of strong leadership. One author has asserted, "Power rested entirely with the Secretary, not only in the technical field of naval construction and equipment but also in the strategic and tactical control of naval operations."[5] Soon after the beginning of the War of 1812, Secretary of the Navy Paul Hamilton, who emphasized squadron, as opposed to single-ship, operations, for "the precious effects which victory will procure," was accused of incompetence.[6] Such charges led to his resignation during the last days of 1812.[7] William Jones was selected as Hamilton's replacement. The new secretary took office during the first days of 1813, and he quickly became known for supporting an oceanic naval strategy that focused on commerce raiding with the object of protracting the war and incurring significant protection costs on the Royal Navy.

President James Madison had selected Jones because his "pursuits and studies have been intimately connected with the objects of the department."[8] Jones had taken up arms against Britain during the American Revolution. Afterward, he commanded merchant ships, became a businessman, and served a term in Congress. In 1801, President Thomas Jefferson offered to appoint him Secretary of the Navy, but Jones declined.[9] This led one of his friends in early 1813 to comment, "I could scarcely believe that you would have been drawn into Public life—knowing how little ambitions [sic] you are in that pursuit."[10] When accepting Madison's offer in January 1813, Jones wrote to the president "that your own and the public confidence far transcends my merit, . . . but the sacred cause in which we are engaged and my confidence indeed attachment to the administration of our Government demands the Sacrifice of every personal consideration."[11] However, Jones noted elsewhere, "the moment peace returns, . . . I shall return to private life and to business."[12]

President Madison, his cabinet, and various members and committees of Congress certainly provided guidance as to what they wished the navy to accomplish, but it was Jones who gave those ideas operational form. The extant correspondence between Jones and Madison is dominated by the situation on the Great Lakes and Lake Champlain, bordering the United States and British Canada.[13] These waters—virtual inland seas, isolated by the rapids on the St. Lawrence River from access by ocean vessels—were the scene of urgent efforts by both sides to build up naval forces. The border between the United

States and British Canada proved the central theater of the war, but Jones saw nonetheless the advantages of protracted oceanic operations. This led Jones to school Madison about what might be accomplished by oceanic operations and how. Jones, for example, explained: "The difference between the Lake and the sea service is that in the former we are compelled to fight them at least man to man and gun to gun while on the ocean five British frigates cannot counteract the depredations of one Sloop of War."[14]

Jones developed a new concept of oceanic operations, which he dispatched to five principal naval officers on 22 February 1813. It predicted that British naval strength on the North American Station would increase during 1813, making it too risky for American warships to operate near the U.S. coast: "Our great inferiority in naval strength, does not permit us to meet them on this ground without hazarding the precious Germ of our national glory." Such a statement served as a warning to American naval officers that battles with the Royal Navy would be discouraged; the dozen operational warships of the U.S. Navy could not defeat a significant portion of the approximately five hundred operational warships of the Royal Navy. Moreover, Jones did not wish to risk the moral advantage obtained from America's 1812 victories over *Guerriere, Macedonian,* and *Java,* as well as the sloops *Frolic* and *Alert.*

Rather than fighting British warships, Jones decided to target maritime commerce: "If any thing can draw the attention of the enemy, from the annoyance of our coast for the protection of his own, rich & exposed Commercial fleets, it will be a course of this nature, & if this effect can be produced, the two fold object, of increasing the pressure upon the enemy & relieving ourselves, will be attained."[15]

To accomplish these objectives, Jones stressed single-ship cruises targeting British commerce while avoiding battle with the Royal Navy. Jones hoped to multiply the strategic effects if the British designed their 1813 operations to counter a continuation of the squadron-sized cruises that had dominated American operations during the first half-year of the war. By defining maritime commerce as Britain's critical vulnerability, Jones planned to use British commerce as bait, forcing the Royal Navy to react in ways both costly and disruptive. In this way, the secretary hoped to husband the strength of the U.S. Navy and disperse that of the Royal Navy.

The Design of British and American Operations

Jones could not have hoped for a more compliant adversary. During the same month he took over as head of the Department of the Navy of the United States, the First Lord of the British Admiralty declared, "It is evident that the Enemy's frigates do not wish to proceed to Sea singly, & we must be prepared accordingly."[16] The British were anticipating squadron-sized American operations similar to those conducted during 1812, and these had to be countered more effectively than in the past. In the words of an Admiralty assessment of December 1812, "The War has now continued some months without any advantage on our parts." The British government demanded from Adm. Sir John Borlase Warren, the commander in chief of the American and West Indian Stations, "more active measures and . . . more successful exertions" against the U.S. Navy. Though the Admiralty was "aware of the great uncertainty of all Naval Operations and of the difficulty of preventing the occasional excursions of an enterprizing Enemy," it expected in the aftermath of America's 1812 naval victories that Warren would restore "the honor of His Majesty's Arms and the preeminence of the Naval Power of the Country."[17]

British naval leaders in London thus saw a need to engage squadron-sized units of the American navy, while Jones planned for single-ship cruises that avoided battles and targeted maritime trade. However, the newly appointed Secretary of the Navy needed his senior leadership to buy into his new strategy. Jones ended a 22 February letter to his principal officers with the following invitation: "Your own ideas of a cruise with this general view will be acceptable to me." This was the beginning of an extensive correspondence with the navy's uniformed leadership. Jones endeavored to work the ideas of senior officers into plans that would support his strategy. For example, Commodore John Rodgers proposed a cruise by his heavy frigate *President* to the Azores and Madeira, since this area served as a rendezvous for scattered British East and West India convoys.[18] *President* would then sail north and operate against commercial shipping around Britain and Ireland before provisioning at a port in Denmark. Subsequently, Rodgers suggested a much farther-ranging cruise, to the East Indies. Jones determined that the first part of the plan would fit his strategy but judged that a cruise into the Indian Ocean was too risky. In the end, he refused to allow Rodgers to sail past the Cape of Good Hope, or at the farthest Mauritius.[19]

Meanwhile, Commodore Stephen Decatur of the heavy frigate *United States* suggested that he sweep up a small British squadron reportedly off the South Carolina coast and then cruise on the route used by British merchant vessels returning to England from the East and West Indies.[20] Jones, however, had intelligence that the Royal Navy squadron off South Carolina had dispersed; instead, therefore, he suggested that Decatur operate solely against British commerce in the West Indies. In contrast to this diplomatic response to a senior officer, Jones had no qualms about ordering the recently promoted Capt. Jacob Jones of the frigate *Macedonian* also to operate in the West Indies and to "compare your ideas [with Decatur's] so as to cruise separately & spread over as great a space as possible."[21]

In another instance, Capt. John Smith of the frigate *Congress* received seven possible cruising options from Jones, who explained, "On the eve of Your departure inform me of the route you contemplate."[22] Smith chose a station along the equator to intercept the

East Indian and South American commerce.[23] As these officers "chose" their cruising grounds, Jones provided more specific orders to his remaining, more junior commanders so as to avoid leaving important regions uncovered. Jones ordered the brig *Argus* to operate around the British Isles, and the commander of the frigate *Chesapeake* received instructions to intercept commerce destined for Canada.[24]

Orders to American naval officers emphasized single-ship cruises that, although sacrificing concentration of force, would compel the British to disperse their warships more widely. As Jones explained, "I have never doubted the effect upon the enemy, would be in proportion to the space covered on the ocean by our cruisers, in those tracks most frequented by his immensely rich, & wide spread commerce."[25] Cruises by American warships into areas the British considered safe from attack could be particularly useful in spurring Britain to costly overreactions.

Commerce raiding by single ships thus promised to increase Britain's wartime expenditures and disrupt other operations, possibly requiring the withdrawal of ships from the U.S. coast. Jones ordered the commanding officer of *Argus* to destroy trade around the British Isles since this "would carry the war home to their direct feelings and interests, and produce an astonishing sensation."[26] For similar reasons he supported operations deep in Canadian waters by the frigate *Chesapeake:* "The enemy will not, in all probability, anticipate our taking this ground [the Gulf of St. Lawrence] with our Public Ships of war."[27]

The targeting of merchant vessels in nearly every corner of the Atlantic had the potential to stretch British naval deployments, but the Royal Navy's strength gave it a considerable degree of elasticity with which to face unanticipated threats. Jones needed additional methods to "in some degree compensate for the great inequality compared with that of the Enemy."[28] One answer was to destroy captured vessels. For financial reasons, naval captains preferred to send captures into friendly ports so that they could be sold as prizes of war, entitling the officers and men of the warship to all or part of the proceeds.[29] Jones's orders explained the drawbacks of this conventional approach: "A Single Cruiser, if ever so successful, can man but a few prizes, and every prize is a serious diminution of her force." By contrast, "a Single Cruiser, destroying every captured Vessel, has . . . the power perhaps, of twenty acting upon pecuniary views alone."[30] Accordingly, "as there is no way of annoying our enemy so effectually as through his Commerce," Jones wrote to one frigate captain, "let devastation be the standing order of your cruize."[31]

The hard fact that the British fleet outnumbered the U.S. Navy by a margin of nearly fifty to one in 1813 made Jones's strategy a risky one, but it was a calculated risk. In the face of such daunting odds, one obvious alternative was to keep the American warships in port and thereby at least tie down the British squadrons that would be needed to guard against a breakout. Yet Jones immediately dismissed such a "fleet in being" strategy: "nothing could more effectually promote his [British] views, than an opportunity of blockading in port our naval force, which one tenth part of the force necessary to watch their motions on the ocean would accomplish."[32] American warships in port, moreover, would be vulnerable to British amphibious operations, as would indeed be demonstrated later in the war. A British operation in Maine involving ground and naval forces during the summer of 1814 resulted in the burning of *Adams,* a flush-decked corvette mounting twenty-seven guns, and the British raid on Washington witnessed the destruction of an American frigate and a ship-sloop. Though the odds against America's warships at sea appeared long, the odds in port were arguably worse.

The Execution of American Strategy

As it was, Jones's single-ship raiding policy soon bore fruit. Two examples occurred during the cruises of *President* and *Congress* during the summer of 1813. The two warships sailed together from Boston, but they soon split up; *President* operated as far north as the Arctic, and *Congress* patrolled along the equator. The frigates' departure from Boston in company confirmed the British assessment that the U.S. Navy would conduct squadron-sized operations, and the Royal Navy reacted accordingly, dispatching forces in pursuit that were larger and costlier than would have been necessary to guard against single-ship commerce raids. Even when it became obvious that the two American warships were operating singly, the British still had to find them, and they disrupted operations from the equator to the Arctic.[33] Of Commodore Rodgers, who commanded *President,* a British newspaper mused sardonically "how flattering it must be to him to learn, that not single ships but squadrons were dispatched after him, and one specifically under the command of *an Admiral.*"[34] Rodgers had every right to assert later that he had caused disruptions to the Royal Navy equivalent to "more than a dozen times the force of a single Frigate."[35]

Broader policy imperatives also drove Jones to accept the risk of oceanic operations so as to obtain moral victories against the stronger power. Overall, the first year of the War of 1812 did not go as well as planned for the United States, but operations at sea resulted in several tactical victories and helped sustain flagging popular support for the conflict. In October 1814 Jones noted that "an increase in force on the Ocean is strongly urged by public writers and by the Legislature."[36] The capture of the British sloop *Epervier* in early 1814 was precisely the sort of achievement Jones thought the country needed: "I like these little events they keep alive the national feeling and produce an effect infinitely beyond their intrinsic importance."[37]

Even so, Jones endeavored to limit the risks to his warships. Writing to one commander in January 1814, he stressed the need to avoid "all unnecessary contact with the Cruisers of the enemy, even with an equal, unless under circumstances that may ensure your triumph without defeating the main object of your Cruise, or jeopardize the safety of the vessel under your Command."[38] Much as a victory over an enemy warship might boost morale, Jones realized that even a successful engagement would almost certainly require the U.S. warship to come into port for repairs. That eventuality would significantly diminish the American presence at sea and thus undercut the larger strategic goal of forcing the British to disperse their effort in the face of widespread commerce raiding operations.

Battles between warships, moreover, were inherently risky; the British might win. Jones cited "the success of the *Argus* [which] . . . in the course of but a few days [of operations against merchant ships in the narrow waters between England and Ireland], was astonishingly great; and had the gallant spirit of Captain [William H.] Allen, but submitted to the restraint of his excellent judgment, he would have rendered more essential service to his country, perhaps, than any single vessel ever did."[39] Instead, Allen had chanced an engagement with a British brig and lost. In another case, Jones wrote of Capt. James Lawrence's decision to bring the frigate *Chesapeake* to action with the British frigate *Shannon*:

> Whilst the gallant spirit and high minded character of our Naval Officers justly excites the national admiration, their zealous devotion to the cause and honour of their country must be tempered by judgment and sound policy. The glory we have acquired is too precious to commit to the wiles of an insidious foe. The just and honorable contest in which we are engaged must be directed to the most effectual annoyance of the enemy, not to Naval Chivalry in which the numbers and force of the respective combatants are unequal by example.[40]

Yet Jones knew some warships would not return. Describing the capture in early 1814 of the American *Frolic*, a ship-sloop mounting twenty-two guns, he wrote that "the loss . . . which though much to be regretted is among the casualties of War."[41]

The U.S. Navy was quite small, and ship losses could not readily be made good. There had been no naval vessels under construction at the outbreak of hostilities;[42] many of the additional vessels built or purchased under wartime programs became operational only after their conclusion. To mitigate attrition, Jones slowed the operational tempo of American warships. During the first six months of the war (before Jones's appointment), eleven frigates sailed on extended cruises, but during the remainder of the conflict American frigates embarked on only ten cruises.

Husbanding of resources in the face of the Royal Navy's dominance and the absence of reinforcements for the U.S. Navy only postponed the nearly inevitable capture or destruction of American warships in chance encounters with superior enemy forces, blockades of ports, and amphibious raids into those ports. The British failed to capture a frigate during all of 1812, but increases in British strength and effective Royal Navy responses to American initiatives made oceanic operations by the U.S. Navy increasingly risky as the war progressed. Beginning in 1813, the British turned back or captured half the U.S. frigates that proceeded to sea.[43] Overall, Jones's time as Secretary of the Navy witnessed the loss of two frigates and five smaller oceangoing warships. In addition, the brig *Enterprize* returned to port without most of its guns, because its crew had heaved them overboard in a desperate bid to escape a pursuing British frigate. Now disarmed, it was withdrawn from oceanic service for the remainder of the war. Moreover, five additional warships could not sail because of the British blockade, and a sixth had been laid up and the crew sent for service on the lakes on the Canadian border by the end of 1814. That left the American oceangoing navy with only four operational warships. Additional vessels procured under wartime programs would not be available for months to come.[44]

Jones, then, could not protract the war indefinitely. In the face of Britain's overwhelming superiority in oceangoing warships, at best he could extend the oceanic struggle for a finite period. Time was not on his side; numbers began to tell, and the British Navy slowly whittled down the U.S. Navy.

Another problem Jones faced was funding. The government's income paid only for a small percentage of its expenditures during the War of 1812. This forced the Treasury Department to raise money through loans and treasury notes, and the results did not meet expectations.[45] In October 1814 Jones lamented, "With respect to money the Department is truly in the most untoward situation. . . . I am destitute of money in all quarters. Seamen remain unpaid and the recruiting Service is at a stand. I have none for the most urgent contingent purposes."[46] Oceanic operations were extremely costly. Without steady funding, Jones's ability to conduct these endeavors became ever more limited as the war progressed, particularly as 1814 drew to a close.

The Limits of Effectiveness

Faced with this bleak picture, Jones decided to resign as Secretary of the Navy. He had warned President Madison of his decision in late April 1814, offering to make the resignation immediate. His stated reason stemmed from the financial embarrassment resulting from personal debt incurred in a failed commercial venture prior to his appointment. Given the state of the U.S. Navy, an alternative argument could be made that Jones felt the naval war had run its course. The financial weakness of the United States, coupled with the attrition of the navy and the potential for further losses because of British power, led Jones to liken his position to "standing upon Gun Powder with a slow match near it."[47] Getting out in 1814 would keep his reputation intact. One thing is certain—Madison did not want Jones to resign. The president wrote of "the gratification I have experienced in the entire fulfillment of my expectations, large as they were, from your talents & exertions." Eventually, the president and his Secretary of the Navy reached a compromise that Jones would serve until 1 December 1814.[48]

In his last days as secretary, Jones crafted a final set of cruising orders. These were only partially implemented, because of funding problems, the strength of British naval deployments, and the looming termination of the war. The overarching target remained British commerce, with the ultimate object of imposing costs on Britain and its navy. As such, *Constitution* and *Congress* received traditional instructions to operate singly.[49] Jones directed the new large frigate *Guerriere,* captured in 1812, to sail from the Delaware River in company with a schooner that could carry extra supplies and serve as a scout.[50] Plans also called for *President* to sail from New York for Asia, in company with the sloops of war *Hornet* and *Peacock,* as well as store ships, so they could operate for a much longer period than in any previous cruise.[51]

The most innovative feature of these orders involved the sailing of small squadrons to the West Indies and the Mediterranean. Each would consist of five vessels resembling privateers.[52] Such commerce raiders were generally procured, fitted, and manned by private citizens as a type of business venture seeking financial gain. To be financially successful, privateers needed to get their captures into friendly ports, but this was proving ever more difficult, given the strength and deployments of the Royal Navy. Jones realized that the destruction of prizes would allow the five-vessel, privateer-like naval squadrons to continue their missions for longer periods and with greater effect than traditional privateers.

Notwithstanding Jones's innovativeness, it was clear that his oceanic strategy had approached the limits of its effectiveness. Meanwhile, *Congress* was not fully manned; *Guerriere*'s departure from the Delaware would be difficult, perhaps impossible, because of a combination of geography and British blockade; and funding problems slowed the creation of the West Indies and Mediterranean privateer-like squadrons, neither of which sailed before the ratification of the peace treaty.

At sea, additional constraints compounded American difficulties. There were only so many bodies of water where commerce raiding could be successfully carried out. Every time U.S. warships did damage in a region, British responses made it more difficult for the Americans to obtain similar effects again. This increasingly limited Jones's options and forced him to exploit new areas, but those regions that remained by late 1814 could be reached only by long and often dangerous passages.

Assessing the Effectiveness of American Commerce Raiding

It would be all too easy to label Jones's strategy a failure. American warships destroyed few vessels in comparison to the immense size of the British merchant fleet. But considered together with privateers, over which Jones had little or no control, they had a much greater quantifiable impact. One author estimates that American warships captured 165 merchant vessels, compared to 1,344 by American privateers.[53] These numbers should be viewed as implying a ratio rather than as precise in themselves; Faye Kert has argued that it is impossible to determine the exact number of British merchant vessels captured and destroyed by the Americans.[54]

In fact, Jones looked beyond the raw numbers of British merchant vessels captured and understood what historian Jan Glete has concluded: "Even a small American fleet... was able to enforce high protection costs on Britain."[55] In a letter to Commodore Rodgers concerning the latter's 1813 summer cruise, Jones explained, "The effects of your Cruize however is not the less felt by the enemy either in his Commercial or Military Marine, for while you have harassed and enhanced the dangers of the one, you provoked the pursuit & abstracted the attention of the other to an extent perhaps equal to the disproportion of our relative forces."[56] As long as the U.S. Navy survived and followed the strategy laid out by Jones, the Royal Navy had to react, expend precious resources, maintain or increase its deployments, and refine a convoy system that was costly to both merchants and the navy.

The American strategy caused the British considerable irritation. In early 1813, the Secretary of the Admiralty asserted to Admiral Warren, the commander in chief of the American and West Indies Stations, that "their Lordships have, not without inconvenience to other Services, placed under your command a force much greater in proportion than the National Navy of the Enemy opposed to you would seem to warrant."[57] That said, the Admiralty continued to reinforce Warren's command, with the object of minimizing or destroying the effectiveness of the U.S. Navy.[58] The First Lord of the Admiralty warned Warren in June 1813 that "any more naval disasters, more especially if they could fairly be ascribed to want of due precaution, would make a strong impression on the public mind in this Country."[59] Warren, for his part, lamented that single American warships at sea "are such small & Difficult Objects to hit—that our chances are few indeed & the good Fortune of these Rascally privateer Frigates makes me almost Despair of ever seeing them."[60]

The British apparent inability to destroy the U.S. Navy led to criticism of Warren's conduct. A letter to the editor in the influential *Naval Chronicle* argued that Warren "sailed from England with the confidence of the nation—that he will possess it on his return, I greatly doubt. . . . I fear they [the Americans] have shewn, that the British lion is *sound asleep*. . . . It is too certain that *little has been done*, certainly nothing great or worthy of this powerful fleet." The writer, however, believed the naval failures went beyond Warren to include "the apathy and supineness of the B[oard] of A[dmiralty]"—he indicted the Admiralty as "novices."[61]

Yet the Admiralty had an unenviable task of balancing deployments. Until early 1814 Britain faced Napoleonic France. This war had continued, with one short respite, since 1793. The French navy was much larger than the American and included numerous ships of the line, of which the Americans had none operational before the termination of the War of 1812. Moreover, the geographic distinctiveness between the American and French theaters of operations added to the difficulty. The War of 1812 forced the Admiralty to alter its worldwide naval deployments. Forces in North American waters multiplied from twenty-three warships in mid-1812 to 120 in late 1814. The number of ships of the line increased, as noted above, from one to twelve, along with two razees, and frigate strength increased from five to twenty-nine.

The Leeward Islands, Jamaica, and South American Stations continued to demand large squadrons, including some of Britain's best warships. Small squadrons routinely patrolled the busy sea-lanes around the Azores, Madeira, and the Canaries searching for American commerce raiders and protecting British convoys. Around the British Isles, the threat of American naval operations forced the Admiralty to maintain significant deployments even after the defeat of Napoleon. However, deployments shifted from the English Channel to its southern approaches, the coast of Ireland, and from Scotland to the north, so as better to cover the arrival of convoys.[62]

Convoys themselves received stronger escorts. In 1812, convoys from the West Indies to England routinely sailed under the escort of single frigates. These single warships often became squadrons later in the War of 1812. For example, the escorts protecting West India convoys generally grew to include a line-of-battle ship, a frigate, and at least two sloops. In May 1814 the Secretary of the Admiralty explained, "Each convoy therefore equaled in force the whole American navy; the consequence of which was, that not a single merchant-ship had been taken which sailed under convoy, and that no convoy had been at all disturbed, except by weather."[63]

British naval deployments thus minimized commercial losses but meant that the Royal Navy had to maintain a large fleet, including many ships of the line and large frigates, even after the termination of hostilities with Napoleonic France in 1814. Had the British been able to destroy the U.S. Navy or force the United States to rely solely on privateers—which were smaller in size than warships and less apt to fight when brought to bay—the Royal Navy could likely have economized more, decommissioning a greater number of ships of the line and frigates, which were both manpower intensive and costly to operate.

Instead, the possibility of facing the powerful frigates and sloops of the U.S. Navy continued to require an expensive commitment for Britain. This can be seen in the operational strength of the British fleet. During 1812 and 1813, the Royal Navy deployed slightly over five hundred warships. The fall of Napoleon in the spring of 1814 should have resulted in a major drawdown. Though operational strength did decline, Royal Navy deployments in late 1814 still totaled approximately 350 warships, of which thirty-three were ships of the line or razees and eighty-three were frigates. To be sure, the British maintained squadrons in places that had little to do with the War of 1812, like the Mediterranean and the East Indies, but operations relating to the conflict with the United States still accounted for more than half of Britain's warships in late 1814.[64]

Conclusions

Beginning in January 1813, Secretary of the Navy William Jones crafted the U.S. Navy's oceanic strategy to include a greater emphasis on commerce raiding. Moreover, Jones was the operational planner who designed the cruising orders for the navy, thus creating the tangible element of the strategy at sea.[65] According to all available evidence, President Madison largely left oceanic strategy and the conduct of operations in the hands of Jones, who had the great advantage of understanding the intricacies and vulnerabilities of the global maritime commercial system. Jones was therefore the driving force in protracting an oceanic naval war that provided reasonable dividends to the United States at a considerable cost to the British navy. This was quite an achievement in the face of overwhelming British maritime power.

Jones's commerce raiding strategy inflicted significant costs on the Royal Navy by creating a festering irritation that could not be eliminated before hostilities concluded. To sustain this irregular naval war, Jones explained, "The species of force called for is undoubtedly well calculated to annoy the enemy and in order to meet the wishes which have been expressed on the subject by the President and in accordance with my ardent desire to employ every possible means of annoyance against the enemy."[66]

Aiming for mere annoyance allowed Jones to create a strategy that could prolong the war and make it increasingly expensive for the British. As long as the war at sea against the Americans wore on, the Royal Navy had to remain on a war footing, and British merchants had to adhere to convoy regulations rather than return to more efficient and less costly peacetime practices. Peace was greatly desired among the British political public, which had endured more than two decades of war with revolutionary and then

Napoleonic France. This put greater diplomatic leverage in American hands than would have been the case with another, lesser, naval strategy.

Notes

An earlier version of this research appeared under the title "Waging Protracted Naval War: The Strategic Leadership of Secretary of the U.S. Navy William Jones in the War of 1812," *Northern Mariner / Le marin du nord* 21 (April 2011), pp. 143–57.

The thoughts and opinions expressed in this publication are those of the author and are not necessarily those of the U.S. government, the U.S. Navy Department, or the Naval War College.

1. The largest ship captured by the Americans after 1 January 1813 was *Cyane*, rated for twenty-two guns but mounting thirty-two. It was normal for warships of the period to mount extra guns. Details for *Cyane, Guerriere, Macedonian,* and *Java*, Admiralty Papers [hereafter ADM] 7/556, Public Record Office / The National Archives, Kew, U.K. [hereafter PRO/TNA]; Brian Lavery, *Nelson's Navy: The Ships, Men and Organization, 1793–1815,* rev. ed. (1989; repr. Annapolis, Md.: Naval Institute Press, 1997), pp. 40, 81–83; William James, *Naval Occurrences of the War of 1812* (1817; repr. London: Conway Maritime, 2004), pp. 11–12.

2. Ships in Sea Pay, 1 July 1812, 1813, ADM 8/100, PRO/TNA; Admiralty Board Minutes, late 1814, ADM 7/266, PRO/TNA. To make these calculations, frigates are warships rated from thirty-two to forty-four guns; ships of the line were sixty-four-gun ships and larger; and "razees" were ships of the line from which part of their armament had been removed. Razees mounted fifty-seven or fifty-eight guns and were a response to America's large frigates, such as *Constitution*.

3. Croker to Warren, 10 February 1813, ADM 2/1376/73–87, PRO/TNA; Croker to the Several Commanders in Chief . . . , 10 July 1813, ADM 2/1377/154–56, PRO/TNA.

4. This chapter is an adaptation of a paper presented at the 2008 Society of Military History Conference in Ogden, Utah. More recently, Stephen Budiansky, who recognized the earlier work of this author, has explored aspects of Secretary Jones's role in the War of 1812 on the high seas; see his "Giant Killer," *Military History Quarterly* (Spring 2009), pp. 50–60.

5. Howard I. Chapelle, *The History of the American Sailing Navy: The Ships and Their Development* (New York: Norton, 1949; repr. n.d.), p. 177. A contradictory interpretation has been posited by Edward K. Eckert, who asserts, "One thing is certain; no case can be made for Jones' [as Secretary of the Navy] directing the nation's naval strategy"; see Eckert, *Navy Department in the War of 1812* (Gainesville: Univ. of Florida Press, 1973), p. 74.

6. Hamilton to Rodgers, 22 June 1812, Letters from Captains to the Secretary of the Navy, No. 125 [hereafter No. 125], reel 23/58, Record Group [hereafter RG] 45, National Archives and Records Administration, Washington, D.C. [hereafter NARA].

7. Christopher McKee, *A Gentlemanly and Honorable Profession: The Creation of the U.S. Naval Officer Corps, 1794–1815* (Annapolis, Md.: Naval Institute Press, 1991), pp. 10–11.

8. Jones to Eleanor Jones (his wife), 23 January 1813, in *The Naval War of 1812: A Documentary History,* ed. William S. Dudley and Michael J. Crawford (Washington, D.C.: Naval Historical Center, 1992), vol. 2, pp. 34–35.

9. Jefferson to Jones, 16 March 1801, and Jones to Jefferson, 20 March 1801, Thomas Jefferson Papers, series 1, Library of Congress, Washington, D.C. [hereafter LC]; John K. Mahon, *The War of 1812* (Gainesville: Univ. of Florida Press, 1972), pp. 103–104. The most complete description of William Jones while Secretary of the Navy can be found in Eckert, *Navy Department in the War of 1812.*

10. Bainbridge to Jones, 1 March 1813, Papers of William Jones, U. C. Smith Collection, Historical Society of Pennsylvania, Philadelphia [hereafter HSP].

11. Jones to Madison, 14 January 1813, James Madison Papers [hereafter JMP], series 1, reel 14, LC.

12. Jones to William Young, 11 April 1813, Papers of William Jones, HSP.

13. Jones to Madison, Madison to Jones, various dates, JMP, series 1, reels 15–16, LC.
14. Jones to Madison, 26 October 1814, JMP, series 1, reel 16, LC.
15. Circular letter, Jones to Rodgers, Decatur, Bainbridge, Stewart, Morris, 22 February 1813, Letters from the Secretary of the Navy to Naval Officers, No. 149 [hereafter No. 149], reel 10/266, p. 77, RG 45, NARA. For the strengths of the two navies, see Ships in Sea Pay, 1 January 1813, ADM 8/100, PRO/TNA; "Ships of the United States Navy, Winter 1811," in *The New American State Papers: Naval Affairs,* ed. K. Jack Bauer (Wilmington, Del.: Scholarly Resources, 1981), vol. 1, p. 71. The 1811 figure for the U.S. Navy works since the Americans failed to commission any oceangoing warships between late 1811 and early 1813. The remainder after removal of the losses sustained in the war's first months—including *Wasp, Nautilus, Vixen,* and *Viper*—provides a snapshot of the navy.
16. Melville to Warren, 9 January 1813, Warren Papers, WAR/82/41–45, National Maritime Museum, Greenwich, U.K. [hereafter NMM].
17. Croker to Warren, 2 December 1812, ADM 2/1107/346–51, PRO/TNA.
18. Henry Veitch (Consul Madeira) to Croker, 2 October 1813, ADM 1/3845, PRO/TNA.
19. Rodgers to Jones, 22 April 1813, No. 125, reel 28/28, NARA; Jones to Rodgers, 29 April 1813, Confidential Letters of the Secretary of the Navy [hereafter CL], p. 14, RG 45, NARA.
20. Decatur to Jones, 10 March 1813, No. 125, reel 27/31, NARA.
21. Jones to Decatur, 15, 17 March 1813, and Jones to Jacob Jones, 17 March 1813, No. 149, reel 10/304–309, NARA.
22. Jones to Smith, 21 March 1813, No. 149, reel 10/314–16, NARA.
23. Letter from an officer of *Congress,* 12 December 1813, printed in *(Boston) Repertory,* 16 December 1813.
24. Jones to Evans, 6 May 1813, and Jones to Allen, 5 June 1813, CL, pp. 19–22, 29–31, NARA.
25. Jones to Smith, 21 March 1813.
26. Jones to Allen, 5 June 1813.
27. Jones to Evans, 6 May 1813.
28. Jones to Parker, 8 December 1813, in *Naval War of 1812,* ed. Dudley and Crawford, vol. 2, pp. 294–96.
29. McKee, *Gentlemanly and Honorable Profession,* p. 341.
30. Jones to Parker, 8 December 1813.
31. Jones to Jacob Jones, 3 May 1813, CL, pp. 16–17, NARA.
32. Jones to Smith, 21 March 1813.
33. Rodgers to Jones, 27 September 1813, No. 125, reel 31/100, NARA; Log of *Congress,* May–December 1813, RG 24, NARA; Admiralty to Charles Paget, 10 July 1813, ADM 2/1377/145–49, PRO/TNA; John Spratt Rainer to Young, 28 August 1813, ADM 1/573/375A, PRO/TNA; Warren to Croker, 16 October 1813, ADM 1/504/223, PRO/TNA; Dixon to Croker, 20 August 1813, ADM 1/21/83, PRO/TNA.
34. *(London) Morning Chronicle,* 13 November 1813.
35. Rodgers to Jones, 27 September 1813.
36. Jones to Madison, 26 October 1814.
37. Jones to Madison, 10 May 1814, JMP, series 1, reel 16, LC.
38. Jones to Joseph Bainbridge, 16 January 1814, CL, pp. 91–93, NARA.
39. Jones to Warrington, 26 February 1814, CL, pp. 102–105, NARA.
40. Jones to the editors of the *National Intelligencer,* 9 June 1813, No. 125, reel 29/12½, NARA.
41. Jones to Joseph Bainbridge, 13 June 1814, No. 149, reel 11/340, NARA. This *Frolic* was an American ship, named to recall the capture in 1812 of the British ship of that name (though it was soon lost again).
42. Peter J. Kastor, "Toward 'the Maritime War Only': The Question of Naval Mobilization, 1811–12," *Journal of Military History* 61 (July 1997), pp. 472–73.
43. This does not include short cruises in 1812, when *Constitution* sailed in July and *Essex* in September. The long cruises of 1812 comprise the following: *President, Congress,* and *United States* sailed in June; *Essex* sailed in July; *Constitution* sailed in August; *President, Congress, United States, Essex,* and *Constitution* sailed in October; and *Chesapeake* sailed in December. The 1813–15 sailings are as follows: *Constellation* in February 1813 (prevented); *President* and *Congress* in April 1813 (to sea);

United States and *Macedonian* in June 1813 (prevented); and *Chesapeake* sailed in June 1813 (captured); *President* and *Constitution* in December 1813 (to sea); *Constitution* in December 1814 (to sea); and *President* in January 1815 (captured).

44. Smith to Jones, 9 June 1814, No. 125, reel 37/42, NARA; Biddle to Jones, 19 November 1814, Letters from Commanders to the Secretary of the Navy, No. 147, reel 5/82, RG 45, NARA; Renshaw to Jones, 18 July 1814, Letters from Officers below the Rank of Commander to the Secretary of the Navy, No. 148, reel 13/22, RG 45, NARA; Jones to Madison, 26 October 1814; Jones, "Report on the State of the Navy, 22 February 1814," in *New American State Papers*, ed. Bauer, vol. 4, pp. 198–201; Paul H. Silverstone, *The Sailing Navy, 1775–1854* (Annapolis, Md.: Naval Institute Press, 2001), pp. 23, 28, 30, 32, 34–36, 39, 46.

45. Donald R. Hickey, *The War of 1812: A Forgotten Conflict* (Urbana: Univ. of Illinois Press, 1990), pp. 247–48.

46. Jones to Madison, 15 October 1814, JMP, series 1, reel 16, LC.

47. Jones to his wife, 6 November 1814, Papers of William Jones, HSP.

48. Madison to Jones, [26] April 1814, and Jones to Madison, 25 April, 11 September 1814, JMP, series 1, reel 16, LC.

49. Jones to Stewart, 29 November 1814, and Jones to Morris, 30 November 1814, CL, pp. 217–20, NARA.

50. Jones to Rodgers, 30 November 1814, CL, pp. 218–19, NARA.

51. Jones to Decatur, 17, 29 November 1814, CL, pp. 210–12, 216–17, NARA.

52. Jones to Perry, Jones to Porter, 30 November 1814, CL, pp. 220–23, NARA.

53. Robert Gardiner, ed., *The Naval War of 1812* (London: Chatham, 1998; repr. London: Caxton, in association with the National Maritime Museum, 2001), p. 28.

54. Faye Kert, "The Fortunes of War: Commercial Warfare and Maritime Risk in the War of 1812," *Northern Mariner* (October 1998), p. 2.

55. Jan Glete, *Navies and Nations: Warships, Navies and State Building in Europe and America, 1500–1860* (Stockholm: Almqvist and Wiksell International, 1993), vol. 2, p. 395.

56. Jones to Rodgers, 4 October 1813, in *Naval War of 1812*, ed. Dudley and Crawford, vol. 2, pp. 254–55.

57. Croker to Warren, 10 February 1813.

58. Ships in Sea Pay, 1 July 1813.

59. Melville to Warren, 4 June 1813, WAR/82/73–77, NMM.

60. Warren to Barrie, 19 January 1814, in *Naval War of 1812*, ed. Dudley and Crawford, vol. 3, pp. 16–17.

61. Letter to the editor of the *Albion*, 16 December 1813, *Naval Chronicle* 31 (January to June 1814), pp. 118–20.

62. Ships in Sea Pay, 1 June 1812, ADM 8/100, PRO/TNA; Admiralty Board Minutes, late 1814, ADM 7/266, PRO/TNA.

63. Statement of Croker, 13 May 1814, House of Commons, in *Parliamentary Debates. From the Year 1803 to the Present Time*, ed. T. C. Hansard (London: HMSO, 1803–20), vol. 27, pp. 869; Croker to Laforey, 9 December 1812, ADM 2/1107/369–71, PRO/TNA.

64. Ships in Sea Pay, 1 July 1812, 1813.

65. This was illustrated when news of the loss of the U.S. brig *Rattlesnake* appeared in the papers. Jones had to inform Madison of the general nature of its cruising order as if this were the first the president had heard of it. Jones to Madison, 30 July 1814, JMP, series 1, reel 16, LC.

66. Jones to Madison, 26 October 1814.

CSS *Alabama* and Confederate Commerce Raiders during the U.S. Civil War

SPENCER C. TUCKER

On 8 February 1861, seven southern states established at Montgomery, Alabama, the Confederate States of America. The American Civil War began on 12 April, when shore batteries at Charleston, South Carolina, opened fire on Union-held Fort Sumter at the entrance to the harbor. At great disadvantage vis-à-vis the North in both population and industrial strength, the South necessarily made the army its military priority. The Confederate president, Jefferson Davis, left the naval war largely to his able Secretary of the Navy, Stephen R. Mallory.

Mallory hoped to secure a few technologically advanced ironclads, break the Union blockade, and attack Northern ports and shipping. He also planned to send out commerce raiders to destroy Union merchant shipping on the high seas. These would not be privateers but regular commissioned naval vessels, operating under established international law. As the war progressed, Mallory shifted to a more defensively oriented approach and increasingly experimented with new methods of warfare, including mines and the submarine, but commerce raiding remained a consistent strategy.

Mallory hoped that a campaign of *guerre de course* would cause serious economic distress in the North, divert U.S. naval assets from the blockade, and pressure Northern businessmen to demand a negotiated end to the war that would grant Southern independence. In the event, while the Confederate campaign against Union commerce did drive up insurance rates and force the North to shift naval assets in an attempt to hunt down the raiders, its principal lasting effect was to initiate the flight of U.S.-flagged merchant ships to foreign registries.

Creating a Confederate Fleet

At the onset of the war, Secretary Mallory had very few ships. Like the Union navy secretary, Gideon Welles, he purchased steamers for conversion into warships; but Mallory had almost no resources available, whether of ships to purchase or facilities for their conversion, let alone for new construction. Only the Tredegar Iron Works (J. R. Anderson & Co.) of Richmond, Virginia, could manufacture entire steam-propulsion systems. Thus the Confederacy never built any cruisers in its ports during the war, although it did manage to construct a great many wooden gunboats and a number of ironclads.

In the months before fighting broke out, Mallory dispatched agents to purchase supplies in the North and in Canada. He also ordered naval representatives to Europe both to purchase ships for conversion to cruisers and to contract for the construction of purpose-built warships. By far the most able of these individuals was James D. Bulloch, a former U.S. Navy lieutenant who had resigned from the service in 1853. Bulloch arrived in Liverpool in June 1861 and by August had placed contracts with British yards for the ships that would become the Confederate cruisers *Alabama* and *Florida*. He and other Confederate agents eventually contracted for eighteen vessels, the best of which were those secured in Britain: *Alabama, Florida, Shenandoah, Chickamauga, Georgia, Rappahannock,* and *Tallahassee*. The other eleven ships became blockade-runners, were sequestered by the British or French governments, or were not completed prior to the end of the war.

With so few ships available at the beginning of the war, the Confederacy first turned to private vessels. On 15 April 1861, following the shelling of Fort Sumter, President Abraham Lincoln declared the existence of an "insurrection" and called for seventy-five thousand Union volunteers. Jefferson Davis responded two days later with a call for letters of marque and reprisal to carry out privateering operations against American merchant shipping. Privateering involved the capture of civilian property by private individuals and thus could not involve destruction of enemy vessels. The Confederate Congress passed, and Davis signed into law on 6 May, a bill recognizing a state of war with the Union and establishing regulations for "letters of marque, prizes, and prize goods" similar to those issued by the United States in the War of 1812.[1]

Mallory had little confidence in privateers, but even modest success in this quarter would force up insurance rates in the North and adversely affect its business sector. Also, even a few privateers could force Welles to shift warships away from the blockade to hunt for them. Davis and other Southern leaders claimed the practice was legal, in that the United States had failed to ratify the 1856 Declaration of Paris, the signatories of which foreswore privateering. In retaliation for the Southern declaration, however, Lincoln proclaimed a naval blockade of the Confederacy and warned that any captured privateers would be treated as pirates.[2] Lincoln also offered to bind the United States to adhere unconditionally to the 1856 Declaration of Paris, but the British secretary of state for foreign affairs, Lord John Russell, pointed out that any European powers signing such a convention with the United States would be bound to treat all Confederate privateers as pirates, which they were unwilling to do.[3]

Meanwhile, on 14 May 1861, the British government issued a proclamation recognizing the Confederate States as a belligerent power, thereupon rejecting the Lincoln administration's contention that Confederate privateers were pirates. At the same time, however, on 1 June the British forbade armed ships of either the Union or the Confederacy to bring prizes into British home or colonial ports; France and the other major European maritime powers promptly followed suit. These decisions were a serious blow to the Confederacy. To be legal, all captured vessels had to be taken into port and there adjudicated by prize courts as legal captures. Without access to prize courts, privateers could be treated as little more than pirates. Not only did a declaration of neutrality prohibit the entry of prize vessels into ports, but it prohibited that nation's citizens from fitting out privateers under the flag of either belligerent. Yet international law held that state warships could legally destroy captured vessels, so the Confederate vessels were not prohibited from sinking Union ships.

Lincoln's threat did not deter applications for letters of marque in the South. On 10 May, the day the Confederate regulations were published, the Confederacy granted its first commission to the thirty-ton schooner *Triton* of Brunswick, Georgia, armed with a single six-pounder. In all, the Confederacy issued letters of marque for fifty-two privateers, most of which operated out of Charleston and New Orleans.[4] The few privateers that got to sea in May found easy pickings. The first success came on 16 May, when *Calhoun* of New Orleans captured the American merchant bark *Ocean Eagle* of Rockland, Maine, off the Mississippi River mouth. *Calhoun* and two other New Orleans privateers took nine other Union ships before the arrival at the end of May of the U.S. Navy screw sloop *Brooklyn* to patrol the area.

Typical of Atlantic coast privateers was the fast schooner *Savannah*, a fifty-three-ton vessel with a crew of twenty men and armed with a single gun. *Savannah* sailed from Charleston on 2 June and soon captured and sent into port the brig *Joseph* of Philadelphia. At dusk that same day *Savannah* spotted a sail and ran toward it, but the ship turned out to be the U.S. Navy brig *Perry*, armed with six thirty-two-pounders. Having lost part of its upper works in a storm the night before, *Savannah* could not outrun its opponent. The outclassed *Savannah* surrendered after a twenty-minute fight.

Sailed to New York, *Savannah* was there condemned and sold. Branded as "pirates" by the Northern press and Federal government, the crew was put on trial and threatened with the death penalty. President Davis promptly warned that if the men were executed, he would hang captured Union officers on a one-for-one basis. Washington backed down. In February 1862 it decided that privateersmen would be classified as prisoners of war.[5]

Union warships soon swept up most of the remaining privateers. Also, as the Union blockade became more effective, it became more difficult to send prizes to the South, and increasing numbers of the latter were recaptured. Many privateer vessels were subsequently converted into blockade-runners, a course that ran counter to Confederate naval strategy.

Maximizing Commerce Raiding

Secretary of the Navy Mallory sought to get Confederate commerce raiders to sea. Given the Confederacy's lack of facilities, Great Britain, the world's most advanced and largest shipbuilder, was the logical source from which to buy the ships, especially as its leadership was sympathetic to the South. Mallory urged Bulloch in England "to get cruising ships . . . afloat with the quickest possible dispatch."[6]

Mallory had decided views on the type of ships required. In his words, such ships should be

> enabled to keep the sea, and to make extended cruises, propellers fast [fixed firmly] under both steam and canvas suggest themselves to us with special favor. Large ships are unnecessary for this service; our policy demands that they shall be no larger than may be sufficient to combine the requisite speed and power, a battery of one or two heavy pivot guns and two or more broadside guns, being sufficient against commerce. By getting small ships we can afford a greater number, an important consideration. The character of our coasts and harbors indicate attention to the draft of water of our vessels. Speed in propeller and the protection of her machinery can not be obtained upon a very light draft, but they should draw as little water as may be compatible with their efficiency otherwise.[7]

Pending foreign construction, Mallory sought to outfit some ships at home. On 17 April 1861 he met with Cdr. Raphael Semmes, a staunch advocate of commerce raiding with a strong hatred of the North. A former career U.S. Navy officer, Semmes had distinguished himself during the Mexican-American War. Interestingly, in a book about his war experiences, *Service Afloat and Ashore during the Mexican War* (1852), Semmes had argued that if Mexico had fitted out privateers against American shipping during that war, Washington should have treated them as pirates.[8]

Following their discussions, Mallory gave Semmes command of the former steamer packet *Habana* at New Orleans. Launched in 1857, this 437-ton vessel had been employed on the route between New Orleans and Havana. Renamed *Sumter* and commissioned on 3 June, it was the first Confederate Navy commerce raider. The ship had a retractable funnel and a screw propeller; there would be no outward means to identify it as a steamer. Its armament consisted of one nine-inch Dahlgren gun and four thirty-two-pounders.[9] *Sumter* escaped to sea from the mouth of the Mississippi on 30 June, outrunning *Brooklyn*, which was caught off station.[10] On 3 July *Sumter* took its first prize, the merchant bark *Golden Rocket*.[11]

Semmes, like other Confederate raider captains, found himself handicapped by the neutrality of the major powers. Semmes tried to talk Spanish officials in Cuba into adjudicating five of his prizes there, but they refused, and these ships were eventually returned to their owners. As a result, Confederate captains routinely burned the Northern merchant ships they captured. Occasionally a ship would also be let go "on bond," because of a large number of passengers or because its cargo belonged to a neutral nation. Bonding meant that a captain signed a paper guaranteeing to pay a set sum to the Confederate government at the end of the war, the amount to be decided by condemnation procedures.

Semmes also discovered that neutrality laws limited the time that cruisers might spend in port and the repairs that might be made there. Large numbers of captured seamen and passengers were both a problem and a danger to a commerce raider. Those captured were routinely sent ashore in their own boats or, if no land was in sight, transferred to neutral ships or to Union merchant ships carrying cargoes belonging to neutral nations.

Semmes cruised the Caribbean and the South American coast to Brazil and back to the West Indies before crossing the Atlantic. With his ship now in poor repair, he put in to Cádiz, but Spanish authorities there would not permit an overhaul of its engine and ordered him to depart. Semmes then proceeded to Gibraltar, but U.S. Navy warships, including the screw sloop *Kearsarge,* soon arrived. Since the repairs his ship needed could not be effected at Gibraltar either, in April 1862 Semmes, having received authorization to do so, laid up the ship. In December it was sold to a British firm and put back into commercial service as *Gibraltar*.[12]

Though *Sumter* had proved both too small and too slow to be an effective commerce raider, Semmes had taken seventeen prizes in just six months, at a cost to the Confederate government of only twenty-eight thousand dollars. This figure was less than that of the least valuable of its prizes. Advanced to captain, Semmes was at Nassau in June 1863, hoping to catch passage on a blockade-runner to the South, when orders arrived from Mallory sending him back to England to take command of a ship nearing completion at Liverpool.[13]

Bulloch, meanwhile, had managed to skirt the Foreign Enlistment Act of 1819, which prohibited British subjects from equipping, furnishing, fitting out, or arming any vessel intended for service by foreign belligerent navies. Liverpool lawyer F. S. Hull advised him that construction of such a ship was not illegal in itself, whatever the intent, and that the offense lay only in the equipping. Bulloch thus saw to it that none of his cruisers went to sea with ordnance, small arms, or military stores. He shipped these in other vessels, and the cruisers were then outfitted in international waters.[14]

The first British-built Confederate raider was *Florida*. Commissioned in August 1862 by Lt. John N. Maffitt, it made two spectacular passages through the blockade of Mobile Bay and captured thirty-three Union merchant ships, causing an estimated $4,051,000 in damages. Expenses of the raider's construction and cruises probably totaled only $400,000, so this was a tenfold return on investment. *Florida* was captured at Bahia, Brazil, on 7 October 1864 by the Union screw sloop *Wachusett* in defiance of international law and sailed to the United States.[15]

Alabama was the second English-built Confederate commerce raider and by far the most successful. On 1 August 1861 Bulloch contracted for the ship with the Birkenhead Ironworks, owned by the firm of John Laird and Sons. Identified on the ways as Hull No. 290, it was launched on 15 May as *Enrica*. Bulloch expected to command it himself, but Mallory decided that command would go to Semmes, now without a ship, while Bulloch continued his important contract and logistics work.[16]

Any trained observer could see that Hull No. 290 was designed for easy conversion into an armed cruiser; the American consul at Liverpool hired a private detective and soon learned more about the vessel. The U.S. minister to Britain, Charles Francis Adams, complained to London and furnished evidence as to *Enrica*'s true nature. Adams also arranged that the U.S. Navy screw sloop *Tuscarora*, then at Southampton, be ordered to intercept *Enrica* should it put to sea.

Warned on 26 July that the British government was about to impound his ship, then undergoing sea trials, Bulloch immediately informed the Lairds that he wanted to carry out an additional trial and brought on board a British master, Mathew J. Butcher, and a skeleton crew. On the morning of 29 July, Bulloch and invited guests set out in *Enrica*, with a steam tug as tender. After lunch, Bulloch informed his guests that the ship would stay at sea that night and took them back with him to Liverpool in the tug. Early the next morning Bulloch returned in the tug with additional crewmen. Learning that *Tuscarora* was at sea searching for *Enrica* toward Queenstown on the southern Irish coast, Bulloch ordered Butcher to head north around Ireland, thence to Terceira Island in the Azores.[17]

Returning to Liverpool, Bulloch sent out another ship, *Agrippina*, with stores, ordnance, ammunition, and 250 tons of coal. On 13 August he and Semmes, who had only just arrived, departed Liverpool aboard *Bahama*. Meanwhile, *Enrica* arrived at Porto Praia da Vitória, on Terceira, on 9 August, followed by *Agrippina* on 18 August, and *Bahama* on 20 August. Semmes ordered the latter two ships to Angra Bay to fit out *Enrica*.[18] On 24 August, in international waters, Semmes commissioned his ship *Alabama*. He also persuaded some eighty seamen from the other two ships to sign on, promising them double standard wages in gold, along with prize money for ships destroyed. Bulloch, meanwhile, returned to Liverpool in *Bahama*.[19]

Alabama was a sleek, three-masted, barkentine-rigged wooden ship, described by Semmes as "a very perfect ship of her class."[20] Some 230 feet long and nine hundred tons burden, it had a single screw propeller powered by two three-hundred-horsepower engines and four boilers, with a retractable funnel. The propeller could be detached from the shaft and lifted into a special well to enable higher speed under sail alone.

Alabama was capable of thirteen knots under steam and sail, ten knots under sail alone. It mounted eight guns: two pivot-mounted guns—a rifled seven-inch (110-pounder) Blakely on the forecastle and a smoothbore sixty-eight-pounder abaft the main mast—as well as six heavy thirty-two-pounders in broadside. The average crew numbered twenty-four officers and 120 men. Designed to keep the sea for long periods, *Alabama* boasted a fully equipped machine shop so that the members of the crew might make all ordinary repairs themselves. It carried sufficient coal for eighteen days' continuous steaming. Semmes used the coal sparingly and made most captures under sail alone. The entire cost of the ship, including outfitting, came to $250,000.[21]

Semmes's first three lieutenants had served with him on *Sumter*. The first lieutenant, John McIntosh Kell, and the fourth lieutenant, Arthur F. Sinclair, both later wrote books about their experiences. While his officers proved capable, however, Semmes had problems with his crew. The vast majority of its members were British seamen, many of them castoffs from Liverpool. Difficulties were especially pronounced in port when alcohol was available. Partly for this reason, Semmes rarely allowed his men ashore, which in turn created morale problems. The large number of non-American crew members also made it more difficult to enforce discipline. The same problems affected other raiders, including *Florida,* whose Spanish and Italian seamen did not get on well together and also had difficulty understanding orders delivered in English.[22]

Alabama took its first prizes, all whalers, in the vicinity of the Azores. The first, on 5 September, was *Ocmulgee* of Massachusetts. As was his practice, Semmes had approached it under a false U.S. flag; Semmes also regularly presented his ship as a British or Dutch vessel—even as a U.S. Navy warship.[23] During two weeks, Semmes decimated the Union whaling fleet in the Azores. After weathering a major storm, *Alabama* proceeded west, arriving off Newfoundland and New England. That October Semmes took eleven vessels, destroying eight and releasing three on bond. Nature intruded, however, in the form of a hurricane, which on 16 October split sails and snapped the main yard, but *Alabama* proved its ability to withstand heavy weather.[24]

More than a dozen Union warships now searched for *Alabama* and the other raiders, but they were always a little late or in the wrong location. Semmes next proceeded to Fort-de-France, Martinique, to receive coal from *Agrippina*. The tender was already in port when *Alabama* arrived on 18 November. Semmes ordered *Agrippina* to Blanquilla Island off Venezuela, and the tender was hardly clear of the port when the U.S. Navy screw frigate *San Jacinto* arrived and took up a position off the harbor. Although it had a much more powerful armament than *Alabama,* the Union warship could make only seven knots, and that same night Semmes took advantage of a squall to escape to Blanquilla.[25]

Alabama Sails to Galveston

Semmes soon learned from newspapers of the U.S. capture of Galveston and that a Union expeditionary force was expected to invade Texas in January 1863. Aware that Galveston Harbor was shallow and that all Union transports would thus have to anchor offshore, Semmes developed a daring plan to sail there and attack the transports. He hoped also to take, en route to Galveston, a steamer from Panama carrying gold transshipped from California.

On 29 November 1862, *Alabama* made the passage between San Domingo and Puerto Rico, the usual route for mail steamers. Semmes took several prizes, among them the large bark-rigged steamer *Ariel* of the Aspinwall Line. Although outward bound and hence not carrying gold, it was Semmes's most important prize. The steamer had more than seven hundred people on board, including some five hundred passengers and 140 U.S. Marines on their way to Pacific Squadron assignments. Semmes disarmed and paroled the Marines, but the large number of prisoners forced him to let *Ariel* proceed under bond.[26]

On 23 December, *Alabama* met *Agrippina* at the Arcas Islands off Yucatan and spent a week there taking on supplies and coal and preparing for the Galveston raid. Semmes planned to arrive there during daylight, reconnoiter, and then return for a night attack. He expected to use *Alabama*'s superior speed to fight or run, as he chose. *Alabama* arrived off Galveston late in the afternoon of 11 January 1863 but found there, instead of a fleet of Federal transports, only five Union warships lobbing shells into Galveston. Semmes correctly concluded that the Confederates had retaken the port; indeed, Galveston had fallen eleven days before, and the Union troops had been diverted to New Orleans.

Lookouts on the Union warships soon spotted *Alabama*. The Union squadron commander, Commodore Henry H. Bell, flew his flag in *Brooklyn,* but since that ship's engine was not functioning, Bell dispatched Lt. Cdr. Homer C. Blake in *Hatteras* to investigate. A former Delaware River excursion side-wheeler, *Hatteras* mounted only four thirty-two-pounders and a 3.67-inch rifle.

Alabama moved slowly along the coast, drawing the Union ship away from the rest of the squadron. As soon as it was dark and the two ships were about twenty miles from the other Federal ships, *Alabama* came about and turned toward *Hatteras* under steam. When *Alabama* came within hailing distance, Blake demanded its identity, only to be told that it was an English vessel. Reassured, Blake demanded and received permission to inspect the ship's registry. After *Hatteras* had lowered a boat, Lieutenant Kell called out, "This is the Confederate States steamer *Alabama* . . . , fire."[27]

Alabama's broadside ripped into *Hatteras* from very close range. Knowing his ship's weakness, Blake tried to ram, but the faster *Alabama* avoided the attempt. His ship on fire and sinking, Blake surrendered after thirteen minutes. Two of his crewmen were dead and five wounded. Hit only five times, *Alabama* had two men wounded. Semmes took the Union crew on board and then sailed for Port Royal, Jamaica, where he paroled his prisoners.[28]

In late January 1863, *Alabama* sailed from Jamaica east through the West Indies to Brazil, arriving on 10 April at Fernando de Noronha, where Semmes coaled from a prize. This was fortunate, because *Agrippina* had been delayed. Semmes then made for Bahia, taking several more prizes en route. There, in mid-May, *Georgia* came in; *Florida* was only a hundred miles north. The only Union warship then in the South Atlantic was the screw sloop *Mohican*. Acting Rear Adm. Charles Wilkes, commander of the West Indian Squadron, created specifically to track down *Alabama* and *Florida,* had been detained in the West Indies with his flagship, the powerful and fast steamer *Vanderbilt*. Had *Vanderbilt* been actively searching with *Mohican,* the career of *Alabama* might have been ended then and there, but Wilkes was more interested in capturing blockade-runners for prize money than in hunting *Alabama;* Welles later relieved him of command for misusing *Vanderbilt.*[29]

On 21 May 1863 *Alabama* sailed from Bahia and cruised off Brazil. *Agrippina* did not arrive at Bahia until 1 June, only to discover the U.S. warships *Mohican* and *Onward* there. Capt. Alexander McQueen of *Agrippina,* fearful that his ship and its contents might be seized by the Union vessels when he left port, sold the coal and took on cargo for Britain. The tender never again encountered *Alabama.*[30]

Between Bahia and Rio, *Alabama* took eight prizes. Of these, five were burned, and two were bonded. The remaining prize, the five-hundred-ton, fast, bark-rigged clipper *Conrad,* was given two twelve-pounders from another prize. Semmes commissioned it as the auxiliary cruiser *Tuscaloosa,* under Lt. John Low. Semmes ordered Low to proceed on his own and rendezvous at Cape Town. Low subsequently took two prizes. After

Tuscaloosa arrived at Cape Town, however, British authorities seized it as an uncondemned prize, eventually turning it over to the American consul.[31]

Semmes, meanwhile, sailed to the Cape of Good Hope to intercept ships homeward bound from the East Indies, but in two months off South Africa he took only one prize. Indeed, of its eventual total of sixty-four prizes, *Alabama* took fifty-two of them in its first ten months at sea. This shift can be attributed to the transfer of American merchant ships to foreign registry, the reliance of American merchants on foreign ships to transport their goods, and the use by merchant skippers of less-frequented trading routes.

Alabama Disrupts U.S. Trade in Asia

On 24 September 1863, aware that *Vanderbilt* was searching for him, Semmes departed Cape Town for the Far East. He took his ship south of Mauritius. Engine problems, meanwhile, forced *Vanderbilt* to return home.[32] Semmes's goal was to cripple the American trade with Asia. During the first half of November, he took four merchantmen. But American ship captains had been warned, and on 21 December, when *Alabama* put in at Singapore, Semmes found twenty-two American merchant ships safely in that harbor. Other U.S. ships had taken refuge at Bangkok, Canton, Shanghai, and Manila. Semmes was also having problems with his crew, and at almost every port, men deserted; fortunately, others usually signed on to replace them.

Alabama was now in need of major overhaul. Its copper plating was coming loose from the wooden hull, and its boilers were so corroded that carrying full steam pressure was dangerous. Learning that the Union screw sloop *Wyoming* was patrolling the Sunda Strait, between Sumatra and Java, Semmes resolved to do battle with this ship, which was not as heavily armed as his own. The two did not meet, however, because *Wyoming* had steamed to Batavia for repairs to its boilers.[33]

Semmes sailed through the Strait of Malacca and took two more U.S. merchant ships before entering the Indian Ocean and briefly calling at Anjengo (now Anchuthengu), on the southwestern Indian coast. He then proceeded west to the Comoro Islands for provisions. The ship departed there on 12 February 1864, retracing its course back to Cape Town, where it arrived on 20 March. On the return trip Semmes took only one prize, and at Cape Town he learned of the seizure of *Tuscaloosa*.

On 25 March 1864, *Alabama* departed for Europe and hoped-for repairs. On 11 June it dropped anchor at Cherbourg. Semmes requested permission from the French authorities there to place his ship in dry dock. They refused, pointing out, as Semmes was well aware, that these facilities were reserved for the French navy and that only Emperor Napoleon III could grant such permission. They suggested that Semmes move his ship to Le Havre or another port with private dockyard facilities, but Semmes declined, expressing confidence that the emperor would approve his request.

Events now moved swiftly. On 12 June, the American minister to Paris, William Dayton, telegraphed news of *Alabama*'s arrival to the Dutch port of Flushing, where Capt. John A. Winslow's screw steam sloop *Kearsarge* was riding at anchor, keeping watch on CSS *Georgia* and *Rappahannock* at Calais. *Kearsarge* was in excellent condition, only two months out of a Dutch dockyard. Having spent a year looking for *Alabama*, Winslow was quickly under way. He arrived off Cherbourg on 14 June and positioned his ship off the breakwater, without anchoring. International law required that if *Kearsarge* anchored in the harbor, *Alabama* would receive a twenty-four-hour head start on departure.

Semmes might have attempted escape. Cherbourg had two channels, and *Kearsarge* could not easily cover both, especially at night. But *Alabama* was in poor condition and could not be kept at sea for much longer. Semmes also might have decided to lay up his ship, as he had *Sumter*. But he elected to fight. To Semmes it was an affair of honor and defense of the flag. Delay would only bring more Union warships. Indeed, Winslow had already telegraphed for reinforcements.

The battle between *Alabama* and *Kearsarge* took place on 19 June, in international waters off the French coast; it was one of the most spectacular naval engagements of the war. Notwithstanding Semmes's later claim that *Kearsarge* had the advantage in size, weight of ordnance, number of guns, and crew, the two ships were in fact closely matched, except that at eleven knots maximum speed, *Kearsarge* was slightly faster. It had four thirty-two-pounder broadside guns, a 4.2-inch rifled gun, a twelve-pounder howitzer, and most importantly, two eleven-inch, pivot-mounted Dahlgren smoothbores, throwing 135-pound shells. *Alabama* would have the edge at long range with its Blakely rifled gun, but advantage would go to *Kearsarge* in medium-to-short-range fire. Both ships could fight only five guns on a side, but *Kearsarge* threw a heavier weight of metal—364 pounds to 274 for *Alabama*.

Semmes expected to use his starboard guns in broadside and shifted a thirty-two-pounder from port to strengthen that side. The movement of weight caused the ship to list about two feet to starboard, but this exposed less of that side to enemy fire. When the two ships were about a mile and a quarter apart, Winslow reversed course and headed for *Alabama*. He too planned to use his starboard battery, so the two ships met going in opposite directions. The battle began at 10:57 AM some six or seven miles offshore and lasted only slightly more than an hour.

The two ships closed. When they were about a mile distant, *Alabama* sheered, turned broadside, and opened fire. The shot went high, probably because the gunners were overcompensating for their ship's starboard list. The two ships now circled each other, starboard to starboard. Because *Kearsarge* was faster and Winslow sought to narrow the range, the circles grew progressively smaller, from a half to a quarter mile in diameter, with each ship firing its starboard battery only, the current gradually carrying both ships westward.

Kearsarge had lengths of chain strung over its vital midships section to protect the engines, boilers, and magazines. An outward sheathing of one-inch wood painted the same color as the rest of the hull concealed this from observation, but the French had informed Semmes of it. *Alabama* had chain in its lockers that might have been used for the same purpose. Semmes later claimed *Kearsarge* had had an unfair advantage as a "concealed ironclad." In his after-action report to Commodore Samuel Barron, he wrote, "The enemy was heavier than myself in ship, battery, and crew, and I did not know until the action was over that she was also ironclad." Semmes convinced himself that he had been tricked into battle and that the chain was the only reason *Alabama* lost. But in fact, whether Semmes knew of the chain or not, Winslow had done nothing untoward. Bulloch himself later observed, "It has never been considered an unworthy ruse for a commander . . . to disguise his strength and to entice a weaker opponent within his reach."[34] Lieutenant Sinclair later criticized Semmes for this very failure, noting that Semmes "knew all about it and could have adopted the same scheme. It was not his election to do so."[35]

As the range narrowed, both sides substituted explosive shell for solid shot. Semmes hoped to close and attempt to board, but Winslow kept to the most effective range for his own guns, able to do so because his ship was both faster and more maneuverable than his opponent's. Repeated hits from *Kearsarge*'s two Dahlgrens tore large holes in *Alabama*'s hull. With *Alabama* taking on water, an eleven-inch shell struck at the waterline and exploded in the engine room, extinguishing the boiler fires. Water now entered the hull at a rate beyond the ability of the pumps to remove it.

At the beginning of the eighth circle, with the two ships about four hundred yards apart, Semmes turned *Alabama* out of the circle, ordering Kell to set all sail in hopes of making the French shore. Semmes also opened fire with his port battery. But *Alabama* was now taking on too much water and was completely at the mercy of *Kearsarge,* whose fire was ever more accurate.

Kell, returning from a check on conditions, reported that *Alabama* could not last ten more minutes, whereupon Semmes ordered him to cease firing, shorten sail, and haul down the colors. Semmes then sent a dinghy to *Kearsarge* to notify Winslow that he was ready to surrender. Semmes and Sinclair both later claimed that *Kearsarge* continued to fire after the colors were struck and a white flag displayed. Winslow asserted that he had ordered fire halted when *Alabama*'s colors came down and a white flag appeared at its stern but that shortly afterward the Confederate ship had fired two port guns; he had then moved his ship into position to rake his antagonist but, seeing the white flag still flying, again held fire. In any case, Semmes ordered all hands to try to save themselves, but only two boats could be used, and most of the crew simply leapt into the sea. Semmes gave his papers to a sailor who was a good swimmer, hurled his sword into the water, and then jumped in himself.[36] *Alabama* suddenly assumed a perpendicular position, bow upward, as its guns and stores shifted aft and then disappeared. Aboard *Kearsarge* there was no elation, only silence.[37]

Surprisingly, during the battle *Alabama* got off twice as many shots as its opponent, 370 rounds to 173, but *Kearsarge* sustained only thirteen hull hits and sixteen in the masts and rigging. Only one shot from *Alabama* caused personnel casualties—a Blakely shell explosion on the quarterdeck wounded three men at the aft pivot gun, one mortally.[38] By contrast, a high percentage of the Union shots struck. Semmes later said that one Union shot alone killed or wounded eighteen men at the after pivot gun. In all, *Alabama* suffered forty-one casualties: nine dead and twenty wounded in action and twelve men drowned.[39]

Winslow was slow to order his men to pick up survivors, partly because most of his own boats had been badly damaged in the exchange of fire. As a result, many of those in the water were taken aboard other ships, especially the English yacht *Deerhound,* which rescued and transported to Southampton forty-two members of the raider's crew, including Semmes and Kell. The British government rejected a demand from Adams to turn them over to American authorities. *Kearsarge* took aboard six officers and sixty-four men, including twenty wounded. Winslow paroled them at Cherbourg. Unfairly, Semmes blamed Winslow for not doing enough to save those in the water, writing, "Ten of my men were permitted to drown."[40]

Semmes was lionized in Britain. After a brief trip to the Continent, he made his way to Havana, then to northern Mexico and overland to Richmond. Promoted to rear admiral in February 1865—and thus second in seniority in the Confederate service only to Franklin Buchanan—Semmes briefly commanded the James River Squadron. Forced to destroy his ships on the night of 2 April 1865 when Confederate forces abandoned Richmond, Semmes formed the men into a naval brigade under his command as a brigadier general, thus becoming the only Confederate to hold flag rank in both the navy and army. Later that month he surrendered his unit in North Carolina with Confederate forces under Gen. Joseph E. Johnston.[41]

Conclusions

The defeat of *Alabama* signaled the beginning of the end for Confederate commerce raiders. Since its commissioning, *Alabama* had sailed seventy-five thousand miles, taken sixty-four prizes, and sent to the bottom a Union warship worth $160,000. In *Sumter* and *Alabama*, Semmes had taken eighty-one Union merchantmen. He later estimated that he had burned $4,613,914 worth of Union shipping and cargoes and bonded others worth $562,250. Another estimate places the total Union loss at nearly six million dollars. Two dozen Union warships had been searching for *Alabama*, another hefty expense. Beyond this, the raider's exploits had been a considerable boost to Southern morale.[42]

Following *Alabama*'s loss, Mallory continued to press the war against Union shipping, and he instructed Bulloch to locate a vessel that might be easily converted to operate in the Pacific against American whalers. (It had to be a conversion, because tightened English neutrality laws now precluded building such a vessel in that country.) In September 1864 Bulloch purchased *Sea King*. The world's first composite auxiliary-screw steamship, it became CSS *Shenandoah* and devastated the U.S. whaling fleet in the Bering Sea. But the war was already over. Finally convinced of the end of the conflict, Lt. Cdr. James Waddell struck *Shenandoah*'s guns below and sailed seventeen thousand miles to Liverpool, arriving there on 6 November 1865, the only Confederate warship to sail around the world. It had taken thirty-eight Union vessels, of which Waddell had burned thirty-two. Damage to Union shipping was estimated at some $1.36 million.[43]

During the Civil War, Confederate commerce raiders took a total of 257 U.S. merchant ships, or only about 5 percent of the total. They hardly disrupted American trade, then, but the cruisers ultimately deployed by the U.S. Navy to hunt down the raiders cost the government some $3,325,000. In fourteen months from January 1863, a total of seventy-seven Union warships and twenty-three chartered vessels were employed in this security effort. The raiders also drove up insurance rates substantially, but a bigger impact was forcing a large number of U.S. vessels into foreign registries. During the Civil War, more than half of the total U.S. merchant fleet was permanently lost to the flag. The cruisers may have burned or sunk 110,000 tons of shipping, therefore, but some 800,000 additional tons were sold to foreign owners—seven hundred ships to British interests alone—and these included some of the best vessels. Legal impediments prevented much of this tonnage from later returning to U.S. ownership.[44]

After the war, the matter of the British government having allowed the fitting-out of a number of Confederate cruisers became a major thorn in Anglo-American relations. U.S. government officials believed, rightly or wrongly, that London's early proclamation of neutrality and then persistent disregard of that neutrality in the early part of the war had heartened the South and prolonged the conflict. There were those in the U.S. government who proposed taking Britain's Western Hemisphere possessions, including Canada, as compensation.

In 1871, when the continental balance of power decisively changed with Prussia's defeat of France, British statesmen concluded that it might be wise to reach some accommodation with the United States against the possibility of a German drive for world hegemony. An international tribunal met in Geneva beginning that December to discuss what became known as the "*Alabama* claims." In September 1872 this tribunal awarded the U.S. government $15,500,500 in damages. This settlement came to be regarded as an important step in the peaceful settlement of international disputes and a victory for the international rule of law.[45]

Notes

Much of the material in this chapter is drawn from Spencer C. Tucker, *Raphael Semmes and the Alabama* (Fort Worth, Tex.: Ryan Place, 1996).

The thoughts and opinions expressed in this publication are those of the author and are not necessarily those of the U.S. government, the U.S. Navy Department, or the Naval War College.

1. William Morrison Robinson, Jr., *The Confederate Privateers* (1928; repr. Columbia: Univ. of South Carolina Press, 1990), pp. 13–17.
2. Ibid., p. 14.
3. U.S. Senate Document No. 332, 64th Cong., Sess. 1, serial No. 6952, p. 19.
4. Robinson, *Confederate Privateers*, p. 30; Paul H. Silverstone, *Civil War Navies, 1855–1883* (Annapolis, Md.: Naval Institute Press, 2001), pp. 193–94.
5. *Official Records of the Union and Confederate Navies in the War of the Rebellion* (Washington, D.C.: U.S. Government Printing Office, 1880–1901) [hereafter *ORN*], ser. I, vol. 5, pp. 692–93, 780; *Trial of the Officers and crew of the privateer Savannah on the charge of piracy, in the United States Circuit Court for the Southern District of New York, Hon. Judges Nelson and Shipman, presiding. Reported by A. F. Warburton, stenographer and corrected by the counsel* (New York: Baker & Godwin, 1862); Robinson, *Confederate Privateers*, pp. 49–57, 133–51.
6. James D. Bulloch, *The Secret Service of the Confederate States in Europe, or How the Confederate Cruisers Were Equipped* (1884; repr. New York: Modern Library, 2001), p. 34.
7. Mallory to Bulloch, 9 May 1861, in U.S. Navy Dept., *Civil War Naval Chronology, 1861–1865* (Washington, D.C.: Naval History Division, 1972), vol. 1, p. 13.
8. David D. Porter, *Naval History of the Civil War* (New York: Sherman, 1886; repr. Secaucus, N.J.: Castle, 1984), p. 602.
9. *ORN*, ser. I, vol. 1, p. 613; Paul H. Silverstone, *Warships of the Civil War Navies* (Annapolis, Md.: Naval Institute Press, 1989), p. 162; Raphael Semmes, *Memoirs of Service Afloat, during the War between the States* (Baltimore: Kelly, Piet, 1869; repr. Secaucus, N.J.: Blue and Grey, 1987), pp. 96, 101–103.
10. Porter, *Naval History of the Civil War*, pp. 605–606.
11. Semmes journal, in *ORN*, ser. I, vol. 1, p. 695.
12. On the cruise of *Sumter*, see Semmes, *Memoirs of Service Afloat*, pp. 108–345; Porter, *Naval History of the Civil War*, pp. 606–20; and John M. Kell, *Recollections of a Naval Life* (Washington, D.C.: Neale, 1900), p. 176.
13. Semmes, *Memoirs of Service Afloat*, pp. 344–53.
14. Philip Van Doren Stern, *The Confederate Navy: A Pictorial History* (Garden City, N.Y.: Doubleday, 1962), p. 36.

15. On *Florida* see Frank L. Owsley, Jr., *The C.S.S. Florida: Her Building and Operations* (Tuscaloosa: Univ. of Alabama Press, 1987).

16. George W. Dalzell, *The Flight from the Flag: The Continuing Effect of the Civil War upon the American Carrying Trade* (Chapel Hill: Univ. of North Carolina Press, 1940), pp. 129–30; Charles M. Robinson III, *Shark of the Confederacy: The Story of the C.S.S. Alabama* (Annapolis, Md.: Naval Institute Press, 1995), p. 20; Stern, *Confederate Navy*, p. 117.

17. Stern, *Confederate Navy*, p. 117; Dalzell, *Flight from the Flag*, pp. 131–36.

18. Robinson, *Shark of the Confederacy*, p. 28; Arthur Sinclair, *Two Years on the Alabama* (Boston: Lee and Shepard, 1895), pp. 11–12.

19. Dalzell, *Flight from the Flag*, pp. 136–47; Robinson, *Shark of the Confederacy*, p. 33.

20. Semmes, *Memoirs of Service Afloat*, p. 402.

21. Ibid., pp. 402–403, 419–20; Sinclair, *Two Years on the Alabama*, p. 3. Most sources differ on statistics; Silverstone and Dalzell have *Alabama*'s displacement at 1,050 tons and dimensions as 220 feet in overall length (211 feet, six inches at the waterline). See Silverstone, *Civil War Navies*, p. 207; and Dalzell, *Flight from the Flag*, pp. 129, 162.

22. Dalzell, *Flight from the Flag*, p. 154; Owsley, *C.S.S. Florida*, p. 102.

23. Semmes, *Memoirs of Service Afloat*, pp. 423–24.

24. Ibid., pp. 456–78.

25. Ibid., pp. 479–519.

26. Ibid., pp. 519–35.

27. Norman C. Delaney, *John McIntosh Kell of the Raider Alabama* (Tuscaloosa: Univ. of Alabama Press, 1973), p. 142.

28. Report of Lt. Cdr. Blake to Welles, 21 January 1863, in *ORN*, ser. I, vol. 2, pp. 18–20; Semmes, *Memoirs of Service Afloat*, pp. 540–50; Porter, *Naval History of the Civil War*, p. 122.

29. Charles Wilkes to Gideon Welles, 11 December 1863, in *ORN*, ser. I, vol. 2, pp. 567–69; Welles to Wilkes, 15 December 1863, in *ORN*, ser. I, vol. 2, pp. 568–71; Robinson, *Shark of the Confederacy*, pp. 97–98.

30. Robinson, *Shark of the Confederacy*, p. 94.

31. Semmes, *Memoirs of Service Afloat*, p. 627; Sinclair, *Two Years on the Alabama*, pp. 124–25; Silverstone, *Civil War Navies*, p. 218.

32. Robinson, *Shark of the Confederacy*, pp. 108–109, 111.

33. Lt. Cdr. D. McDougal to Welles, 9 December 1863, in *ORN*, ser. I, vol. 2, pp. 560–61; Dalzell, *Flight from the Flag*, pp. 154–56; Robinson, *Shark of the Confederacy*, pp. 115, 120–21.

34. Semmes journal, 15 June 1864, in *ORN*, ser. I, vol. 3, p. 677; Semmes to Samuel Barron, 21 June 1864, in *ORN*, ser. I, vol. 3, p. 651; Semmes, *Memoirs of Service Afloat*, pp. 759–62; Dalzell, *Flight from the Flag*, p. 163; Bulloch, *Secret Service of the Confederate States in Europe*, p. 201.

35. Sinclair, *Two Years on the Alabama*, p. 263.

36. John Winslow to Gideon Welles, 19 June and 30 July 1864, in *ORN*, ser. I, vol. 3, pp. 59, 79–81; Sinclair, *Two Years on the Alabama*, pp. 258–60; Semmes, *Memoirs of Service Afloat*, p. 757; William Marvel, *The Alabama & the Kearsarge: The Sailor's Civil War* (Chapel Hill: Univ. of North Carolina Press, 1996), pp. 250–58.

37. Stern, *Confederate Navy*, p. 194.

38. Dalzell, *Flight from the Flag*, p. 160; Frank M. Bennett, *The Monitor and the Navy under Steam* (Boston: Houghton Mifflin, 1900), p. 187.

39. Sinclair, *Two Years on the Alabama*, p. 281; Porter, *Naval History of the Civil War*, pp. 653–54.

40. Sinclair, *Two Years on the Alabama*, p. 270; Semmes, *Memoirs of Service Afloat*, p. 759; Dalzell, *Flight from the Flag*, p. 160; Bennett, *Monitor and the Navy under Steam*, p. 187; Stern, *Confederate Navy*, p. 193.

41. Semmes, *Memoirs of Service Afloat*, pp. 789–92.

42. *ORN*, ser. I, vol. 3, pp. 677–81; U.S. Navy Dept., *Civil War Naval Chronology*, vol. 6, p. 192; Robinson, *Shark of the Confederacy*, p. 194.

43. Bulloch, *Secret Service of the Confederate States in Europe*, pp. 407–29.

44. Dalzell, *Flight from the Flag*, pp. 237–62; Bennett, *Monitor and the Navy under Steam*, p. 184.

45. Dalzell, *Flight from the Flag*, pp. 231–36.

Two Sides of the Same Coin
German and French Maritime Strategies in the Late Nineteenth Century

DAVID H. OLIVIER

For the last thirty years of the nineteenth century, both France and the newly formed German Empire faced similar dilemmas in a potential war—both were likely to fight against superior naval powers. As history had shown, the weaker fleet could rarely wrest control of the sea from its opponent. Thus, the traditional strategy of the weaker power at sea had been to conduct *guerre de course,* hoping to cause enough damage to the foe by attacking its commerce to impair its war effort. However, a number of circumstances had changed by 1871, leaving navies uncertain whether commerce raiding could be conducted at all, let alone with any degree of effectiveness. Changes in maritime law threatened to eliminate a state's ability to hamper the flow of maritime commerce to its rivals. The technological fruits of the Industrial Revolution created uncertainty in naval planning and construction policies. Finally, the most recent major wars featuring some form of commerce raiding—the American Civil War and the Franco-Prussian War—indicated that modern *guerre de course* would have to be very different from what it had been in the Age of Sail.

The responses by the French and German navies to these circumstances were quite different, even if German commentators occasionally borrowed the language of their French counterparts. French naval theorists created an entirely new philosophy of war at sea, the *Jeune École* (the Young School), designed to maximize the threat posed by new naval technology to the economic well-being of the enemy. This philosophy was part and parcel of a greater dispute in the navy over construction policies, promotions, and even political philosophies, and it is arguable that at times the conflict between supporters and opponents was even detrimental to the navy's functioning. The Jeune École sought to create economic chaos in its projected enemy, Great Britain. Financial injury created by French attacks would lead British commercial and business interests to pressure the government for peace. It was not the sinking of ships or the destruction of cargo that was the hoped-for end result but the ensuing panic.

In contrast, commerce raiding never assumed any level of importance in German naval strategy. Instead, it was almost taken for granted, a duty to be expected of ships stationed overseas, which would otherwise have no means of contributing to the war's real focus, which was in European waters. Unlike the French, the Germans never developed a coherent philosophy to support commerce raiding. Greater concern lay in commerce protection and coastal defense, and the navy was expected to support the army's drive against a continental foe. The goal of German commerce raiding was far simpler in nature—to deny the enemy access to overseas raw materials and finished goods. The more ships sunk, the less the enemy had with which to prosecute the war. This was an end that served the needs of the army, by weakening the enemy's ability to continue to fight.

Historical Background

Before 1856, war at sea was often as much a conflict between belligerents and neutrals as it was between the belligerents themselves. Maritime commerce in wartime was affected by the relative strength of the belligerents and the neutrals. In a war in which there were no significant neutrals, such as the Seven Years' War, the most powerful naval state (in that case, Great Britain) was free to do essentially what it pleased to prevent the flow of goods to and from, say, France. However, in a war like that of American independence, involving one or more powerful neutrals, the scope of action was severely limited by the threat of a league of armed neutral states.[1]

In 1856 the major powers agreed to abide by the terms of the Declaration of Paris, which made significant amendments to the law of war at sea. The changes affected blockade, neutral ships and neutral goods, and the practice of privateering—the issuing of licenses to private vessels to make war on enemy commerce. All of these new rules would have a major impact on the conduct of war at sea. Neutral vessels carrying enemy goods were now exempt from capture, and enemy ships carrying neutral goods were also exempt. This meant that far more stringent search procedures would have to be followed by boarding parties.

The change that produced the greatest criticisms—and that prompted the United States, Spain, and Mexico to refuse to ratify the declaration—was the complete abolition of privateering. The ability to charter vessels from a country's merchant marine to serve as commerce raiders was viewed as of paramount importance for such countries as the United States, which in the 1850s possessed a comparatively large merchant fleet but a very small navy. In the Americans' view, the abolition of privateering merely reinforced the maritime dominance of countries with large standing navies, such as Great Britain.[2]

It was not fully realized, however, in the United States or among other naval powers that the Declaration of Paris did not completely eliminate war against maritime commerce. What it required was that such a war be carried out solely by a country's navy instead of by private interests operating under license. The biggest impact of these new rules would be the elimination of the profit motive for attacking enemy trade. Without this motive, the need to keep confiscated cargoes became less important than merely denying them to the enemy, destroying what could not be taken. However, only actual wartime conditions would show how these changes in maritime law would affect the conduct of a *guerre de course*.

In addition, though the innovations of the Industrial Revolution had a significant impact on war at sea in the nineteenth century, that impact was uneven and took time to translate into effective, permanent change. As a result, many advances could only be applied partially, and for most of the second half of the century warships and merchant vessels were a mix of traditional wooden construction and sail power, on the one hand, and modern iron or steel construction and steam power, on the other. In fact, the majority of merchant vessels, even as late as the beginning of the twentieth century, remained sail powered, at the mercy of the winds and tides.[3] A becalmed merchant vessel could go nowhere. Meanwhile, a commerce raider equipped with both sails and a steam engine could use its sails for long-range cruising, then switch to steam propulsion to overtake becalmed or slower-moving sail-powered merchantmen. This provided a considerable advantage to the raider.

The other significant technological advance in the nineteenth century that had an effect on commerce raiding was the self-propelled torpedo. The torpedo was first developed in 1868 but took nearly a decade to become a useful and reliable weapon.[4] The torpedo provided two advantages to the attacker. First, it could be used at a greater range than gunnery, thus avoiding the possibility of return fire or ramming by the merchant vessel. Second, being a compact weapon, it could be carried and used by a much smaller ship. This opened the possibility of conducting a war on enemy commerce using an inexpensive fleet of small vessels. Taken together, these two innovations—steam power and torpedoes—gave commerce raiders greater range and mobility, as well as a punch that made them a threat to both merchant vessels and pursuing warships.

The first two wars to be fought under the new laws of war at sea were the American Civil War and the Franco-Prussian War (1870–71). The use of commerce raiding in the Franco-Prussian War was far more limited in scope than in the Civil War. But France imposed a limited blockade of the coasts of the Germanic states in the opening months of the war, hindered only by its inability to capture neutral British merchant vessels conducting trade into and out of these ports. The reverses on land suffered by the French armies made the blockade more and more superfluous; in addition, the blockading ships were hampered by increasingly bad weather and the need for regular supplies of coal.[5] What caught the attention of French navalists was the exploits of a single Prussian raider, SMS *Augusta,* in January 1871. In its brief career, *Augusta* captured a total of three French merchant vessels—hardly comparable to the two hundred–plus Germanic ships taken by the French—before being forced to seek refuge in a neutral Spanish port.[6] But the panic caused along the French Atlantic coast made a great impression on French observers.

To the Germans, there were entirely different lessons to be learned from the Franco-Prussian War. When Prussian armies had triumphed at the battle of Sedan at the beginning of September 1870, it had been expected that France would soon sue for peace. Instead, a new French government called for continued resistance and raised new armies to replace those lost in the opening campaign. These armies required weapons, and many of those arms were purchased from manufacturers in Great Britain and the United States. An effective *guerre de course* might have hampered the French efforts to prolong the war, but the North German Confederation navy was unable to stem the flow of foreign weapons. It speaks of the attitude toward the navy's contribution to victory that the celebratory postwar parade in Berlin featured a grand total of twenty-two officers and men from that service. In the eyes of many German observers, a prime opportunity had been lost.[7]

The lessons learned from the American Civil War and the Franco-Prussian War helped shape the way French and German naval thinkers saw their navies contributing to future wars. Successes and failures alike were magnified in significance, depending on the kinds of arguments being made. To many French observers, the economic and psychological elements seemed paramount; to the Germans, however, the physical effect itself was the most important. These divergent views influenced French and German thinking about future wars.

France and the Origins of the Jeune École

The French navy had a long tradition of *guerre de course* in its nation's many wars with Great Britain. France seemed unable to defeat the British in naval battles, but it enjoyed much more success when individual ships, either warships or independent privateers, preyed on British merchant vessels. There were sound commercial reasons for the French to remain focused on British trade: vessels could be seized and cargoes deemed prize goods and resold, and money taken out of British pockets went directly to those of the French.

In theory, the Declaration of Paris and the end of privateering should also have ended all attacks on maritime trade. The key to traditional *guerre de course* had always been its profitability; outlawing privateering removed the financial incentive for private individuals and corporations. Furthermore, forcing governments to assume the responsibility for naval warfare required any efforts at attacking enemy trade to be funded out of the government's own purse. Finally, technology appeared to give the merchant ship the means to elude any attempts at commerce raiding, because steam power allowed ships to sail in any direction at any time.

However, such was not the case. In particular, naval strategists in France still believed that it remained essential to attack British shipping in the event of a war. An early proponent was Capt. Richild Grivel. In 1869 Grivel wrote that it was a waste of time and resources for France to continue to build a large navy of battleships for use against Britain; it would be easily countered by Britain's superiority in factories and resources.[8] France could produce a navy large enough to dominate smaller opponents, such as Italy or the North German Confederation, but to build a battle fleet for war against Britain was folly.

Instead, as shown by historical example, France should continue to pressure Britain's merchant fleet. First, the vessels needed by France to engage in such a war were much less expensive to build and maintain than a large battle fleet. Second, this type of war went to the very heart of Britain's success in previous wars—its financial resources. Attacks on British commerce would surely drive up the rates of insurance for British merchant vessels. In turn, this would force the owners to charge more for carrying cargo, eventually so much that no one would be able to afford to ship their goods in British merchant ships. Grivel believed such an effect could be achieved in perhaps two or three years of continuous commerce raiding. Considering the average length of Anglo-French wars, this seemed to him a not unreasonable estimate.[9]

Grivel's theories were neither highly regarded nor eagerly subscribed to at the time. This was due to political considerations as much as anything else; within two years, France had been defeated by the newly formed German Empire, had lost two valuable provinces, and had overthrown the Second Empire and returned to a republican form of government. The navy, having played a minimal role during the war, was of secondary importance to the army, and spending on naval arms was reduced accordingly.

Grivel's theories may not have produced immediate results, but they clearly left their mark on another naval officer, Adm. Hyacinthe-Laurent-Théophile Aube. Since Aube had spent much of his career on overseas duty, geopolitical considerations, especially the growing colonial rivalries between France and Britain, dominated his view of the position of the French navy. It was also influenced by his growing conviction that unless France was willing and able to create a battle fleet at least equal to Britain's, it could never win a fleet-on-fleet encounter. Consequently, Aube was interested in creating a navy that at home would protect France from British invasion and blockade and overseas would safeguard France's colonies and trade while harassing Britain's. This kind of navy did not require large, expensive battleships; instead, a fleet of less expensive cruisers, gunboats, and torpedo boats would suffice.[10]

The torpedo boat was a new weapon of naval warfare, one that changed how the *guerre de course* would be conducted. It would perform three vital tasks: it would reduce the effectiveness of a British blockade of French ports, and it would attack British commerce in vital narrow waters, such as the English Channel or the straits between Tunisia and Malta. Torpedoes would also powerfully deter British warships from cruising too close to French ports. By coordinating attacks from several ports at once by means of telegraphic communication, Aube argued, the French would be able to lift any British blockade, from one port or several, long enough to allow raiders large and small to sally forth to wreak havoc all around the world.[11] Thus, his concern over a British blockade was focused on the inability to send out commerce raiders, rather than over the economic effects of the severance of France's maritime trade.

The French journalist Gabriel Charmes believed that the torpedo boat could play an even greater role in commerce raiding than Aube believed. Charmes argued that fast cruisers and torpedo boats could break any British blockade. Once out to sea, they were to pursue a ruthless war on enemy commerce, while avoiding contact with superior British naval forces—as Charmes put it, "to fall on the weak without pity and to flee the strong at full speed without false shame."[12]

However, small torpedo boats would pose no significant threat to merchant vessels if they observed the traditional rules governing commerce raiding. Therefore, Aube and Charmes advocated a radical departure from the normal stop-and-search procedures, in part to avoid the weakness of the tiny torpedo boat against a much larger merchant vessel, and in part because they viewed warfare as something other than a gentlemanly sport. As Aube wrote in an article in 1885,

> war is the negation of law. It . . . is the recourse to force—the ruler of the world—of an entire people in the incessant and universal struggle for existence. Everything is therefore not only permissible but legitimate against the enemy.
>
> . . . Tomorrow, war breaks out; an autonomous torpedo boat—two officers, a dozen men—meets one of these liners carrying a cargo richer than that of the richest galleons of Spain and a crew and passengers of many hundreds; will the torpedo boat signify to the captain of the liner that it is there, that it is watching him, that it could sink him, and that consequently it makes him prisoner—him, his crew, his passengers—in a word that he has platonically been made a prize and should proceed to the nearest French port? To this declaration . . . the captain of the liner would respond with a well-aimed shell that would send to the bottom the torpedo boat, its crew, and its chivalrous captain, and tranquilly he would continue on his momentarily interrupted voyage. Therefore the torpedo boat will follow from afar, invisible, the liner it has met; and, once night has fallen, perfectly silently and tranquilly it will send into the abyss liner, cargo, crew, passengers; and, his soul not only at rest but fully satisfied, the captain of the torpedo boat will continue his cruise.[13]

The French assault on British shipping was not designed merely to whittle down the British merchant marine but instead formed part of a campaign to cause tremendous harm to the British economy. Aube argued that sufficient economic unrest in Britain, such as would be caused by disrupting its trade with India and other colonies, would provoke social discontent. Charmes insisted that because public wealth was based on the success of private enterprise, no distinction should be made between public and private property when attacking British commerce. Here was the real goal of this new school of naval thought, the Jeune École.[14]

The British shipping industry had prospered in the nineteenth century, and in the last quarter of that century world trade increased dramatically. British firms owned more merchant vessels than the rest of the world combined. As a result of this predominance, Britain also led the world in other economic elements connected to world maritime trade: international banking, finance, and insurance. These financial elements appeared to rest on each other's stability and security.

Germany and Cruiser Warfare

While the Imperial German Navy (IGN) was a new creation, coming into existence in 1871, it had historical foundations in its predecessors, the Prussian and—from 1867—the North German Confederation navies. This meant that the IGN had ships already stationed around the world and also that it had limited experience in commerce warfare.

Unlike that of France, German naval policy until 1888 was under the control of army officers. Thus, the first priority of the navy was coastal defense, its second was the protection of overseas German commerce, and its third was political and diplomatic support for German foreign policy.[15] No serious thought was given during the navy's early years to any coherent theory of commerce raiding—the protection of German commerce was always considered more important.

The limited experience of German commerce raiders in the Wars of Unification, coupled with the example of the Confederacy, served as the template for any German thoughts on *guerre de course*. In 1864, during the Danish War, a lone Prussian corvette took a few Danish prizes in the Far East, operating strictly under prize code. The same was true during the Franco-Prussian War, with the modest success achieved by *Augusta*. This was meager experience on which to base any sort of naval policy.[16]

Nevertheless, the IGN essentially fell by default into a strategy of commerce raiding in the event of war. This situation arose because the navy kept a number of warships on station overseas in the defense of German interests. This was a deliberate naval strategy in the 1870s—to be able to apply timely pressure where needed.[17] The vessels built for overseas service were ideally suited for conducting commerce raiding at the time:

reasonably fast, with sufficient armament, and equipped with dual propulsion systems of both sails and a steam engine.

The biggest drawback to any concerted attempt by the IGN to implement commerce raiding in a war was the lack of overseas naval bases. This was a point made clear by a number of German naval officers, especially those serving overseas. Regular correspondence went from the captains of these warships back to their superiors in Berlin describing the virtues of particular ports and what good bases they would make for German warships.[18] The response was always the same, at least until 1884: Germany had neither interest in nor the need for overseas bases, or colonies of any kind. As a result, the IGN had a number of warships scattered around the globe without any stations of its own to support them.

This was a key point, one that would be a fatal weakness in wartime. There were two reasons why overseas bases were important for the effective conduct of worldwide commerce raiding. First, steam-powered vessels had limitations that their sail-powered counterparts did not. Steam engines required regular replenishment of coal, and boilers needed fresh water; even using sails for cruising and engines only for overhauling becalmed merchantmen, the cruisers would eventually use up these essential materials. This meant that a raider needed to put into port on a regular basis.

The second reason underscored the first. Thanks to events during the American Civil War, the laws of war regarding the use of neutral ports by belligerents had been significantly tightened. The Confederate raiders had benefited from liberal use of neutral ports for supplies and rest. Now belligerents were allowed to use a neutral port only once every three months. German commerce raiders, without ports of their own, would not be able to survive for long without access to the supplies needed.[19] This had been driven home to the Germans during the Franco-Prussian War, when one of their small warships in the Caribbean received no help from the pro-French Spanish, much less from the Americans, after the land war turned in favor of Prussia.[20]

The idea of using *guerre de course* in a potential war was not ignored, but it was not part of significant discussions within the navy's command until a new head of the Admiralty was appointed in 1883. Leo von Caprivi was another army general parachuted into command of the navy, and his first priority was always the navy's contributions to the army in the event of a war in Europe. Nevertheless, he soon became aware of the effects of war on world trade. In October 1883 a predecessor, Eduard Jachmann, submitted a memorandum to Caprivi that called for German warships overseas to form into small squadrons to attack enemy-held bases and enemy commerce.[21]

A more significant opportunity came early in 1884, when Caprivi called an Admiralty Council, an advisory board that had met only once under his predecessor. One of the questions posed to the council concerned the feasibility of cruiser warfare for the IGN. The council believed that commerce raiding had a role to play, especially in denying the flow of overseas supplies to the enemy. The council's belief was based on events in the Franco-Prussian War, when French resistance had been strengthened by the flow of arms purchased in Britain and the United States. This conclusion was stated explicitly in the council's final report:

> A striking proof for [the possible contribution of cruiser warfare] is provided by our last war, in which the need of France soared within months. The amounts of deliveries of weapons, etc.[,] for the newly created armies, were enormous. For example, a single steamer, one of many, had brought in not fewer than 140 cannons and thirty thousand rifles, worth around five million marks, from New York to Bordeaux. Without these colossal overseas supplies the continuation of the war would have been impossible and in this case one can rightly state that an emphatic and successfully conducted cruiser war, while it might not have directly brought about the decision, it would have considerably expedited it.[22]

Caprivi appreciated the efforts of the council, but he did not necessarily share its conclusions. He felt that cruisers would be required that were designed more specifically for the task, and believed that steam-powered merchant ships made commerce warfare less effective, as the cruisers would be outrun by their prey. He also emphasized more immediate needs in his construction policies, such as torpedo boats for coastal defense. More resources allocated to cruisers meant fewer resources for coastal defense. Ironically, political support for Caprivi's budget proposals came in some small measure because politicians believed, just as the Jeune École argued, that the torpedo boat was the low-cost replacement for expensive and obsolete battleships.[23]

If Caprivi was unwilling officially to endorse commerce raiding as a specific strategy for the IGN, however, he did not dismiss it outright. Moreover, the continuing dispatch of German cruisers on overseas duties set conditions whereby a significant portion of the navy would be thousands of miles away from Germany in the event of hostilities. What purpose would those ships serve in war? This question linked ongoing German and French debates on *guerre de course,* in particular after the development of torpedo boats.

The Impact of the Jeune École

France had certainly used *guerre de course* against Britain, and while it had been profitable for the French, it had never won them any wars. What was so different about the theories of the Jeune École that would bring success? First, the new technologies provided hope that British blockades would be less effective than during the sailing-ship era. British frigates had been able to remain on close station on a regular basis in previous wars, providing warning of any attempt by the French to leave port. The switch from wind power to steam allowed British ships to keep station outside French ports regardless of the weather, but this was counterbalanced by the increasing need to coal vessels. Coaling at sea was a difficult and dangerous business; it made more sense to have ships return to nearby ports for refueling. This meant many more ships were required to impose an effective blockade, as at any given time some would be on station, some heading back for coaling, some in harbor coaling, and some returning to station. The French themselves had learned the difficulties of imposing a blockade with steam-powered warships during the Franco-Prussian War.[24] Without overwhelming superior numbers at every port, the British could not blockade the French successfully.

Second, France held a key geostrategic position. Its Channel ports threatened British trade in the English Channel and the North Sea, its Atlantic ports threatened global trade routes, and its Mediterranean ports threatened British supply routes to Egypt and India. Raiders could cause serious harm to the British at any point.

Third, this would be a *guerre de course* unlike any previous in history. Until 1856, commerce raiding had been conducted as warfare for profit. Ships had been seized and brought to friendly ports, where prize courts had ruled on whether they and their cargoes were fair spoils of war. The crews, financial backers, and governments all made money, from the sale either of licenses to become privateers or of confiscated goods. After the Declaration of Paris outlawed privateering, only governments could engage in naval warfare. This reduced profits, as did the increasing difficulty in bringing captured merchant vessels into port to be judged as fair prizes. An enemy blockade changed this, as seen in the American Civil War (during which Confederate raiders, prevented by the blockade from bringing prizes into port at all, burned them, after removing crew, passengers, and any cargo they needed themselves). The size of modern cargo ships was another issue; it was more and more difficult to sort out cargoes that could be seized legitimately from those not susceptible to capture.

One method that had been tried during the American Civil War by Confederate commerce raiders was to capture a ship, evacuate the crew, and then set fire to and sink the merchant vessel. This reduced the number of ships in the Union's merchant fleet, but it posed two problems. One, wooden merchant ships would burn and sink readily, but modern steel-hulled steamers would be tougher to destroy. Two, there was the issue of the captives. If a raider was at sea for an extended period or had a productive spell of action, it could become quickly overrun with captives, posing security and health risks.

The Jeune École's solution was both simple and drastic. French torpedo boats would simply sink their prey, without warning. This eliminated all the problems attendant on commerce raiding under the restrictions of international law. Furthermore, it would create difficulties for the British: loss of vessels, loss of cargoes, loss of experienced crews, and an implied threat to shipowners, shippers, and merchant sailors.

However, it was likely that world opinion would be aghast at the French tactics. Such complete disregard for international law would affect trade in general. Traditionally, goods had been classified as contraband (goods to be used in war) and noncontraband. While there had been disputes over certain items that could be used for either war-related or peaceful purposes, others, such as foodstuffs, had been exempt from the definition of contraband. The Declaration of Paris had failed to clear up ambiguities in this definition, in expectation that the very practice of commerce raiding would slowly disappear. Instead, the Jeune École chose to disregard the question of contraband and noncontraband altogether. By sinking any ship, the French threatened any and all goods coming to Britain, regardless of whether they were raw materials or manufactured goods, were intended for military or civilian use, or were food supplies of any kind. This was the method to be used by France in the event of war with Great Britain.

Aube became minister of marine in 1886, igniting two decades of infighting between factions supporting the Jeune École and those who refused to adopt its precepts. The war at sea was to be directed against British shipping, with the direct goal of creating chaos in the financial heart of Britain. The advantages appeared to the Jeune École to be fairly straightforward. The biggest economic advantage lay in the financial savings such a navy would provide for the French government. For the price of a large battleship, scores of torpedo boats—and even gunboats using the same basic hull design—could be produced. Furthermore, this would please many junior officers, who had been stifled in their career ambitions because there were too few command positions in the navy. Finally, the Jeune École promised victory over Britain using methods that harkened back to French naval tradition—the *guerre de course*.

However, there were a number of problems in the Jeune École's strategy. The first was its overestimation of the ability of the torpedo boat to bear up under ocean conditions. An attempt to send a squadron of torpedo boats from Atlantic ports to Toulon in February 1886 revealed the pitfalls of deploying small vessels in rough waters. The crews were unable to eat or sleep regularly and fell victim to seasickness. Had they been on a raiding cruise, it is doubtful they could have performed effectively.[25]

The second was the dismissiveness toward international law shown by Aube, Charmes, and their supporters. They came under harsh criticism from Adm. Siméon Bourgois, a leading French expert on torpedoes and also a student of international law. Bourgois believed that indiscriminate attacks on merchant ships destined for Britain would cause legal problems. The opinion of neutral nations would be crucial for France, and the Jeune École's strategy would only serve to upset them. This could lead neutrals eventually to become enemies, whereas France's history of war with Britain showed that France fared best when Britain had no Continental allies. Furthermore, Bourgois argued that such callous disregard of human life was morally unacceptable: "The admission that commercial warfare conducted by torpedo boats could lead to such excesses is the most forceful of possible condemnations of this use of the new weapon."[26]

The effects of the Jeune École were felt both within the French navy and in navies around the world, as factions in favor of and opposed to its doctrines, strategy, morality, and construction policies sprang up. In France, the revolving-door nature of cabinet politics meant constant swerves back and forth in strategy and construction policies as pro– and anti–Jeune École ministers of marine came and went. One minister would favor renewed battleship construction; the next would halt that program and instead place orders for torpedo boats and gunboats. The dispute captured public attention and was fought out as much in popular journals as in cabinet rooms and naval offices.

The Importance of National Honor in Germany

German naval thought turned toward the contribution of far-flung cruisers in the event of a European war. Part of the answer to this quandary came in an imperial order issued in March 1885 by Kaiser Wilhelm I, "Concerning the Duties of the Commanders of Overseas Ships." This order made it clear that however ship captains were to act in the event of war, their first priority was to serve "the honor of the flag": "In this context, the commander will have to observe above all that, from now on, it is his first duty to inflict as much damage as possible on the enemy. Whether it is better for this purpose for him to turn against enemy warships or to seek through cruiser warfare to damage the enemy's sea-commerce or coastal places of enemy territory is a decision for him alone."[27]

By using the concept of honor, vitally important to a German officer—and especially one aware of the navy's poor showing in its last war—this order ensured that captains overseas at the outbreak of war would seize opportunities for individual action in commerce raiding if no suitable enemy warships presented themselves. Since these German ships would not likely survive long without bases to fall back on, they were to inflict as much damage as possible on the enemy before the inevitable outcome.[28]

Some did not see any prospect of success for the IGN in commerce raiding. German commerce raiding was closer than what the Jeune École envisioned to that intended by international law: stop, search, and seize. Retired admiral Carl Ferdinand Batsch believed that commerce raiding would not succeed, because it had become a more complicated venture than ever before. Furthermore, like Caprivi, he believed that technology had changed the nature of the game. In his opinion, upon the outbreak of war sailing merchant ships would all tie up safely in harbor, leaving the carriage of commerce to the steam-powered merchant ships capable of outrunning their pursuers.[29]

An even more telling series of observations came from Capt. Alfred Stenzel, one of the first teachers at the Marine-Akademie, a school for German naval officers opened in

1872. Stenzel was the first true German naval theorist, and his ideas shaped the thinking of students who took his courses. Stenzel had been the instigator of the calling of an Admiralty Council in late 1883 and had served on the council when it debated the question of commerce raiding.[30] Stenzel's lectures at the Marine-Akademie were published posthumously.[31] In them he did not differentiate between state or private property in his advocacy of commerce raiding. In this he broke with sections of international law that made a clear distinction between the two; it also meant he disagreed with his own government's declaration at the beginning of the Franco-Prussian War in which the Prussian king had stated, "I make war on French soldiers and not on French citizens."[32]

Stenzel contended that *Kleinekrieg*—meaning "little war," or *guerre de course*—had three aims. The first was to stop the flow of arms to the enemy. In language similar to that used in the Admiralty Council's final report, Stenzel's lectures reminded his students of how in 1870–71 the North German Confederation navy had been helpless to prevent American and British arms from reinforcing France. The second purpose was to attack enemy shipping as a source of wealth to the enemy. Unlike the Jeune École's all-out economic warfare, however, *Kleinekrieg* was designed to deny the enemy the means to purchase arms or raw materials on the open markets. The third goal was to destroy ships and cargoes as a way to hurt the enemy through property damage.[33]

Stenzel also made it clear that in order to be successful German commerce raiders needed safe ports from which to operate. This could be seen as yet another call for Germany to acquire its own chain of bases around the world. Stenzel argued that such bases would support German cruisers operating alone or in squadrons, provide ports into which prizes could be brought, and act as places of refuge for merchant vessels. Clearly, thinking of the potential for prizes as he did, Stenzel did not envision a ruthless, Jeune École–style strategy. Rather Stenzel saw commerce raiding as "an important, perhaps the most important branch of the offensive for a small fleet in a war with a superior."[34]

Commerce raiding remained in the minds of many Germans the one useful function that could be performed by its ships and crews spread around the globe. While it was never officially adopted as a viable strategy by the navy, practically everything the navy said and did—construction policy, general orders, stationing of ships and squadrons—implied that commerce raiding was exactly the role to be played. This apparent contradiction even appeared in official planning. In 1889, war plans were created in the event of hostilities with France and Russia. In the first paragraph of the section on cruisers is the statement "Cruiser warfare is not of decisive influence upon warfare." Yet the very next paragraph ends with "Successful cruiser warfare is therefore particularly suitable, lending lustre to a young navy."[35]

Evidently there was no less confusion surrounding the role of the *guerre de course* in the IGN than there was in the French. It was a type of warfare to be pursued, but to what end and how were much less clear. Such questions would continue to plague naval theorists well into the next century.

Conclusions

The Jeune École was a more fully realized theory than anything on that line coming out of Germany at the same time. The driving force behind the Jeune École was the new technology of naval warfare, especially the automatic torpedo. The torpedo was seen as a cure-all—cheap, easy to deploy, and likely to produce enormous results, out of proportion to its expenses. It was also a weapon that worked equally well against both enemy warships and merchant vessels. France's geographical position astride Britain's main commerce routes meant ready accessibility to these routes by swarms of small, swift vessels.

The Jeune École appeared to solve several of France's naval problems at the same time: it provided a credible threat for far less money than Napoleon III's ironclads of the 1850s and 1860s, it promised victory over Britain, and it provided many commands for junior officers stuck in lengthy waits for commands in a small navy of a few large ships. The notion of creating panic in Britain's financial sector, and possibly even in the general public itself, worried about unemployment or the next food shipment, spoke of a war that could be won with minimal exertion and loss of life.

The drawbacks to the Jeune École were first and foremost its reliance on new technology. Charmes's hopes for oceangoing torpedo boats, in particular, proved to be ill founded. Too much of the technology needed by the Jeune École was either too new, insufficiently tested, or simply incapable of what was desired. The other significant drawback to the theory was its callous disregard of the laws of war at sea, especially in terms of morality. The concept of sinking enemy merchant vessels without any warning went against all previous practice, and in the late nineteenth century such immorality was unacceptable. Furthermore, the Jeune École failed to take into account neutral reaction. France could not afford to have Continental foes at the same time as it was fighting Britain, and this policy of sinking any vessel carrying on trade with Britain might push some nations into war.

Finally, the very theory itself became part of the greater chaos of French politics of the period. Cabinets changed very quickly, and successive naval ministers seemed to be pro- or anti–Jeune École based on their spending priorities—battleships or torpedo boats, whales or minnows. The last fifteen years of the century saw far less cohesion within the navy than was needed for an effective buildup to face a potential war. This was especially apparent in 1898, when the possibility of war with Britain increased and France was forced to give way because it was not ready for a major war.

German ideas on commerce raiding wrapped in the rubric of "national honor" would have been equally difficult to implement in the event of war with France. The Imperial German Navy was prepared to fight a more traditional war on commerce than were the French, observing prize rules whenever possible and focusing on material losses rather than economic or psychological pressure. In German warships stationed overseas on political and diplomatic duties, the nucleus of a far-flung *guerre de course* already existed. Since these vessels were equipped for long-range cruising, they were technologically appropriate for the mission.

Like the Jeune École, the German way of commerce raiding had flaws. As maritime technology changed, ships stationed overseas were less and less capable of conducting a *guerre de course* strategy. Already by 1890 navies had begun to realize that dual-propulsion sail and steam warships were no longer practical and had begun to phase them out. Dual-propulsion vessels were less and less likely to catch up to the superior merchant ships being brought into service.

More importantly, the switch to all-steam propulsion revealed the weakness in Germany's world geostrategic position. Without overseas bases to support operations, its cruisers would run out of essential supplies—coal and water—within a matter of weeks. As the laws of neutrality also tightened, warfare in the style of the Union versus the Confederacy—relying on the kindness of sympathetic neutrals—became less and less possible. In the end, German captains were left to fall honorably on their swords, inflicting as much damage on the enemy as possible before being sunk, interning themselves in a neutral port, or making a frantic dash for home waters.

As a result, neither the French nor the German method of commerce raiding appeared to hold out much chance for success. However, the boldness and daring of the Jeune École captured the attention of naval theorists and interested publics outside France. Some sought to emulate French thinking; others sought ways to counter it. Even members of the IGN, while preparing to fight a different kind of *guerre de course,* used the language of the Jeune École to make their points. These ongoing debates had a particularly important impact on future wars, not just in Europe, as the various naval theorists naturally assumed, but also in Asia.

Notes

The thoughts and opinions expressed in this publication are those of the author and are not necessarily those of the U.S. government, the U.S. Navy Department, or the Naval War College.

1. David H. Olivier, *German Naval Strategy, 1856–1888: Forerunners of Tirpitz* (London: Frank Cass / Routledge, 2004), p. 12; Rolf Hobson, *Imperialism at Sea: Naval Strategic Thought, the Ideology of Sea Power, and the Tirpitz Plan, 1875–1914* (Boston: Brill, 2002), pp. 67–68.

2. Bryan Ranft, "Restraints on War at Sea before 1945," in *Restraints on War: Studies in the Limitation of Armed Conflict,* ed. Michael Howard (Oxford, U.K.: Oxford Univ. Press, 1979), p. 45.

3. Arne Røksund, *The Jeune École: The Strategy of the Weak* (Leiden: Brill, 2007), p. 123.

4. Theodore Ropp, *The Development of a Modern Navy: French Naval Policy 1871–1904,* ed. Stephen S. Roberts (Annapolis, Md.: Naval Institute Press, 1987), p. 128.

5. Ibid., p. 23.

6. Lawrence Sondhaus, *Preparing for Weltpolitik: German Sea Power before the Tirpitz Era* (Annapolis, Md.: Naval Institute Press, 1997), p. 96.

7. Ibid.; Olivier, *German Naval Strategy,* p. 68.

8. Richild Grivel, *De la guerre maritime avant et depuis les nouvelles inventions* (Paris: Arthus Bertrand, 1869), p. 278.

9. Ibid., pp. 259–61; Røksund, *Jeune École,* pp. 4–5.

10. Ropp, *Development of a Modern Navy,* p. 157.

11. Hobson, *Imperialism at Sea,* p. 101.

12. Ranft, "Restraints on War at Sea before 1945," p. 51.

13. As quoted in Ropp, *Development of a Modern Navy,* p. 165.

14. Røksund, *Jeune École,* pp. 10–12.

15. Lawrence Sondhaus, *Naval Warfare, 1815–1914* (London: Routledge, 2001), pp. 120–21.

16. Sondhaus, *Preparing for Weltpolitik,* p. 96; Olivier, *German Naval Strategy,* p. 56.

17. Hobson, *Imperialism at Sea,* p. 114.

18. Olivier, *German Naval Strategy,* pp. 105–106.

19. C. John Colombos, *The International Law of the Sea,* 6th rev. ed. (London: Longman, 1967), pp. 648–49.

20. As noted in the memoirs of Adm. Eduard Knorr, who commanded SMS *Meteor* in 1870; N578–Nachlass Knorr, Bundesarchiv-Militärarchiv, Freiburg [hereafter BA-MA].

21. Olivier, *German Naval Strategy,* pp. 137–38. The original is in "Die Entstehung der Preußisch Deutschen Flotte," RM1/1845, BA-MA.

22. Sondhaus, *Preparing for Weltpolitik,* p. 152; Olivier, *German Naval Strategy,* pp. 138–41. Original of quotation found in "Gutachten über die Frage," 21 January 1884, RM1/2795, BA-MA.

23. Sondhaus, *Preparing for Weltpolitik,* pp. 158–59; Hobson, *Imperialism at Sea,* p. 118.

24. Ropp, *Development of a Modern Navy,* pp. 23–24.

25. Ibid., p. 176; Røksund, *Jeune École,* pp. 64–68, disputes some of Ropp's conclusions on this issue.

26. Ropp, *Development of a Modern Navy,* pp. 168–69 (quotation from Bourgois on p. 169); Røksund, *Jeune École,* pp. 26–27.

27. As quoted in Rolf Güth, *Von Revolution zu Revolution: Entwicklungen und Führungsprobleme der deutschen Marine 1848–1918* (Herford, Ger.: E. S. Mittler, 1978), p. 74.

28. Olivier, *German Naval Strategy,* p. 160. For a definition of honor as interpreted by the imperial German military, see Isabel V. Hull, *Absolute Destruction: Military Culture and the Practices of War in Imperial Germany* (Ithaca, N.Y.: Cornell Univ. Press, 2005), p. 317.

29. Vice Adm. [Carl Ferdinand] Batsch, "The British Sea War-Game," *Royal United Service Institution Journal* 34 (1890–91), p. 529, translated from the original German published in *Deutsche Revue* 4 (November–December 1889).

30. Hobson, *Imperialism at Sea,* pp. 136–38; Olivier, *German Naval Strategy,* p. 138.

31. Alfred Stenzel, *Kriegführung zur See. Lehre vom Seekriege,* ed. Hermann Kirchoff (Hannover and Leipzig, Ger.: Hahnsche Buchhandlung, 1913); Hobson, *Imperialism at Sea,* p. 138 and note 107, explains why this work can be taken as primarily Stenzel's and not the editor's.

32. Stenzel, *Kriegführung zur See,* p. 247; Olivier, *German Naval Strategy,* p. 65.

33. Stenzel, *Kriegführung zur See,* p. 247.

34. Ibid., pp. 247–48.

35. "Promemoria betreffend die Kriegführung der Marine gegen Frankreich, gegen Rußland und gegen Frankreich und Rußland zusammen," 16 October 1889, RM1/1656, BA-MA.

CHAPTER SEVEN

Missed Opportunities in the First Sino-Japanese War, 1894–1895
S. C. M. PAINE

The Sino-Japanese War overturned the East Asian balance of power, transforming Japan into the dominant regional power for a century to come. Japan initially mobilized well over a hundred thousand troops.[1] But in the first phase of the war, naval force played the key role. If Japan could not secure command of the sea, it could not safely transport its troops to the mainland. By contrast, if China could not command the sea or at least deny Japan access to the Korean and Manchuria littoral, it would have to depend on its antiquated and fragmented land forces to fight the Japanese on shore.

Japan's military strategy required rapid seizure of control of the sea so that its navy could transport soldiers at will to the mainland. Once it landed troops in Korea, it planned to expel China. After the occupation of Korea, Japan would take the Chinese naval base at Weihaiwei on the southern shore of the broad Gulf of Bo Hai, providing naval access to Beijing. It would follow with an invasion of the metropolitan province of Zhili. If this strategy proved unfeasible, Japan would simply push all Chinese troops out of Korea. However, if disaster struck and China, not Japan, took command of the seas, Japan would focus on its own coastal defenses to prevent a Chinese invasion.[2]

Commerce raiding occurred exactly once during the First Sino-Japanese War, when the Imperial Japanese Navy sank the British-owned steamship *Kowshing* (高陞), killing many of the 1,100 Chinese troops on board and making headlines around the globe. Given the fine navies possessed by both belligerents, the stakes at risk, and the reliance by both on seaborne trade, the absence of commerce raiding is surprising. Had China followed Japan's lead and targeted logistical lines and particularly troop transports, it might have won the war or, at the very least, managed to negotiate more favorable terms at the end of the conflict.

Japan's Strengths

Japanese leaders leveraged their country's Westernized institutions, powerful nationalism, and extraordinary leadership to defeat China. These institutions included a Westernized financial sector that allowed Japan to raise funds internally and externally.[3] Japan also had a six-division national army, based on universal male conscription and organized on the Prussian model, complete with a general staff. The artillery and soldiers' kit were all standardized. Men served for three years in the regular army, followed by four years in the reserves. They were uniformly armed with the Japanese-made single-shot Murata breechloader rifle, while elite units had the new five-round, clip-fed model. During wartime, commanders were given precise objectives but with the freedom to decide how to reach them.[4] In 1892, the army conducted comprehensive war games that left a French observer with "the best impression."[5]

Compared to China, by 1894, Japan had an extensive industrial base to support its military. In 1890 Japan had over 1,700 miles of railways, which grew to 2,100 miles by 1895; China had just 175 miles.[6] By 1880 the Japanese government had opened three shipyards, ten mines, and five munitions factories. Two private shipyards opened in the 1880s. In 1892 Japan manufactured its first locomotive.[7] By 1880 telegraph lines linked virtually all major cities.[8] In contrast, at the beginning of the Sino-Japanese War, China's Self-Strengthening Movement of the preceding thirty years had managed to produce only one arsenal, one shipyard, thirteen mines, and six munitions factories.[9] Whereas the Japanese government had all military production and the country's entire infrastructure at its disposal, in China provincial governors were personally responsible for the defense of their provinces and did not necessarily cooperate with other provinces.

Part of the institutional reforms in Japan concerned public education, which was compulsory. By the early 1890s, there was a large literate public eager to read about the latest battle reports. Once Japanese armies started winning in the field, they received strong public support. In Japan, with higher literacy rates than China, press coverage flourished, with feature stories on common-man heroes and color prints depicting battles. Newspaper readership became national for the first time in Japanese history, and the press coverage was highly jingoistic. In China, a mass press did not develop until after the war, and the public displayed no meaningful support for the Qing armies.

The Japanese creators of these institutions were members of a brilliant generation of leaders who had traveled the world in search of best practices to emulate at home. None of China's top leaders had left the country, let alone conducted extended tours abroad. In addition, Japan's primary civil and military leaders all knew each other. Their generation had overthrown the Tokugawa shogunate and worked together to modernize Japan by introducing a complete array of Westernized institutions. The Meiji generation is known for the ability to match national capabilities to the goal of transforming Japan into a great power capable of defending its national security.[10]

The Japanese pitted their strengths—their Westernized civil and military institutions, their intense nationalism, and their brilliant leadership—against China's weaknesses. These included Manchu minority rule, China's fractured traditional institutional structure, and the compromised telegraphic communications of the man in charge of the Chinese forces, Li Hongzhang. All of these weaknesses combined to produce a best-case military scenario for Japan.

China's Weaknesses

China had bitterly resisted Westernizing its institutions. Its military remained capable of putting down internal unrest and defeating technologically inferior nomad invaders on the empire's inland borders but was no match for Westernized forces, as its defeats in the Opium Wars (1839–42 and 1856–60) and the Sino-French War (1884–85) had shown. In 1892, when a reporter remarked in surprise at seeing Chinese soldiers "actually armed with bows and arrows and firearms of the most antique pattern," the inspector general of the Chinese Imperial Maritime Customs since 1861, Sir Robert Hart, replied, "Most people are surprised, for incredible as it may appear, while possessing as she does some of the finest types of modern warships, the Chinese army is still in many respects absolutely what it was three hundred years ago—merely an armed undisciplined horde. There seems as yet no signs of her waking up from this lethargy."[11]

The Chinese army consisted of "banner" forces, segregated under Muslim, Manchu, Mongol, and Han Chinese banners; the Green Standard Army; "braves," or hired mercenaries; and the foreign-drilled army. The Qing dynasty had ethnically segregated its original military forces, the banners, in order to preserve its leadership in an empire of which Manchus constituted perhaps 2 percent of the population.[12] The banners remained the praetorian guard of the dynasty.[13]

Han Chinese mainly served in the Green Standard Army, as well as in the various provincial armies. By the late Qing period, the Green Standard Army, originally the Ming dynasty army, had been broken down throughout the empire into small, relatively independent units used for constabulary, not combat, duty. Contrary to the name of the Green Standard Army, it was actually trained less for war than for civilian police duties.

The braves, or hired mercenaries, were organized by province to serve collectively as a national guard to be sent to hot spots in the empire. Their units were descendants of the Ever-Victorious Army, which had been trained by foreigners to help put down the mid-nineteenth-century Taiping Rebellion, but most of the foreign advisers had long since departed. The foreign-drilled army simply consisted of those braves who had received some Europeanized military training. It formed a microscopic percentage of the total Chinese force.

Of the Chinese armed forces, the Green Standard troops were the most numerous, followed by bannermen. Braves and trained recruits together represented perhaps 10 percent of the total. But this is just a guess, since the Chinese government did not maintain accurate records. The Chinese army lacked an organized engineer corps, a commissariat, transport services, and a medical division.[14] Even within (let alone among) provinces, guns and ammunition were not standardized, greatly impeding wartime logistics.[15] The Chinese army, and even its comparatively state-of-the-art navy, was also fractured by command, organization, province, race, and training. The Qing deliberately prevented unified command, out of not stupidity but a rational fear that a unified Han army would make short work of Manchu minority rule—fears that would be borne out in the 1911 Revolution, when the recently Westernized Han military toppled the dynasty.

When war broke out in 1894, local inhabitants in Manchuria, far from supporting their troops, lived in dread of their arrival, since Chinese armies acquired their provisions at gunpoint from civilians.[16] In the nineteenth century a succession of huge peasant rebellions, ethnic secession movements, and defeats in foreign wars had cost the Qing their reputation as great leaders. Loyalties were not to the dynasty, much less to the state, but to native place in a society where nationalism had yet to develop. Unlike the members of the Japanese ruling house, the Manchus could not count on the loyalty of their subjects.

To make matters worse, China failed to maintain secrecy. China's key strategist, the seventy-one-year-old viceroy Li Hongzhang, the governor of the key province of Zhili (containing Beijing), did much of his communication by telegraph, which the Japanese apparently tapped. The Japanese, who excelled at cryptography, had broken the Chinese code by June 1894, before the outbreak of hostilities. They made a point of reading Li's messages, with the result that he inadvertently alerted Japan of his evolving plans.[17]

The Manchu division of power by province exacerbated all of these problems. Troops, arsenals, and munitions were the property of their provinces of origin and could not be counted on to support other provinces in the event of hostilities. According to a contemporary German press account, each of the provincial armies was the personal creation of that province's governor. "It is naturally in the interest of each [provincial] Viceroy to retain the fruit of his exertions for himself; in no case is he inclined to come to the assistance of a neighbour who is worse provided, and incur the danger of denuding his own province, for whose safety he is responsible with his head." The same system of individual responsibility applied down through the military ranks. It squelched initiative and promoted defensive rather than preemptive action. "By this system, common action is virtually excluded."[18]

For instance, during the Sino-French War, Li Hongzhang, who controlled the Beiyang Squadron, which had the most modern elements of the Chinese navy, had declined to heed a call for help from the south. The Beiyang Squadron had remained in North China. Ten years later, during the First Sino-Japanese War, the southern fleet would pay him back by ignoring his calls for help.[19] The failure to coordinate (on any level) greatly facilitated the victories of China's enemies.

The Japanese Navy's Decision to Sink *Kowshing*

As war loomed between China and Japan over which country would dominate the Korean Peninsula, China desperately sought third-party intervention, while Japan applied agile public relations and deft diplomacy to discourage the participation of additional belligerents. As Japan's government wished to demonstrate to the industrialized world that the country had become a modern power, the battle for world public opinion became a key theater. The Japanese courted the American and European press and, according to a contemporary report, sought to "capture the European press" by demonstrating that they were "engaged in a crusade against darkness and barbarism, and were spreading light with which they had themselves been illumined by Christendom."[20]

Japan took care to notify Britain days before the outbreak of hostilities that Shanghai lay outside the sphere of hostilities, since commerce raiding had the potential to widen the war by affecting neutral trade.[21] Excluding Shanghai was a costly decision, since it was the location of the main Chinese arsenal, and ships transported its munitions unmolested for the duration of the war.[22] But Shanghai was also a key center of British commerce, which the Japanese did not want to disrupt, for fear that Britain would certainly retaliate. Although in terms of capital ships the Imperial Japanese Navy and the Chinese navy ranked, respectively, third and fourth globally, the Royal Navy ranked number one, as ever, and by a wide margin.[23] So cutting off Chinese arms deliveries through commerce raiding was a nonstarter for Japan. The same was true, for the same reason, of Chinese commerce raiding against Japan's even more important imports.

But a gray area remained, that of targeting merchantmen carrying troops to the theater, a strategy Japan initially pursued. The Korean and Chinese road systems were both deplorable, making the deployment of troops by sea far more efficient than by land. There could be no war if Japan could not ferry its troops to Korea. China, although connected to the theater by land, was actually in a similar position. It possessed only one short railway line from Tianjin to the coast and then north to Shanhaiguan, where the Great

Wall meets the sea. The line was good as far as it went, but it ended a long way from the Korean border. China had no railway lines beyond the Great Wall.[24]

On 20 July 1894, nearly two weeks before the formal declarations of war, the Chinese began massing troops in Korea in preparation for a possible pincer movement on Seoul. They deployed troops two hundred kilometers to the north of the capital, to P'yŏngyang, and forty kilometers to the south, to Asan.[25] Those sent nearer Seoul came by sea along a predictable route, since large ships require large ports. The Japanese military realized that these troops would be easier to dispose of at sea before they dug in on land. So the Japanese navy went after troopships en route to Asan. As tensions continued to escalate, the Chinese commander in Asan requested reinforcements.[26] On 22 July, eight transports set off from Dagu, a fortified city at the mouth of the Hai River, linking Tianjin and Beijing. Two of these transports headed for Asan, where they arrived safely forty-eight hours later.[27]

That day news reached Asan that the Japanese had occupied the royal palace in Seoul. In response, China ordered its warships in Korean waters to return home, to avoid any chance of a confrontation. On 25 July, returning from Asan, two of these warships encountered three Japanese cruisers in the vicinity of Feng Island, on the sea approaches to Inch'ŏn and Asan.[28] The Japanese disabled one Chinese vessel and damaged the other. While pursuing the damaged ship, the Japanese sighted a troop transport, the Chinese-leased but British-owned steamship *Kowshing*. According to its master, Thomas Ryder Galsworthy, the steamer had picked up 1,100 Chinese troops and officers at Dagu. On 23 July, it had left for Asan.

The three Japanese men-of-war now intercepted *Kowshing*. When Chinese officers on board refused to heed the Japanese order to follow them to port, the crew of the ship mutinied and demanded to be returned to Dagu. During several hours of fruitless negotiations, Chinese officers refused either to allow the Europeans to leave the ship or to heed their advice to follow Japanese orders. The Japanese commander, Tōgō Heihachirō, who had studied naval science in Great Britain and was probably aware of the strategies favored by the *Jeune École* and its followers, carefully examined the relevant provisions of international law. The Jeune École had emphasized the great value of attacking such soft targets as troop transports. Upon finding the law to be on his side, Tōgō sank *Kowshing*.[29] In doing so, he strictly followed international law, which gave him the right to sink the transport if it failed to follow his orders.

Unlike most of the Chinese, many of the Europeans could swim and so jumped overboard as *Kowshing* started to go down. The Chinese responded by opening fire on their own foreign advisers.[30] The Japanese made an effort to pick up the Europeans, but not the many drowning Chinese. On the contrary, some Japanese opened fire on them, perhaps because significant numbers of Chinese troops were still armed and putting up a struggle. The Japanese then damaged a Chinese cruiser, captured one gunboat, and sank another, in addition to those damaged earlier, while two other naval ships escaped.[31] Although accounts differ, approximately half of the Chinese troops went down with the ship, were drowned, or were shot. Many observers considered the Chinese troops lost on *Kowshing* to have been the best in the land.[32]

The owners of the sunken *Kowshing*—the British firm of Jardine, Matheson and Company, of opium-trade fame—aggressively sought compensation from the Japanese government. For a time it had the support of members of the British press. But on 26 February 1895 the British government exonerated the Japanese of any wrongdoing: since both China and Japan had been on a war footing as of the hostilities at Feng Island, since *Kowshing* had been carrying troops, and since it had refused to follow reasonable Japanese orders, the British government considered the Japanese to have been justified in sinking it as a hostile ship.[33]

Renowned British international legal experts came out in support of Tōgō.[34] The legal digests concluded

> that at the time of the sinking of the *Kowshing* a state of war de facto existed between China and Japan; that the *Kowshing*, as a neutral ship engaged in the transport service of a belligerent, was liable to be visited and taken in for adjudication, with the use of so much force as might be necessary; that, as one of a fleet of transports and men-of-war engaged in carrying reinforcements to the Chinese troops on the mainland, she was clearly part of a hostile expedition, or one which might be treated as hostile, which the Japanese were entitled by all needful force to arrest; that the force used did not appear to be excessive, either for the capture of an enemy's neutral transport or for barring the progress of a hostile expedition, and that, as the rescued officers were duly set at liberty, no apology was due to the British Government and no indemnity to any person.[35]

This was a victory, both legal and moral, for Japan. Owing in part to these complications—the legal niceties of the *Kowshing* sinking were not settled until two weeks after the destruction of the Chinese navy—neither Japan nor China again engaged in commerce raiding of any variety during the war.

Japan's Winning Strategy

The hostilities at Feng Island and the sinking of *Kowshing* precipitated formal declarations of war on 1 August.[36] Within the first week of the war, the Japanese army settled its operational strategy. It divided its forces into two armies. The First Army, under General Yamagata Aritomo, would invade Korea and enter Manchuria from the north, while the Second Army, under the minister of war, General Ōyama Iwao, would invade Manchuria from the south to take the Lüshun (Port Arthur) naval base, on the Liaodong Peninsula, and, once the two armies met, leave for Shandong Province to take the naval base at Weihaiwei.[37] Possession of Lüshun and Weihaiwei would put Japan in control of the mouth of the Bo Hai and sea access to Beijing.[38]

On 28 July the Japanese defeated Chinese troops at Sŏnghwan, taking the town of Asan with ease the next day.[39] During a three-day period in mid-September 1894, following a month-and-a-half hiatus in hostilities, Japan trounced China on land and sea. On 16 September, Japan overran China's prepared positions at P'yŏngyang after China failed to contest a dangerous river crossing or attack vulnerable supply lines. Chinese forces then retreated all the way to the Chinese bank of the Yalu River bordering Manchuria, ceding Korea to Japan. A month and a half later, the Chinese forces again failed to contest a Japanese river crossing, when Japan pursued over the Yalu. China thereby ceded to Japan the latter's original war objective, which was the expulsion of China from Korea.

Meanwhile, on 17 September, the Japanese navy sought out China's Beiyang Squadron, sinking four Chinese vessels near the mouth of the Yalu River without losing any ships. Again Chinese forces all too easily gave up the initiative. They never again crossed the Yalu–Weihaiwei line, an imaginary line running from the river mouth to the naval base, effectively ceding to Japan command of the sea, allowing it to deploy and supply at will.

A second pair of key battles took place over the winter of 1894–95, at Lüshun and Weihaiwei. China failed to contest the landing of Japanese troops at Huayuankou, on the southern coast of the Liaodong Peninsula about a hundred miles northeast of Lüshun, which had the only facilities adequate to repair capital ships. On 21–22 November Lüshun fell to the ensuing landward attack. Japan then attacked China's second naval base, at Weihaiwei, again by land, first blockading the Chinese fleet in port. On 12 February 1895 Weihaiwei fell, and Japan destroyed the trapped fleet and captured its most modern warships, thus ending Chinese naval power for over a century. Japan threatened to make Beijing the next target, with regime-change implications that forced the Manchus to capitulate. In the resulting Treaty of Shimonoseki Japan gained full control over Korea and valuable rights in Manchuria, and it annexed Taiwan and the Pescadores.

For the Manchus, virtually any outcome would have been better than this one. They had lost every battle and the entire Beiyang Squadron, not in battle but at anchor. The fleet, which had consumed so much of the government's scarce funds, had largely sat out the war. The foreign powers responded to Chinese incompetence in the field by dividing up the failing empire into a welter of spheres of influence—the so-called scramble for concessions portrayed in Chinese history texts as the "Era of Humiliations." The Manchus tried to use Russia as a postwar counterbalance to Japan. In practice, this meant that neighboring Russia, rather than overseas Japan, occupied Manchuria. Japan and Russia soon, in 1904–1905, fought a war on Chinese territory over which would dominate Manchuria. The Qing dynasty limped along for another six years before collapsing in the face of a mutiny of the army it had Westernized in response to its defeat by Japan in 1895. All of these events might have been avoided had China adopted a more aggressive *guerre de course* campaign aimed at interdicting Japan's troop transports and supply ships.

China's Alternative Strategy

In 1894–95 China actually had considerable strengths and Japan notable weaknesses, particularly its vulnerable logistical lines. China had manpower, resources, strategic depth, and interior lines that Japan could not match. If the Manchus had deliberately pulled the fighting inland, Japan would have quickly run out of manpower. Japan could ill afford a protracted war, not only because of its manpower limitations but also because of its financial ones. Japan was not a rich country. Its rural population bore the burden of industrialization, through taxes that allowed the government to invest in industry. The Meiji reforms that had Westernized Japanese political, economic, military, legal, and educational institutions remained deeply unpopular, as indicated by the antagonistic relations between the powerless but obstreperous Diet and the so-called oligarchs who actually ran the country.[40] Had the fighting gone badly, popular anger might have focused on the government, with potential for domestic unrest.

Few realize that at the beginning of the Sino-Japanese War China had a navy that was the equal of and in some ways superior to the Imperial Japanese Navy. Although it was divided into four autonomous squadrons—the Beiyang (northern), Nanyang (southern), Fujian, and Guangdong Squadrons—which did not cooperate, the Beiyang Squadron by itself had a force structure on a par with that of the Imperial Japanese Navy. The other three squadrons together had twenty-one destroyers and six cruisers, generally older.[41]

If Chinese land forces were outdated, the Beiyang Squadron ranked among the top navies of its day. Both the Beiyang Squadron and the Imperial Japanese Navy had state-of-the-art equipment; prior to the war there was no consensus on which was superior.[42] China's best ships were larger and carried bigger guns, while Japan had faster ships overall and an advantage in quick-firing guns.[43] In May 1894 Viceroy Li had made a three-week triennial inspection of the northern coastal defenses. One foreign correspondent present openly admired the "powerful forts, dock yards, work shops, armouries, piers, store-rooms, colleges, hospitals, etc., etc.," at Lüshun, Dalian, and Weihaiwei.[44]

Despite the well-orchestrated naval maneuvers of that May, other commentators took a less sanguine view, noting peculation, inefficiency, lax discipline, insufficient stores,

Chinese and Japanese Naval Force Structure Comparison, 1894

| | JAPAN | | | | CHINA | | | |
| | NAVY | | COAST GUARD | | BEIYANG SQUADRON | | NANYANG, FUJIAN, GUANGDONG SQUADRONS | |
	SHIPS	TONNAGE	SHIPS	TONNAGE	SHIPS	TONNAGE	SHIPS	TONNAGE
Battleships	3	12,840			4	20,480		
Cruisers	8	25,570			9	15,720	6	13,400
Destroyers	12	12,640	6	5,970	10	5,560	21	15,700
Torpedo boats	6	540	18	990	7	763	12	450
Small craft	29	51,590	24	6,960	30	42,523	39	29,550

Source: Tanaka, *Illustrated Volume on the Meiji Navy*, p. 35.

insufficient coal, and "gross nepotism."[45] Although the Beiyang Squadron had China's most modern ships, its guns and ammunition were not standardized. The gunpowder was locally manufactured and not of the appropriate grade for the ships' imported guns. This greatly complicated supplying the correct ammunition in adequate quantities. Moreover, the supply system was ad hoc, with the result that ships were grossly undersupplied with ordnance. Foreign employees had complained about these problems for at least a year prior to the war.[46] China could have addressed them, particularly as war loomed.

Instead of withdrawing behind the Yalu–Weihaiwei line after the battle of the Yalu, China should have waged a *guerre de course* against Japan's troop transports. China still had two naval bases in the vicinity of the anticipated fighting, Lüshun and Weihaiwei, to support commerce raiding, while Japan had none.[47] The Japanese government had long recognized the importance of the ability to conduct rapid and massive troop deployments. It had subsidized the country's large steamship companies, to make their vessels available in times of war. The Japan Mail Steamship Company supplied almost ninety steamships for transport service during the war.[48] Even so, Japan required the services of foreign merchant ships to help supply its troops on the Asian mainland.[49] Japan could not have easily compensated for merchantman sinkings or for troop losses at sea.

Also, Japan's choices for landing points were limited and predictable. Korea's finest harbors lay at Pusan, Wŏnsan, and Inch'ŏn. Each was a long march from the theater, which China had determined would be P'yŏngyang, spending the month and a half between the declarations of war and the battle of P'yŏngyang improving the city's fortifications. In the event, the initial Japanese landings of elements of the First Army took place at Pusan on 19 August 1894 at the southern tip of the Korean Peninsula, nearest Japan, followed by a second landing on 27 August 1894 of a detachment of the First Army in

Wŏnsan, on the central part of the eastern coast, as far away as possible from China's two naval bases.[50] These deployments entailed arduous marches to Seoul. They confirm deep Japanese concerns that Chinese forces might have attacked.

So long as the Beiyang Squadron remained intact, the Japanese military had to be very cautious about transporting these troops, which were the most vulnerable to attack when en route to Korea in transport ships. In late August, after the Japanese had defeated the Chinese at Sŏnghwan and the Chinese had fled northward to P'yŏngyang, the Japanese army concluded that considerations of disease, fatigue, and time made landing at Inch'ŏn worth the risk.[51] Japan had already made three landings there prior to the outbreak of hostilities—completed on 12 June, 16 June, and 28 June, respectively—of a detachment followed by the 9th Brigade, in two groups. The fourth landing took place at the end of August, of the 5th Division, followed by a fifth and final unopposed landing at Inch'ŏn, completed on 1 October, of the 3rd Division.[52]

China could have easily tried to interfere with the final two landings, which were also the largest. Three divisions made up a field army, and Japan had only two. Given the right circumstances China could have crippled two-thirds of an army at Inch'ŏn. The port is a notoriously difficult place for an opposed landing, given its huge tides, constrained approaches, endless mudflats at low tide, and steep coastal embankment. China also had many small ships that could have operated in the narrow approaches to transform the troop landings into suicide runs. It could have posted troops on the high embankments and, if able to delay follow-on landings, eliminated the mud-bound troops below.

Between 24 October and 2 November 1894 Japan undertook an even riskier landing of the Second Army at Huayuankou, about a hundred miles up the southern coast of the Liaodong Peninsula from Lüshun, on the far side of the Biliu River.[53] In theory, this Japanese army landed in the heartland of Manchu territory. China would have totally upset Japanese plans had it contested the landing of the entire Second Army, which in the event would capture the naval base of Lüshun by land and then repeat the maneuver at Weihaiwei. Had the Chinese patrolled the waters along the Liaodong Peninsula and put spotters on the coast, they could have contested the risky landing at Huayuankou, which took a whole week to complete. The Beiyang Squadron had nine cruisers to patrol the 175-mile-long southern coast of the Liaodong Peninsula. It had the entire local population, including fishermen, at its disposal to report on Japanese movements. The vulnerable troop transports, which were merchantmen not designed to withstand attack, could have been sunk by a wide variety of warships; the Beiyang Squadron had seven torpedo boats and ten destroyers fit for the task. Moreover, Japan had no counterparts for China's two largest German-built battleships, which were virtually unsinkable, given the thickness of their hulls and the caliber of ordnance available to the Japanese.[54] Contested landings would have forced a deployment of the Imperial Japanese Navy to cover the transports, some of which China would surely have sunk anyway, particularly if it had used its battleships to great advantage. China could have added to the Japanese losses if it had used mines more extensively. A joint defense from land and sea would have proved costly and perhaps lethal to Japan.

The risk equation also favored China. Japan would lose the war if it lost its navy, because its army could not get to or remain in the theater without the navy. After the defeat in Korea, the loss of the Chinese navy, however, had no effect, other than in morale, on China's armies. Therefore, China could afford to risk its navy in a way that Japan could not. China should have hunted the Japanese transport ships and sunk them.

A *guerre de course* targeting Japanese troop transports by itself would not necessarily have delivered a Chinese victory, but in combination with other feasible adjustments, China might have imposed costs on Japan severe enough to reverse the outcome. China could have also contested Japan's difficult river crossings at P'yŏngyang and the Yalu, causing further attrition of limited Japanese manpower. China could have then drawn Japan inland toward the historical Manchu capital at Mukden, where the Japanese initially headed, awaited the coming of winter, and then targeted Japan's logistical lines. The Chinese could have protected their key naval bases more carefully. Foreign observers did not understand why the defenders abandoned their positions so rapidly. How the Chinese fleet allowed itself to be blockaded at Weihaiwei remains a mystery to this day.

Conclusions

Viceroy Li Hongzhang, whose military expertise had come from land warfare during the Taiping and Nian Rebellions thirty years prior, demonstrated his lack of understanding of naval or joint warfare. Rather than seizing the initiative and choosing the time and place of attack to suit Chinese interests, he fought on Japanese terms. Apparently he wished to minimize the risk to his modern fleet, in order to save his two ironclad battleships to deter attack on the coast and then to use his fleet for convoy duty, to protect Chinese troop deployments, rather than to target those of Japan.[55] He focused on a prevent-defeat strategy—that is, on a strategy to preserve his modern navy intact to fight Japan another day. He must have believed time was on his side, that an attrition strategy by Chinese land forces would yield victory.[56] The viceroy may have assumed that if hostilities dragged on into the winter months, the bitter Korean weather would take its toll on the Japanese troops and their long logistical lines.[57] He did not seem to perceive the possibility of destroying Japan's land forces at sea, where they were most concentrated and vulnerable.

Li apparently thought of naval forces in terms of fleet-on-fleet engagements and convoy duties. Capital ships, in his view, could either target enemy naval vessels or protect

Chinese merchantmen. He did not turn the equation around to see that a naval ship could also attack enemy merchantmen and troop transports, creating a very favorable situation in which the naval ship could sequester or, if necessary, sink a defenseless vessel.

China should have applied its strength—its state-of-the-art navy, which it could afford to put at risk—against Japan's vulnerable troop transports and navy, which Japan could ill afford to lose. Japan could lose the war on land or sea, whereas China had to be defeated on land. Loss at sea would have been financially costly for either but fatal only for Japan. As an island nation, Japan had to keep the sea lines of communication open to deliver and support its troops, but China did not. The Chinese could have responded to the sinking of *Kowshing* by adopting an aggressive *guerre de course* campaign. Japan was far more vulnerable than China to the cost-efficient strategy of disposing of armies at sea rather than fighting them on land.

Notes

The thoughts and opinions expressed in this publication are those of the author and are not necessarily those of the U.S. government, the U.S. Navy Department, or the Naval War College.

1. "Telegraphs in Korea," *Japan Weekly Mail* (Yokohama), 4 August 1894, p. 122.

2. Stewart Lone, *Japan's First Modern War: Army and Society in the Conflict with China, 1894–95* (London: St. Martin's, 1994), p. 33; Takahashi Hidenao [高橋秀直], 日清戦争の道 [The Road to the Sino-Japanese War] (Tokyo: 東京創社, 1995), pp. 484–85.

3. Kenneth B. Pyle, *The Making of Modern Japan*, 2nd ed. (Lexington, Mass.: D. C. Heath, 1996), pp. 107–108.

4. Vladimir [Zenone Volpicelli], *The China-Japan War Compiled from Japanese, Chinese, and Foreign Sources* (Kansas City, Mo.: Franklin Hudson, 1905), pp. 56–62; Japan, Imperial General Staff, *A History of the War between Japan and China*, trans. Major Jikemura and Arthur Lloyd (Tokyo: Kinkodo, 1904), vol. 1, pp. 33–37; "Extracts from an Epitome of the Chino-Japanese War," in *Chino-Japanese War, 1894–95*, ed. N. W. H. Du Boulay (London: typescript, ca. 1903), pp. 4–6; Edward J. Drea, *Japan's Imperial Army: Its Rise and Fall, 1853–1945* (Lawrence: Univ. Press of Kansas, 2009), p. 74.

5. "L'Armée japonaise aux grandes manœuvres de 1892" [The Japanese Army in the Major Maneuvers of 1892], *Le Journal des débats politiques et littéraires* [Journal of Political and Literary Debates] (Paris), 8 August 1894, morning edition, pp. 1–2.

6. Janet E. Hunter, comp., *Concise Dictionary of Modern Japanese History* (Berkeley: Univ. of California Press, 1984), p. 173; W. G. Beasley, *A Modern History of Japan*, 2nd ed. (New York: Praeger, 1974), p. 146; Stanley Spector, *Li Hung-chang and the Huai Army: A Study in Nineteenth-Century Chinese Regionalism* (Seattle: Univ. of Washington Press, 1964), p. 234.

7. Beasley, *Modern History of Japan*, pp. 147–49.

8. Pyle, *Making of Modern Japan*, p. 107.

9. See Bruce A. Elleman and S. C. M. Paine, *Modern China: Continuity and Change 1644 to the Present* (Boston: Prentice Hall, 2010), p. 166, table 10.1, "Self-Strengthening Projects."

10. See S. C. M. Paine, *The Sino-Japanese War of 1894–1895: Perceptions, Power, and Primacy* (Cambridge, U.K.: Cambridge Univ. Press, 2003), pp. 77–87.

11. Julius M. Price, *World*, quoted in "A Talk with Sir Robert Hart at Peking," *North-China Herald* (Shanghai), 21 September 1894, p. 500.

12. Robert H. G. Lee, *The Manchurian Frontier in Ch'ing History* (Cambridge, Mass.: Harvard Univ. Press, 1970), p. 62; Joseph Fletcher, "Ch'ing Inner Asia c. 1800," in *Cambridge History of China*, ed. John King Fairbank (Cambridge, U.K.: Cambridge Univ. Press, 1978), vol. 10, p. 39.

13. "An Epitome of the Chino-Japanese War," pp. 1–3.

14. "China's Armies," *North-China Herald* (Shanghai), 27 July 1894, p. 150.

15. John L. Rawlinson, *China's Struggle for Naval Development 1839–1895* (Cambridge, Mass.: Harvard Univ. Press, 1967), pp. 148–50.

16. Robert Coltman, Jr., *The Chinese, Their Present and Future: Medical, Political, and Social* (Philadelphia: F. A. Davis, 1891), pp. 209–10.

17. Jen Hwa Chow, *China and Japan: The History of Chinese Diplomatic Missions in Japan 1877–1911* (Singapore: Chopmen, 1975), p. 106; Mutsu Munemitsu, *Kenkenryoku: A Diplomatic Record of the Sino-Japanese War, 1894–95*, trans. Gordon Mark Berger (Princeton, N.J.: Princeton Univ. Press, 1982), pp. 266 note 2, 58, 109, 172, 191; Japan, Imperial General Staff, *History of the War between Japan and China*, p. 46.

18. Citing *Der Ostasiatische* in "China's Armies," p. 150.

19. Rawlinson, *China's Struggle for Naval Development*, pp. 81–82, 94, 120–21.

20. "Japanese Embroglio," *Blackwood's Edinburgh Magazine* 158 (September 1895), p. 313.

21. "Shanghai to Be Respected," *North-China Herald* (Shanghai), 27 July 1894, p. 131.

22. Hosea Ballou Morse, *The International Relations of the Chinese Empire* (Shanghai: Kelly and Walsh, 1918), vol. 3, p. 31.

23. Benjamin Franklin Cooling, *Gray Steel and Blue Water Navy: The Formative Years of America's Military-Industrial Complex 1881–1917* (Hamden, Conn.: Archon Books, 1979), p. 136; Tanaka Ken'ichi [田中健一], comp., 図説東郷平八郎目でみる明治の海軍 [Illustrated Volume on the Meiji Navy through the Eyes of Tōgō Heihachirō] (Tokyo: 東郷神社・東郷会, 1996), p. 35.

24. Japan, Imperial General Staff, *History of the War between Japan and China*, p. 21.

25. John Robert Russell, "The Development of a 'Modern' Army in Nineteenth Century Japan" (master's thesis, Columbia Univ., New York, 1957), p. 90.

26. Japan, Imperial General Staff, *History of the War between Japan and China*, p. 48; Bonnie Bongwan Oh, "The Background of Chinese Policy Formation in the Sino-Japanese War of 1894–1895" (PhD diss., Univ. of Chicago, 1974), p. 306; Sun Kefu [孙克复] and Guan Jie [关捷], 甲午中日陆战史 [A History of Land Engagements of the Sino-Japanese War] (Harbin, PRC: 黑龙江人民出版社, 1984), p. 122.

27. U.S. Adjutant-General's Office, Military Information Division, *Notes on the War between China and Japan* (Washington, D.C.: U.S. Government Printing Office [hereafter GPO], 1896), p. 12.

28. Japan, Imperial General Staff, *History of the War between Japan and China*, p. 47.

29. "The 'Kowshing' Affair: Captain Galsworthy's Report," *Japan Weekly Mail* (Yokohama), 18 August 1894, p. 199; George A. Ballard, *The Influence of the Sea on the Political History of Japan* (New York: E. P. Dutton, 1921), p. 142; David C. Evans and Mark R. Peattie, *Kaigun: Strategy, Tactics, and Technology in the Imperial Japanese Navy, 1887–1941* (Annapolis, Md.: Naval Institute Press, 1997), pp. 82–83. A decade later, Tōgō would command the Japanese fleet in the Russo-Japanese War and either sink or capture most of the Imperial Russian Navy in the battle of Tsushima.

30. "'Kowshing' Affair," p. 199.

31. "The 'Kowshing' Affair," *Japan Weekly Mail* (Yokohama), 11 August 1894, p. 163; "Japanese Victory in Korea," *Japan Weekly Mail* (Yokohama), 4 August 1894, p. 124; "Korean Affairs," *Japan Weekly Mail* (Yokohama), 4 August 1894, pp. 125–26; Kencho Suyematsu, "The Report Relating to the 'Kowshing' Affair," *Japan Weekly Mail* (Yokohama), 18 August 1894, pp. 205–208; Evans and Peattie, *Kaigun*, p. 41. See Bruce A. Elleman, *Modern Chinese Warfare, 1795–1989* (London: Routledge, 2001), p. 98.

32. Allen Fung, "Testing Self-Strengthening: The Chinese Army in the Sino-Japanese War of 1894–1895," *Modern Asian Studies* 40, no. 4 (1966), p. 1015.

33. Earl of Kimberley to O'Conor, 26 February 1895, in *British Documents on Foreign Affairs: Reports and Papers from the Foreign Office Confidential Print*, ed. Ian Nish (Bethesda, Md.: University Publications of America,

1989), doc. 137, part 1, series E, vol. 5, pp. 93–94; "The Kowshing Inquiry," *North-China Herald* (Shanghai), 31 August 1894, p. 344; Ballard, *Influence of the Sea on the Political History of Japan*, p. 142.

34. For example, Professors J. Westlake and T. E. Holland wrote letters of support in the London *Times* on 3 and 7 August 1894, respectively; see Samuel Wells Williams and Frederick Wells Williams, *A History of China: Being the Historical Chapters from "The Middle Kingdom"* (London: Sampson Low, Marston, 1897), p. 443.

35. Legal opinion cited in John Bassett Moore and Francis Wharton, *A Digest of International Law, as Embodied in Diplomatic Discussions . . .* (Washington, D.C.: GPO, 1906), pp. 414–15.

36. Vladimir, *China-Japan War*, pp. 245–47.

37. "Movements of the Headquarters of the Japanese Armies" and "A Summary of the Port Arthur Campaign," in *Chino-Japanese War*, ed. Du Boulay; Zhang Qiyun [張其昀], comp., 中國歷史地圖 [A Historical Atlas of China] (Taipei: Chinese Culture Univ. Press, 1980), vol. 2, p. 117.

38. Takahashi, *Road to the Sino-Japanese War*, p. 484.

39. Warrington Eastlake and Yamada Yoshi-aki, *Heroic Japan: A History of the War between China & Japan* (1897; repr. Washington, D.C.: University Publications of America, 1979), pp. 16–21; Russell, "Development of a 'Modern' Army in Nineteenth Century Japan," p. 90.

40. Paine, *Sino-Japanese War*, pp. 88–89.

41. Japan, Imperial General Staff, *History of the War between Japan and China*, p. 32; Tanaka, *Illustrated Volume on the Meiji Navy*, p. 35.

42. Paine, *Sino-Japanese War*, pp. 154, 156–57.

43. Evans and Peattie, *Kaigun*, pp. 38–39; Vladimir, *China-Japan War*, p. 63; "An Epitome of the Chino-Japanese War," p. 4.

44. "The Viceroy Li's Inspection," *North-China Herald* (Shanghai), 8 June 1894, p. 883; "Li Hung-chang's Tour of Inspection," *North-China Herald* (Shanghai), 15 June 1894, p. 925.

45. Ballard, *Influence of the Sea on the Political History of Japan*, p. 136; Alexander Michie, *The Englishman in China during the Victorian Era* (Edinburgh, U.K.: William Blackwood and Sons, 1900), vol. 2, p. 412; Lee McGiffin, *Yankee of the Yalu: Philo Norton McGiffin, American Captain in the Chinese Navy (1885–1895)* (New York: E. P. Dutton, 1968), pp. 107–11; "Finding Fault," *North-China Herald* (Shanghai), 4 May 1894, p. 685; "Li Hung-chang's Tour of Inspection," pp. 925–26; "The Chinese Navy," *Japan Weekly Mail* (Yokohama), 8 September 1894, pp. 294–95; "Shanghai News," *Japan Weekly Mail* (Yokohama), 14 July 1894, p. 42; "Foreigners and the War," *North-China Herald* (Shanghai), 24 August 1894, p. 293.

46. Rawlinson, *China's Struggle for Naval Development*, pp. 148–40.

47. Evans and Peattie, *Kaigun*, p. 38

48. "Compilers' note, Mil. Inf. Div., War Dept. on 'Notes on the Japan-China War' General Inf. Series, No. XIV, Off. of Naval Intelligence," in *Chino-Japanese War*, ed. Du Boulay, p. 1; Eastlake and Yamada, *Heroic Japan*, pp. 434–45.

49. Russell, "Development of a 'Modern' Army in Nineteenth Century Japan," pp. 86–87.

50. Tanaka, *Illustrated Volume on the Meiji Navy*, pp. 34; 117; Zhang Qiyun, comp., *Historical Atlas of China*, p. 117.

51. Lone, *Japan's First Modern War*, p. 25; Sun Kefu and Guan Jie, *History of Land Engagements of the Sino-Japanese War*, p. 119.

52. Okumura Fusao [奥村房夫] and Kuwada Etsu [桑田 悦], eds., 近代日本戦争史 [Modern Japanese Military History], vol. 1, 日清・日露戦争 [The Japanese-Qing and Russo-Japanese Wars] (Tokyo: 同台経済懇話会, 1995), map.

53. Museum of the Chinese People's Revolutionary Military Affairs [中国人民革命军事博物馆], ed., 中国战争史地图集 [Map Collection of Chinese Military History] (Beijing: 星球地图出版社, 2007), p. 205; Okumura and Kuwada, *Modern Japanese Military History*, map.

54. Tanaka, *Illustrated Volume on the Meiji Navy*, p. 35; Evans and Peattie, *Kaigun*, p. 38.

55. Samuel C. Chu, "The Sino-Japanese War of 1894: A Preliminary Assessment from U.S.A.," 近代史研究所集刊 [Proceedings of the Institute for Modern History] 14 (June 1985), p. 366; Ballard, *Influence of the Sea on the Political History of Japan*, pp. 144–46.

56. Fung, "Testing Self-Strengthening," pp. 1019–20.

57. "The Japanese in Corea," *North-China Herald* (Shanghai), 10 August 1894, p. 240; "Peking," *North-China Herald* (Shanghai), 9 November 1894, p. 774.

Chinese Neutrality and Russian Commerce Raiding during the Russo-Japanese War, 1904–1905
BRUCE A. ELLEMAN

During the Russo-Japanese War, the Russian squadron based at Vladivostok conducted commerce raiding against Japanese inbound and outbound transport ships and merchantmen, even as Russian warships and "volunteer" cruisers stopped and searched commercial shipping far to the west in the Red Sea. While the campaign was initially quite effective, a truly successful Russian *guerre de course* operation would have required more numerous naval bases. Japan profited, therefore, from China's declaration of neutrality, which excluded the Russian navy from its ports. Although the ground portion of the conflict took place on Chinese territory—mainly in Manchuria—the Chinese government and its citizens did not join the war on behalf of either side.

Chinese neutrality made Beijing legally responsible for closing Chinese ports to Russian commerce raiders, interning enemy ships, and monitoring the behavior of any Russian sailors and soldiers on parole. Both the Russian and Japanese governments at times accused China of not acting as a truly neutral power. But China's policy of neutrality was left intentionally vague from the very beginning to give the greatest range of action to all of the interested powers. Often the undefined nature of China's neutrality led to divergent interpretations of international law and diplomatic practice.

This chapter will begin by examining Russia's commerce-raiding operations in East Asia and in the Red Sea. It will then turn to the international diplomacy surrounding China's declaration of neutrality, Japan's assistance to China to eliminate Russian extraterritoriality, and the ongoing Russo-Japanese disputes over the "incomplete" neutrality of Chinese ports. Despite all of the problems, neutrality allowed China to avoid becoming embroiled as a belligerent in the Russo-Japanese War, and Beijing eventually halted Russian commerce raiding from Chinese ports.

Russian Commerce Raiding Based in Vladivostok

Soon after war broke out between Russia and Japan on 8 February 1904, ships from the Russian naval squadron based in Vladivostok carried out a successful commerce-raiding strategy. This effort was conducted by three heavy cruisers, *Riurik, Rossiia,* and *Gromoboi,* supported by a single armed merchantman and a number of torpedo boats.[1] On the basis of orders that had been issued by Vice Adm. Oskar Victorovich Stark on 9 January 1904, any "valuable prizes captured at no great distance from Vladivostok may be sent to that port; all the remainder must be sent to the bottom without considerations of pity and without hesitation."[2] This order not only followed the teachings of the *Jeune École* but took account of the highly publicized sinking of *Kowshing* during the Sino-Japanese War.

The Japanese navy and the Russian Pacific Fleet were approximately equal in numbers of battleships, with seven apiece, but Japan had six armored cruisers to Russia's four, plus Japan had an additional eighteen protected cruisers and ten small cruisers. As for destroyers, Russia had twenty-five to Japan's nineteen, but Japan's eighty-five torpedo boats could easily outmatch Russia's twenty-five. Given these uneven numbers, a fleet-on-fleet engagement would have been very one-sided, so Russia wanted to avoid one. Good strategic sense dictated that the Russian forces in the Pacific avoid battle until the arrival of the Baltic Fleet. In the meantime, they adopted a strategy emphasizing commerce raiding. During the first year of the war, Russian cruisers attacked transport ships moving troops and supplies from Japan to Korea; they also transited the Tsugaru Strait and operated along the east coast of Japan, at one point cruising as far south as the mouth of Tokyo Bay and beyond.

At the beginning of the conflict the Japanese merchant marine was approximately 50 percent larger than that of Russia, an estimated 979,000 tons compared to Russia's 679,000 tons.[3] The primary strategic goal of the Russian squadron based at Vladivostok was to stop Japanese military supplies being shipped to Korea. Secondary goals included interfering actively with neutral shipping to Japan and by so doing raising the insurance rates paid by commercial shippers. In fact, over the course of several months dozens of Japanese transport ships, as well as many "neutral" British, German, and American commercial ships, were detained or sunk, interrupting the flow of food, guns, ammunition, and vital railway supplies to Manchuria.

The Russian *guerre de course* campaign had a significant impact on Japanese operations in Manchuria. On 15 June 1904, for example, *Gromoboi* sank a three-thousand-ton commercial vessel carrying Japanese wounded back from Port Arthur, as well as *Hitachi Maru,* a six-thousand-ton transport. The sinking of *Hitachi Maru* destroyed eighteen eleven-inch Krupp siege guns en route to Port Arthur: "The loss of arms, locomotives, and Krupp siege guns intended for the bombardment of the Port Arthur fortress was a serious blow for the Japanese." Without these guns the Japanese assault on Port Arthur was delayed by "many months," and only in early December were replacements delivered and emplaced.[4] During this period Japan lost the equivalent of a quarter of its force surrounding Port Arthur. This delay therefore helped Russia prolong its control over this crucial port. Port Arthur fell almost exactly one month, on New Year's Day, after the guns arrived.

By contrast, Russian losses during these *guerre de course* operations were comparatively minor. For example, the protected cruiser *Bogatyr* was permanently disabled in May 1904, but by hitting a rock, not through enemy action. During this early period of the conflict, therefore, Russia's naval operations in East Asia appeared to be highly successful. These attacks proved to be a "sober reminder that the fate of the Japanese army depended on [Admiral] Togo's ability to command the sea and its supply lines to the home islands."[5] At the same time, Russia was conducting an apparently equally effective commerce-raiding operation to the west, in the Red Sea.

Russian Commerce Raiding in the Red Sea

Almost immediately after war broke out, the Russian navy adopted a *guerre de course* strategy in the Red Sea, to halt the shipment by neutral countries of "contraband" to Japan. The list of Russian contraband items was long, including not only obvious military supplies like guns and ammunition but also any technological products for use in constructing telegraphs, telephones, or railways. In addition, all food—including staples like rice—was prohibited, as were all "fuel" products.[6] Over time it became clear that this latter term included "coal, naphtha, alcohol, and similar materials."[7] This contraband list was potentially of concern to the British, major exporters of coal to Japan.

The best place to intercept the coal shipments was in fact not in East Asia, where Japanese warships might be able to interfere, but closer to the source. One highly strategic choke point was in the Red Sea, just south of the Suez Canal. On 20 February 1904, a group of five Russian cruisers and three torpedo boats stopped the P&O steamship *Mongolia* in the middle of the Red Sea, at latitude 18° north, longitude 39° east. The search of *Mongolia* only took eight minutes. When no contraband items were found, the captain of the Russian cruiser politely signaled, "I beg to be excused," and released the ship.[8]

However, other British merchant ships were not so lucky. On 21 February 1904 Russian warships approximately twenty miles south of Suez, on the "high seas," seized and retained for several days SS *Frankby*. After being released, the ship had to return to Port Said, at the southern exit of the Suez Canal, to be reloaded, leaving again on 29 February after a delay of eight days.[9] The owners of SS *Ardova,* which was detained on 17 July and released a week later on 25 July, later claimed from the Russian government a total of £3,218 sterling in compensation for "the seizure and detention of the vessel."[10]

British shipping companies and insurers were outraged at the Russian operations, since Britain remained neutral in the Russo-Japanese War. Insurance rates began to increase to offset the risk of being stopped and detained. The insurers were particularly upset, because many of these commercial ships had departed Britain prior to the outbreak of war, carrying what had been at that point not contraband but "peaceful" cargoes.[11] Their worst fears were realized in early March 1904, when the Russian government announced the creation of four prize courts, two in Europe, at Libau (in Latvia) and Sevastopol, and two in Asia, at Vladivostok and Port Arthur. The creation of European prize courts suggested that neutral ships in the Red Sea caught transporting contraband could be confiscated by Russia and treated as war prizes.

The British protested, noting that in December 1884 the Russian government had firmly declared that coal should never be considered a contraband item.[12] In early March 1904 the British government even considered sending Royal Navy ships to the Red Sea to protect British commercial shipping. However, concern about being drawn into the war militated against such a policy, since sending warships "would likely lead to increased friction, suspicion and unrest."[13] Instead, the British took advice given on 2 March 1904 by Capt. Edmond J. W. Slade, RN, Senior Officer, Red Sea Division, who pointed out that the Russian cruisers were "using an anchorage in neutral waters as a base from which to exercise the belligerent right of search."[14] Rather than trying to pressure the Russian government directly, therefore, the British urged the Egyptian government to assert its rights as a neutral state and deny the use of its ports to Russian warships. Almost at precisely the same time, similar concerns about Russian warships making use of neutral ports had convinced the Chinese government as well to declare its neutrality.

The Chinese Declaration of Neutrality

From almost the very beginning of the war, both Russia and Japan had urged China to declare its neutrality. There were many reasons for this, including Russian concern about a Sino-Japanese alliance, should China decide to back Japan. There were also fears that France and Great Britain might get dragged into the war if it embroiled China. But even more importantly, Tokyo was concerned that Chinese ports might be used by Russian vessels to launch raids on troop transports and supply ships. One way to eliminate this possibility would be to convince China to remain strictly neutral, which would close its ports to Russian commerce raiders.

During February 1904 Japan worked hard to secure China's neutrality. According to one account, obtaining China's promise to remain neutral was the "most hazardous aspect of Japan's diplomatic struggle."[15] The Japanese feared that a too-public Sino-Japanese alliance against Russia might be proclaimed by St. Petersburg as proof of the oft-cited "yellow peril." Not only would this claim gain sympathy from other European states, especially Germany, but it might inhibit Great Britain, Japan's ally since the 1902 Anglo-Japanese alliance, from assisting Tokyo. Therefore, Japan wanted to make clear that the Russo-Japanese War did not represent warfare of East against West.

A second factor for the Japanese, which was discussed by the cabinet as early as 28 December 1903, was the fear that the war might escalate to include other European countries:

> If we were to let China enter the war, the situation might become difficult and we could not be sure that complications might not take place. In view of the Franco-Russian declaration [of March 1902] which was issued as a counterpoise to the Anglo-Japanese Alliance, it might be that, if a third country like China were to enter the war against Russia alongside Japan, France would have no alternative but to come to Russia's aid. If France were to help Russia, Britain would also be required to support Japan, thus leading ultimately to the involvement of all world powers. We therefore believe that it would be most opportune if China and all other countries stayed neutral and thus restricted the scope of the war to Russia and Japan.

As a result, "any Chinese suggestions for taking part in the war had to be avoided like the plague."[16]

From the beginning, St. Petersburg was more lukewarm than Tokyo about China's neutrality. Some thought Beijing might be tempted to join Russia's side. However, considering that St. Petersburg had violated its promises to withdraw troops from Manchuria, there was more reason to believe that China would support Japan. Therefore, following the advice of Germany, France, Great Britain, and the United States, the Russian government agreed to accept China's declaration of neutrality, while simultaneously waiting for China to commit a breach of neutrality that would work to Russia's advantage. In the meantime, St. Petersburg undoubtedly hoped that its commerce-raiding operations in East Asia would have a significant impact on the war.

Russian Commerce Raiding in East Asia

During the initial months of the war, Russian warships operating out of Vladivostok conducted a highly disruptive commerce-raiding operation. On 19 July 1904 the Russian ships sortied again from Vladivostok for a two-week cruise, returning on 1 August. Passing through the Tsugaru Strait, the Russian ships cruised along Japan's eastern coast, sinking or capturing a total of eight merchant vessels. Before returning north they passed the mouth of Tokyo Bay, causing panic in Tokyo.[17] Since they never encountered any Japanese warships, the total Russian losses were negligible—one torpedo boat grounded and abandoned.[18]

Under the command of Adm. Kamimura Hikonojō, the Japanese navy force protecting Tokyo Harbor wished to intercept the Russian ships, but to no avail. Kamimura's force consisted of four armored cruisers, five protected cruisers, and two flotillas of torpedo boats. Since he had to keep his ships ready for action in case the main Russian fleet at Port Arthur broke out of the blockade and attempted to flee to Vladivostok, he could not afford to chase the Russian commerce raiders. While this gave the Japanese public the impression that Kamimura was ineffectual, his decision actually followed orders given by Adm. Tōgō Heihachirō, whose top priority was dealing with the Russian ships trapped at Port Arthur. Tōgō admitted that "there is no special plan to deal with the large cruisers" working out of Vladivostok.[19]

On 11 August the Vladivostok squadron again left port and sailed for Korea. Admiral Kamimura sortied his fleet and encountered the three Russian armored cruisers on 14 August about forty miles off Pusan. He was now in command of four armored cruisers, two protected cruisers, and a number of torpedo boats, although the latter had not been able to keep up with the faster vessels and so played no role in the ensuing battle. The slowest Russian vessel, *Riurik,* was quickly damaged; its commanding officer eventually scuttled the ship so it would not be taken by the Japanese. Meanwhile, *Rossiia* and *Gromoboi* were both badly damaged. These two cruisers returned to Vladivostok and were eventually repaired, but they rarely left port during the remainder of the war. For all intents and purposes, this battle put an end to Russian commerce raiding from Vladivostok.

Up to this point, the Russian ships working from Vladivostok had sunk a total of fifteen ships and captured another three.[20] Because most of the Russian Pacific Fleet remained blockaded at Port Arthur, the "only practicable role for the weaker Russian squadron was the *guerre de course* against enemy communications," and for a time these ships were "moderately successful."[21] Julian S. Corbett, in his two-volume study of the Russo-Japanese War, points out that this campaign was useful, acknowledging that the Japanese navy could not "ensure absolute immunity for sea communications, so long as any fragments of the enemy's fleet remain[ed] in the theatre of operations."[22] This observation also largely describes British efforts to convince Egypt to enforce its neutrality in the Red Sea, thereby halting Russian commerce raiding there.

The Diplomatic Consequences of Egyptian Neutrality

On the one hand, the British government wanted to halt Russian patrols in the Red Sea, but on the other hand, it did not want to become an active belligerent. As a compromise solution, it pressured the Egyptian government to enforce Egyptian neutrality in its Red Sea ports. As noted by the North of England Protecting and Indemnity Association in a 17 June 1904 letter of protest, the Russian naval patrols were actively infringing on "international law" when they made "use of neutral ports from which to search neutral vessels."[23]

Beginning in July 1904, the Russian navy shifted from using conventional warships to relying on Russian "volunteer ships" to enforce its *guerre de course* strategy in the Red Sea. On 13 July, for example, it was reported that the Russian volunteer ship *Petersburg* had stopped and searched *Menelaus* and *Crewe Hall* two days before, at latitude 18° north, longitude 40° east.[24] Five days later, a second Russian volunteer ship, *Smolensk,* stopped SS *Persia*. Of even greater concern, in early July the Russians stopped a German ship, *Prinz Heinrich,* and removed thirty-one sacks of mail and twenty-four sacks of parcels addressed to Japan. On 18 July, presumably after searching these sacks for anything of interest, *Persia* was ordered to deliver these mail sacks to Japan. The British considered such actions to "usurp the functions of His Majesty's Post-Master General, detaining mail steamers and causing them to carry mails."[25]

After these reports, it became clear that the two volunteer ships *Petersburg* and *Smolensk* had transited the Dardanelles and Suez as commercial ships and only afterward "transformed into cruisers," a situation that the British government argued violated the 18 March 1856 Treaty of Paris, which denied Russian warships access from the Black Sea to the Mediterranean.[26] A "Very Confidential" report from Charles Hardinge, a British Foreign Office official stationed in St. Petersburg, stated that "on emerging into the Red Sea [the Russian volunteer ships] had thrown away the mask, hoisted the naval flag, mounted guns, and initiated a crusade against the merchant shipping in those waters."[27] According to R. Tupper, captain of HMS *Venus,* these Russian cruisers were being "mothered by the German S.S. *Holsatia*" and if allowed to refuel at will at local ports could operate indefinitely in the Red Sea.[28]

The British government loudly protested to St. Petersburg the decision to use volunteer ships to conduct a formal naval operation in the Red Sea. In particular, it argued that the Russian ships should not be allowed to buy coal in neutral ports in the Red Sea. In July 1904 British-controlled ports, including Aden, were ordered not to allow the Russian ships to purchase coal.[29] Meanwhile, at Britain's urging the Egyptian government had already announced in late March 1904 that Russian ships stopping at Port Said could purchase only enough coal "to take them to the nearest port on their direct route."[30]

In response to these protests, the Russian government finally agreed to recall *Petersburg* and *Smolensk* and replace them with two regular warships, *Don* and *Vral*. However, rather than going north through the Suez Canal back to Russia, the two volunteer ships proceeded southward. On 22 August *Smolensk* stopped and searched SS *Comedian* just eighty miles from East London, South Africa. St. Petersburg disingenuously claimed that it was unable to communicate with its ships to recall them and so agreed to allow British ships to transfer a message telling *Petersburg* and *Smolensk* to "cease to act as a Cruiser."[31] This message was finally delivered by HMS *Forte* on 5 September 1904 off the coast of Zanzibar.[32]

Only after a British warship delivered to the two volunteer ships notice of their recall did they stop their search efforts and agree to return to Russia, passing Gibraltar on 4 October and arriving in Libau in mid-October. It was clear to the British that St. Petersburg's failure to recall the ships promptly was "proof of bad faith and lack of good will"; Charles Hardinge commented that it was a "pity that the Russian Government had not made a similar effort six weeks earlier."[33] The Foreign Office called the Russian conduct "shifty."[34] This commerce-raiding strategy in the Red Sea helped set the pattern in East Asia, where Russian commerce-raiding ships attempted to base their efforts in neutral Chinese harbors.

Chinese Internment of Russian Ships

In line with China's standing as a neutral country, Beijing officials promised to monitor any belligerent ships claiming succor. As Japanese fleets progressively dominated the seas, more and more Russian ships sought safety in Chinese ports. In August 1904, for example, the Russian warship *Reshitelny* took refuge in the northern Shandong port of Qifu (Chefoo). The next day Japan seized this Russian ship, disregarding China's neutrality, and towed it out of the port. In the face of numerous protests, the Japanese argued that China's neutrality was imperfect—since China was clearly incapable of enforcing its duties as a neutral, Japan had no choice but to act alone. This legal interpretation was to serve Japan's needs quite well during the 1904–1905 conflict.

According to international law, Chinese neutrality included jurisdiction over ships that sought refuge in its ports. On 10 August 1904 *Reshitelny,* under the command of Lieutenant Roshchakovskii, entered Qifu, a neutral port in Shandong Province. Roshchakovskii reportedly told the Chinese naval authorities at Qifu that he intended to disarm, but he did not immediately do so. There were legitimate fears in Tokyo that after resupplying, the ship might leave to conduct commerce raiding. For that reason, on the following day, 11 August 1904, a Japanese party under the command of Lieutenant Terashima boarded the ship. After determining that it had not yet disarmed, he ordered *Reshitelny* "either to get out into the open sea for a fight or prepare to be towed out."[35]

In response to Japan's ultimatum, Roshchakovskii ordered that charges be set to blow up his ship, which would be technically a violation of his request for refuge in a neutral port. Tension quickly increased. Roshchakovskii struck Terashima, and the two fell overboard. Terashima was immediately pulled out, but Roshchakovskii reportedly remained in the water for almost an hour; several nearby Chinese ships evidently refused to allow him on board. Fighting between the Japanese party and the Russian crew resulted, in which one Japanese was killed and fourteen were wounded. *Reshitelny*'s magazine exploded, apparently in an attempt to scuttle the ship, but it remained afloat. *Reshitelny* was soon towed out of port by a Japanese ship, after which Japan claimed the ship as a spoil of war.

This incident caused quite a stir in China. The Chinese government was represented at Qifu by Vice Adm. Sa Zhenping, who had personally agreed to give the Russian ship refuge. During this dispute, Sa protested Japan's action, but to no effect. Evidently Sa was "so deeply hurt that he handed over the command of his squadron to one of his captains."[36] Later Sa petitioned the empress dowager to relieve him from duty and punish him, but she declined.[37]

The U.S. government was particularly concerned by Japan's apparent failure to respect Chinese neutrality. On 17 August 1904, President Roosevelt met with the Japanese ambassador Takahira Kogorō to "urge the surrender of the Russian ship to Chinese jurisdiction." On instructions from Tokyo, Takahira explained that Russian ships were making only a "pretense" of disarming and were still "in a position to take to the sea." In other words, he thought they were planning to use China's neutral ports as bases to conduct *guerre de course* operations.[38] To offset this possibility, Japan was "helping" China to enforce its neutrality against Russian ships.

For its part, the Russian government, in a parallel strategy to its successful commerce raiding in the Red Sea, was clearly trying to use neutral ports in Asia to carry out further attacks on Japanese shipping. In fact, by late summer 1904 there were already a fairly large number of Russian naval vessels docked in Chinese ports, not far from the war zone in southern Manchuria. Therefore, the concern in Tokyo that these Russian ships might continue to operate as independent commerce raiders to attack Japanese transports and logistical lines was legitimate. Convincing China to enforce its neutrality against the Russian crews became an important diplomatic objective, since only with Beijing's active cooperation could Tokyo stop Russian commerce raiders from using its ports. To do this effectively Beijing had to abolish, or at the very least modify, Russia's right of extraterritoriality, or diplomatic exemption from local law.

Chinese Attempts to Abolish Russian Extraterritoriality

A second important development limiting Russian commerce raiding in East Asia was China's promise to detain the Russian crews and enforce their paroles. As part of

China's decision to declare neutrality, Beijing assumed responsibility for any interned belligerents, including all Russian crews. The Chinese government soon published a guide entitled *Manual of Neutral Public Law,* which stated, "Any belligerent warship, which has entered the area of neutral territories in consequence of her defeat in battle, shall be disarmed and detained until the end of the war." Furthermore, the captain, officers, and crew were required to give paroles, thereby promising not to fight again during the remainder of the conflict.[39]

In 1904 China agreed to intern paroled Russian sailors as part of its responsibility as a neutral power. However, it was virtually impossible to regulate the behavior of the interned Russian sailors so long as extraterritoriality existed. After receiving parole and promising to remain in China, many Russian officers and their crews simply broke their promises and, after returning to Russia, rejoined the Russian fleet. This undermined the whole reason for paroling them in the first place. With the tacit support of Japan, therefore, China determined that extraterritoriality could not possibly apply in these cases, thus abrogating the right of Russians to this privilege. Such a legal determination worked in 1904–1905 to the advantage of Japan, which wanted the maximum administrative control over the interned Russian sailors.

However, many Russian officers and sailors still breached parole and rejoined the Russian navy. The cases of *Variag* and *Koreets* were perhaps the most famous in this connection. After the beginning of hostilities, these two ships left the Korean port of Inch'ŏn to engage the Japanese fleet. Damaged beyond repair, *Variag* was destroyed by fire, while *Koreets* was dynamited. Their officers and crews were taken to Shanghai and released after pledging "not to come again to the north of Shanghai during the war." However, many Russian officers soon left Shanghai and returned to Russia. When the Russian naval reserves were called to service, the London *Times* reported "that both officers and men of the *Koreetz* [sic] and the *Variag* had taken service again."[40]

Japan protested such Russian violations of parole on many occasions. In particular, on 21 January 1905, a memorandum was addressed to Beijing denouncing the recent violation of parole of both the captain and first officer of the Russian torpedo-destroyer *Rastoropny*. This protest forcefully concluded, "I believe the present event must have arisen from the imperfectness of your internment of the Russians, therefore, I must advise Your Excellency to put them under strict guard, and not to repeat such a troublesome occurrence."[41]

The Chinese officials were at a loss how to stop the Russians, however, since they did not have any legal authority over them. China's declaration of neutrality was in direct conflict with its treaty obligations to uphold the extraterritorial legal privileges of foreigners. Prompted by protests from Japan, however, Chinese officials determined that these circumstances demanded that China become the "constable of International Law." Therefore "it was obviously requisite for her to have the corresponding high sovereign powers necessary to enable her to fulfill those duties." In practice, this meant that "extraterritoriality could not be allowed to interfere with her neutral functions."[42]

As a result of this legal reinterpretation, China determined that its neutrality outweighed the Russian sailors' extraterritorial privileges. When Russian sailors refused to "give parole," Chinese officials were authorized to retain them "as prisoners on board a Chinese man-of-war without reference to the Russian Consulate."[43] One case that potentially could have tested China's new legal interpretation was known as the "Shanghai murder case." On 15 December 1904, a sailor from *Askold* named Terente Ageef killed a Chinese civilian. Instead of being handed over to Chinese authorities, Ageef was retained by the Russian consul at Shanghai. Over Chinese protests, a special court composed of Russians tried and sentenced Ageef to four years' hard labor, although he was imprisoned in Shanghai's French concession.

The Chinese Foreign Ministry sent an official protest via the Chinese minister in St. Petersburg, demanding the extradition of Ageef. China also asked Russia to agree that in all future cases "China shall, without the interference of the Russian authorities, have power to try those Russians who may, in violation of China's neutrality, attempt to escape and are arrested."[44] The Russian naval authorities eventually began to extradite criminals to China, but it was too late to retry Ageef. The Chinese government stated that in any future case in "which the prisoner was in their hands the case should be tried either at the Consular Court with a Chinese officer on the bench or on board a Chinese man-of-war by the Consul."[45] This Chinese decision embodied a new interpretation of neutral rights and responsibilities.

A New Interpretation of Neutral Rights and Responsibilities

Faced with Japan's insistence that all Russian ships observe China's neutrality and that Russian crews honor their paroles, the Russian government accused the Chinese government of "complicity" with Japan. Pointing specifically to Vice Admiral Sa's presence in Qifu when *Reshitelny* had been taken by force, St. Petersburg demanded "punishment of the Chinese Admiral, and the restoration of the ship."[46] On 15 August 1904 the "Russian Minister P. M. Lessar accused the Chinese government of not fulfilling its responsibilities for the protection of neutral vessels."[47]

In response to these Russian criticisms, however, the Japanese government argued that "the neutrality of China is imperfect and conditional." Tokyo concluded, "Russia cannot escape the consequences of an unsuccessful war by moving her army or navy into those portions of China which have by arrangement been made conditionally neutral."

According to Japan's main argument, "Experience has shown that China will take no adequate steps to enforce her neutrality laws."⁴⁸ In short, if the Chinese government was unable to enforce its neutrality, the Japanese government would.

Japan's arguments embodied a new interpretation of neutrality laws. According to one legal analysis, this controversy was likely to "open up a new chapter in the history of warfare." This was especially the case because "at present, everything seems to depend upon the capacity of the nation whose neutrality is thus affected to maintain that neutrality, if necessary by force of arms."⁴⁹ If this interpretation were generally accepted, only strong countries could afford to declare their neutrality; weak countries could not enforce it.

Later, following the defeat of the Baltic Fleet in the battle of Tsushima, Russian ships fleeing to the German port at Qingdao were quickly interned and their crews paroled. One Russian ship, *Lena,* was ordered to sea to continue fruitless *guerre de course* operations, but its crew quickly mutinied; the ship arrived in San Francisco on 12 September 1904, where it was interned by American authorities.⁵⁰ However, when several Russian ships appeared in Shanghai and refused to disarm, it appeared that Japan might be forced to attack again. The American Secretary of State, John Hay, was so disturbed by this possibility that he "was ready to throw up in despair the whole business of China's neutrality in her ports" and instead make the ports "spheres of hostility," in which Japan and Russia would be free to fight it out. Fortunately, Russia backed down and agreed to disarm and intern these ships.⁵¹

During the Russo-Japanese War, Great Britain and Japan were generally able to enforce their interpretations of the obligations of a neutral country. Russia's attempt to conduct commerce-raiding operations from neutral ports in the Red Sea and to keep Japanese ships busy guarding neutral Chinese ports in which its ships had taken refuge ultimately failed. In 1905 one scholar concluded that Russia intended to waste Japanese resources and so there was a "probability that Russia has been cynically using the uncertainty which prevails as to the treatment of refugee ships to assist her materially in her warlike operations."⁵² Japan's tactics, however, proved more effective, especially after the resolution of the *Reshitelny* incident, when it became clear that Japan would not allow Russian warships to use Chinese neutral ports for commerce raiding.

Conclusions

The Russian navy's policy of using volunteer ships in the Red Sea largely backfired, even while its Pacific squadron based in Vladivostok carried out a successful, albeit short-lived, *guerre de course* operation. According to British statistics, forty-four British, seven German, and eight other neutral ships—a total of fifty-nine vessels—were directly impacted.⁵³ But this was a relatively small percentage of the total ships trading with Japan.

As one historian of the war concludes, "With the defeat of the final attempt at a naval sortie from Port Arthur and with the destruction of the Vladivostok squadron, both occurring in August [1904], the Japanese had acquired total sea control in the theater of war."[54] Japan met any further Russian attempts to continue *guerre de course* from Chinese ports with firm resistance. As a result, the overall impact of Russian commerce raiders proved insufficient to alter the outcome of the war.

Without a doubt, China's neutrality during the Russo-Japanese War kept Russian commerce raiding to a minimum and thereby ultimately helped Japan win the war. Initially, it was unclear to the belligerents whether Chinese neutrality would help or hurt their respective military situations. During the war, neither side was satisfied: "China, due to her ignorance of international law and the inexperience of her officials, could not help making blunders in her endeavor to maintain a strict neutrality. Consequently, she was accused by both belligerents as having favored one party against the other while the war was going on."[55]

But taken as a whole, China's neutrality played an important role in Japan's victory, since it ensured that Russian commerce-raiding operations could not be conducted from Chinese ports. More importantly, no Russian naval vessel, once interned in Chinese ports, was ever allowed to leave again, and crews were eventually subject to Chinese legal regulations. To accomplish this, China used its declaration of neutrality to undermine the extraterritorial privileges of interned Russian sailors in Shanghai. The Chinese government may have failed to achieve its goal of canceling Russian extraterritoriality altogether, but the diplomatic precedents created during this earlier period finally began to bear fruit during World War I, when Chinese neutrality resulted in not only the complete elimination of German and Austrian extraterritorial rights but also the eventual return of their territorial concessions to China.

Notes

The thoughts and opinions expressed in this publication are those of the author and are not necessarily those of the U.S. government, the U.S. Navy Department, or the Naval War College.

1. Denis Warner and Peggy Warner, *The Tide at Sunrise: A History of the Russo-Japanese War, 1904–1905* (New York: Charterhouse, 1974), p. 284.
2. Julian S. Corbett, *Maritime Operations in the Russo-Japanese War: 1904–1905* (Annapolis, Md.: Naval Institute Press, 1994), vol. 2, p. 404.
3. Donald W. Mitchell, *A History of Russian and Soviet Sea Power* (New York: Macmillan, 1974), p. 206.
4. Warner and Warner, *Tide at Sunrise*, pp. 284–85.
5. Pertii Luntinen and Bruce W. Menning, "The Russian Navy at War, 1904–05," in *The Russo-Japanese War in Global Perspective: World War Zero*, ed. John W. Steinberg et al. (Leiden: Brill, 2005), p. 239.
6. Russian Declaration of Contraband Goods, *London Gazette,* 11 March 1904, FO881/8101,

Public Record Office / The National Archives, Kew, U.K. [hereafter PRO/TNA].

7. Russian Contraband List, 15 February 1904, p. 85, FO46/621, PRO/TNA.
8. Report from Aden, 22 February 1904, p. 44, FO46/620, PRO/TNA.
9. FO46/625, pp. 7–8, PRO/TNA.
10. Fall 1905, pp. 137–46, FO46/641, PRO/TNA.
11. Letter of Protest by the North of England Protecting and Indemnity Association, 7 March 1904, p. 48, FO46/621, PRO/TNA.
12. FO46/621, p. 38, PRO/TNA.
13. 3 March 1904, p. 454, FO46/620, PRO/TNA.
14. Letter from Captain Slade to the Earl of Cromer, 2 March 1904, pp. 310–13, FO46/621, PRO/TNA.
15. John White, *The Diplomacy of the Russo-Japanese War* (Princeton, N.J.: Princeton Univ. Press, 1964), p. 167.
16. Ian Nish, *The Origins of the Russo-Japanese War* (New York: Longman, 1985), pp. 200–201.
17. Luntinen and Menning, "Russian Navy at War, 1904–05," p. 239.
18. Mitchell, *History of Russian and Soviet Sea Power*, p. 224.
19. Aizawa Kiyoshi, "Differences Regarding Togo's Surprise Attack on Port Arthur," in *Russo-Japanese War in Global Perspective*, ed. Steinberg et al., p. 89.
20. Mitchell, *History of Russian and Soviet Sea Power*, pp. 228–29.
21. Ibid., p. 223.
22. Corbett, *Maritime Operations in the Russo-Japanese War*, p. 379.
23. Protest by the North of England Protecting and Indemnity Association, 17 June 1904, pp. 97–99, FO46/625, PRO/TNA.
24. 13 July 1904, p. 118, FO46/626, PRO/TNA.
25. 18 July 1904, p. 164, FO46/626, PRO/TNA.
26. 20 July 1904, p. 212, FO46/626, PRO/TNA.
27. 19 July 1904, p. 367, FO46/626, PRO/TNA.
28. 18 August 1904, pp. 184–85, FO46/630, PRO/TNA.
29. July 1904, p. 135, FO46/626, PRO/TNA.
30. 20 March 1904, pp. 86–87, FO46/622, PRO/TNA.
31. 24 August 1904, p. 142, FO46/629, PRO/TNA.
32. 6 September 1904, p. 408, FO46/630, PRO/TNA.
33. September 1904, pp. 329–330, FO46/630, PRO/TNA.
34. 1 September 1904, p. 270, FO46/630, PRO/TNA.
35. *Cassell's History of the Russo-Japanese War* (New York: Cassell, 1905), vol. 2, pp. 58–59.
36. Ibid., p. 59.
37. Kuo Ting-yee, *Sino-Japanese Relations, 1862–1927* (New York: Columbia Univ. Press, 1965), p. 92.
38. Raymond A. Esthus, *Theodore Roosevelt and Japan* (Seattle: Univ. of Washington Press, 1967), pp. 50–51.
39. Sakuye Takahashi, *International Law Applied to the Russo-Japanese War* (London: Stevens and Sons, 1908), p. 470.
40. Ibid., pp. 463–67.
41. Ibid., p. 468.
42. Hosea Ballou Morse, *The International Relations of the Chinese Empire* (Shanghai: Kelly and Walsh, 1918), vol. 3, p. 481.
43. Ibid.
44. Takahashi, *International Law Applied to the Russo-Japanese War*, pp. 476–82.
45. Morse, *International Relations of the Chinese Empire*, p. 481.
46. *Cassell's History of the Russo-Japanese War*, p. 59.
47. Kuo Ting-yee, *Sino-Japanese Relations*, p. 92.
48. *Cassell's History of the Russo-Japanese War*, pp. 61–62.
49. Ibid., p. 66.
50. Mitchell, *History of Russian and Soviet Sea Power*, p. 229.
51. Esthus, *Theodore Roosevelt and Japan*, pp. 52–53.
52. *Cassell's History of the Russo-Japanese War*, p. 66.
53. FO46/630, pp. 244–45, PRO/TNA.
54. William C. Fuller, Jr., *Strategy and Power in Russia, 1600–1914* (New York: Free Press, 1992), p. 401.
55. Ken Shen Weigh, *Russo-Chinese Diplomacy* (Shanghai: Commercial, 1928), p. 117.

"Handelskrieg mit U-Booten"
The German Submarine Offensive in World War I
PAUL G. HALPERN

The German navy faced a difficult geographic problem when examining the prospect of a maritime war against Great Britain. The British Isles have been compared to the stopper in a bottle, hindering Germany's access to the open sea. Warships attempting to attack British trade would have either to proceed far to the north around Scotland or to risk the narrow waters of the Dover Straits. Furthermore, in comparison to the British Empire, Germany at the turn of the twentieth century possessed relatively few overseas bases where ships could obtain fuel or a safe refuge. This seemed to make the traditional recourse of a weaker navy—*guerre de course*—impractical.

This geographic accident was in fact one of the major arguments used by the state secretary of the Imperial Navy Office (Reichsmarineamt), Adm. Alfred von Tirpitz, for the view that Germany had no option but to develop a powerful battle fleet, one strong enough to deter the British, if only by inflicting losses heavy enough to imperil Britain's worldwide position.[1] This entailed the crippling of the Royal Navy in the face of such possible British enemies as Japan, France, and Russia. A battle fleet could also, however, increase Germany's value as a potential ally.[2]

Nevertheless, the German fleet would have to pass developmentally through a danger zone before it was strong enough to constitute a true deterrent. Tirpitz succeeded in getting two major naval laws through the Reichstag in 1898 and 1900. The Germans thereby embarked on a steady building program that produced the second-largest fleet in the world. At this early stage, U-boats were not intended to play a major role.

The Challenge of German Expansion of Its Surface Fleet

Despite the technical excellence of the German navy, Tirpitz's strategy of threatening the Royal Navy failed. The British responded by increasing their own naval construction so that the danger zone kept lengthening. Moreover, they reached defensive agreements with major or potential rivals, notably the Japanese in 1902, the French in 1904, and the Russians in 1907. It was Germany and not Great Britain that appeared to be isolated, with only one reliable ally, Austria-Hungary, which, however powerful in the Adriatic, was hardly a world naval power.

Furthermore, confident German assumptions that technical superiority—in the form of submarines, torpedoes, and mines off the German coast—would whittle down the Royal Navy to the point where the smaller German navy might engage it with a good chance of success also proved to be false. The British adapted to the new technologies and abandoned the concept of a close blockade in favor of a distant one. They would not obligingly place their fleet in a potential trap. This problem was apparent even before the war, when Tirpitz asked the commander of the High Seas Fleet what he would do if the British did not come.[3] The Germans never evolved a satisfactory answer to this conundrum.

Most importantly, Tirpitz failed to obtain a supplementary naval law to increase his building program in the year before the outbreak of war. The money was devoted instead to an increase in the army.[4] Moreover, the German navy was assigned no role in the German army's plan for a decisive and quick victory against France. After less than a month of hostilities a British raid deep into German waters resulted in the loss of three German cruisers and a destroyer in the Heligoland Bight before the bulk of the High Seas Fleet could intervene. The kaiser and the naval high command, however, opposed risking the fleet in operations close to the British Isles. Even Tirpitz, despite subsequent claims to the contrary, apparently opposed seeking battle more than a hundred miles from the German coast.[5] The most risk the high command would accept was offensive minelaying or sporadic raids by fast battle cruisers on British coastal towns. This caution, however, involved political risks. Tirpitz pointed out that if the war ended "without the fleet having bled and worked, we shall get nothing more for the fleet, and all the scanty money that there may be will be spent on the army."[6]

Given these constraints on offensive operations, what could the German navy do? What were the prospects for an alternative *guerre de course*? The German government had in fact prepared for a certain amount of commerce warfare before the war. It was to be executed for the most part by "auxiliary cruisers," converted merchant ships. There were thirteen of these auxiliary cruisers in existence at the beginning of the war.[7] The Germans also organized the *Etappe* system, whereby the oceans were divided into zones linked to specific communications centers, each under a naval officer charged with transmitting messages to German ships in his zone in the event of war. These officers, usually located in cities where the Germans maintained diplomatic or consular representatives, were, in turn, linked to Germany by, wherever possible, wireless communication.

The important shipping lines had equipped their ships with wireless units, and the German government had issued sealed orders to be opened in the event of war referring captains to the relevant *Etappe*. The Germans also sent out a large number of colliers and supply ships to provision cruisers at prearranged rendezvous, often remote or obscure anchorages. The system was particularly noteworthy in South America, where the Germans counted on benevolent interpretations of neutrality.[8]

There were furthermore a number of regular German cruisers at sea at the outbreak of war, the most notable group being the Far East Squadron, under Adm. Maximilian von Spee. This squadron, however, would be quickly eliminated in the action off the Falkland Islands in December 1914. One of Spee's cruisers, *Emden,* detached early in the war, enjoyed a spectacular career against commerce in the Indian Ocean before being sunk.

The auxiliary cruisers at the beginning of the war were on the whole less successful than *Emden*. Contemporary ships required great quantities of coal, which was difficult to transfer at sea. Wireless transmissions could quickly reveal the presence of a warship in neutral ports, although by the same token careless transmissions by British warships sometimes warned the Germans of their proximity. The *Etappe* system was a diminishing asset. The vulnerable colliers were gradually sunk, captured, or interned.

Throughout the war, the Germans sent out surface raiders, generally converted freighters that were relatively economical in their consumption of coal. Some, such as the raider *Wolf* in 1917, had considerable success; *Wolf* itself even managed to return safely to German waters, after a voyage of 444 days.[9] But the losses these surface raiders inflicted are, given the scale of British and allied shipping, best described as pinpricks. A *guerre de course* waged by surface raiders could not have a significant effect on the war.

A Gradual Shift to U-boats

Surface ships proved inadequate to conduct *guerre de course* effectively. The potential danger that submarines posed to surface warships was not ignored before the war. However, more attention seemed directed at the threat to warships than to merchant shipping. The Germans began the war with but twenty-eight U-boats, of which four were suitable only for training and fourteen had unreliable, smoke-producing engines that required ventilation pipes that had to be lowered and stowed before the boat could submerge.[10]

During the early stages of the war this small number of U-boats scored a number of spectacular successes against British warships. However, these were sometimes against older ships, handled in a foolish fashion. The dreadnoughts of the Grand Fleet, the major striking force of the Royal Navy, were carefully screened by destroyers, and no dreadnought of the Grand Fleet would be sunk by a submarine during the entire war (one would be sunk by a mine laid by an auxiliary minelayer). The commander of the Grand Fleet, Adm. John Jellicoe, was so conscious of the threat from German U-boats that he temporarily shifted his fleet to the west coast of Scotland, and later northern Ireland, until the main base at Scapa Flow could be properly protected. Consequently, even an offensive use of U-boats in British waters was unlikely to reduce the Grand Fleet to such an extent that the German fleet could chance a battle.[11]

On 8 October 1914 a proposal to commence commerce raiding by U-boats along the British coast was made by Korvettenkapitän Hermann Bauer, commander of the German U-boat flotilla in the North Sea (Führer der Unterseeboote, or F.d.U.). This action was ostensibly to be taken in retaliation for the British laying of a minefield in the approaches to the English Channel east of the Dover–Calais line, which Bauer considered a violation of international law.[12] (Similar claims, that German actions were retaliations for British violations of international law, would be made throughout the war, but a proper evaluation of the rights and wrongs on both sides would require a volume of its own.) Bauer submitted a further memorandum at the end of December 1914 claiming that there would be sufficient U-boats in service to commence a campaign against commerce by the end of January 1915.

The first British merchant ship to become a victim of a U-boat is believed to have been the small steamer *Glitra* (866 tons), off the Norwegian coast on 20 October 1914. The commanding officer of *U.17* allowed the crew ten minutes to abandon ship before sinking it and then towed the lifeboats in the general direction of the coast before casting them off, giving them general directions toward land. This action, while relatively humane and in most respects conforming to the accepted rules of "cruiser warfare," also demonstrated the potential difficulties of using U-boats against merchant shipping. There would not normally be sufficient men to form a prize crew to bring the ship into port, and U-boats were too small to accommodate the crews of sunken ships. This meant that under the best circumstances a crew would be left to the mercy of the sea in small boats; in other words, the safety of passengers and crew could not be assured.

Mistakes were also possible. Less than a week after the *Glitra* affair, another German U-boat torpedoed what its commander assumed was a troopship off Cape Gris-Nez in the Strait of Dover. The ship was actually a French liner carrying about 2,500 Belgian refugees; between thirty and forty people were killed, although the ship was not sunk. Fortunately for the Germans, the explosion was at first attributed to a mine rather than a submarine, but the potential for embarrassing charges of atrocities was obvious.

When these atrocities were committed on neutrals, diplomatic difficulties arose for the German government.

In January 1915 the Germans attempted another raid with battle cruisers against British fishing vessels assumed to be acting as scouts on the Dogger Bank in the North Sea. Thanks to their ability to decode intercepted German wireless messages, the British were forewarned, and in the ensuing action the Germans lost an armored cruiser and were lucky not to suffer even greater losses. The fiasco strengthened the kaiser's determination not to risk the fleet again, led to a change of command in the High Seas Fleet, and convinced Tirpitz that a *guerre de course* involving not only cruisers but also a submarine blockade against British trade offered the best chance of success. He also advocated airship attacks against the docks and warehouses of London, as well as aggressive raiding and minelaying in the Thames, to be executed from German bases in occupied Belgium.[13]

German Support for Submarine Warfare

By the end of January 1915 there was substantial support in Germany for U-boat warfare. Within the navy, the commander of the High Seas Fleet, Adm. Hugo von Pohl, as well as Vice Adm. Reinhard Scheer, commander of the 2nd Battle Squadron, favored increasing submarine warfare. Meanwhile, outside of the navy, an interview given by Tirpitz in November 1914 to an American journalist had been published in the German press and aroused widespread interest and support among German professors, financial experts, and segments of the public and press. There was a tendency to regard U-boats as a wonder weapon, capable of deciding the war.

During early 1915 there developed what has been termed a "U-Boat party," whose noisy claims for "unrestricted" submarine warfare frequently went even beyond what Tirpitz at the time advocated.[14] At the beginning of February a reluctant Chancellor Theobald von Bethmann Hollweg, despite the reservations of the German Foreign Office, accepted the navy's proposals for the U-boat offensive against commerce. On 15 February 1915 the waters around Great Britain and Ireland, including the entire English Channel, were designated by the German Admiralty Staff to be a "war zone" in which any enemy merchant vessels could be sunk.

Most importantly, this destruction could occur even when it was not possible to ensure the safety of passengers and crew. Further, because the Germans believed the British had misused neutral flags, it was declared that neutral ships entering the zone might also be attacked if mistaken for enemy ships. As a minor concession to the neutral Dutch and Scandinavians, a strip thirty miles wide along the Dutch coast was exempted, as were waters north of the Shetland Islands and in the eastern part of the North Sea.[15]

The Germans began what has been termed the "first unrestricted submarine campaign" with a relatively small number of U-boats. While published accounts differ as to exact numbers, it is possible the Germans never had available for this offensive more than twenty-five U-boats, eight of them elderly with unreliable engines. Of this number, only one-third could be assumed to be in their operational areas at a given time, and not all U-boats had deck guns. A further seventeen boats were under construction when the war began, and orders placed after the outbreak of hostilities produced another seventeen during the December 1915–December 1916 period.[16]

The German navy also took over five U-boats under construction for the Austrian navy in German yards, as well as one for Norway. There was initially a tendency to hold back on new contracts, on the assumption that war would likely be over before orders were completed. Small coastal U-boats of the UB.I-class and UC.I-class submarine minelayers could be built much more quickly, and the German seizure of the Belgian coast gave proximity to British waters that could be exploited by these classes. Consequently, the navy ordered fifteen UB.I boats, plus two for Austria, and fifteen UC.Is in October and November 1914.[17] The majority of these were destined for the newly constituted Flanders Submarine Flotilla.

As the war progressed, however, the Germans developed an inland submarine base at Bruges, linked by canal to outlets to the North Sea at Zeebrugge and Ostende. Bombproof shelters were eventually constructed at Bruges, and the Belgian coast would be heavily fortified with batteries of large-caliber guns. The area was held by a combined-arms force of corps strength, designated the Marinekorps Flandern, under a vice admiral directly responsible to the kaiser.[18]

The number of U-boats available in February 1915 to conduct commerce raiding was obviously very small. Tirpitz claimed that the declaration of unrestricted submarine warfare had been premature, although this did not stop him from proclaiming on 16 February that the British would be forced to give in within six weeks.[19] The campaign had no sooner been declared than it ran into strong opposition from the United States, specifically a warning that should American ships or lives be lost, the German government would be held strictly accountable. This prompted Chancellor Bethmann Hollweg and the Foreign Office to assure the Americans that neutral shipping would not be harmed if recognized as such, thereby antagonizing German naval leaders, who hoped a ruthless campaign would frighten neutrals, whom they believed carried approximately 25 percent of traffic to the British Isles.

With this development, a three-way tension developed among the chancellor and the Foreign Office, naval leaders, and the U.S. government that would persist for the following two years. On 18 February 1915, U-boat commanders were ordered to spare ships flying neutral flags unless clearly identified as enemy, although they would not be held responsible for mistakes. Furthermore, ships of the Belgian Relief Commission were not to be attacked, nor were hospital ships, unless obviously transporting troops.[20]

Despite grumbling at such restrictions by naval leaders, in the first three months of the campaign, from March to May 1915, the Germans sank 115 ships (approximately twenty-two of them neutral) representing 255,000 tons, for a loss of five U-boats. This represented what a British admiral later termed an "exchange ratio" of twenty to one, meaning about twenty ships sunk for every U-boat lost. These spectacular results were obtained even though there was a daily average of only six U-boats actually at sea. British and French countermeasures were obviously ineffective. However, the specifically British losses were only about fifty ships total, which were more than replaced by new construction and by the capture of German and Austrian merchant ships. Neutrals seeking profit were not deterred from trading with the British Isles, and the amount of tonnage available to the Entente rose, not fell, during this period.[21]

The Impact of German U-boats on Neutral Shipping

The perhaps inevitable incidents between German U-boats and neutrals occurred. These episodes involved Norwegian-, Swedish-, and Greek-flagged ships. A particularly troublesome incident involved the Dutch *Katwijk,* torpedoed and sunk without warning while en route from Rotterdam to Baltimore, Maryland, in an area designated safe by the Germans. This sinking involved a neutral ship trading between two neutral ports, which made it seem particularly outrageous to some. After a strong Dutch protest, the Germans offered to pay compensation—as they did in a few other cases as well—if it was determined that a U-boat had been involved. On 18 April 1915 the German government once again ordered U-boat commanders not to attack neutrals. However, after a U-boat was erroneously believed to have been lost through British misuse of neutral flags, those same U-boat commanders had received orders on 2 April no longer to surface to verify the identity of neutrals.

These incidents against neutral ships culminated in the sinking of the British liner *Lusitania* on 7 May 1915. This disaster resulted in the death of 128 American citizens among the 1,201 lost. Regardless of the legal questions involved, whether the ship had been carrying ammunition, or whether the Germans had warned civilians against embarking, the incident was a public-relations disaster. It turned American public opinion against Germany. In the face of President Woodrow Wilson's strong protests and demands to cease submarine attacks against commerce, Bethmann Hollweg managed to obtain the kaiser's consent to order U-boat commanders not to attack large passenger liners even if they flew the enemy flag. This sharpened the division between the civilian government and the naval leaders and led Tirpitz and the chief of the Admiralty Staff to attempt to resign. Tirpitz's resignation was not accepted, but that of the chief of the Admiralty Staff was.[22]

After these new restrictions were adopted, the amount of tonnage sunk dipped slightly for two months but then climbed above pre-restriction levels, reaching 182,772 tons in August and 136,048 tons in September. The Germans were able to keep a larger number of U-boats at sea. They commissioned fifteen U-boats in this period, ten of them UC-class coastal minelayers. This offset the ten U-boats lost during this period, as well as two provided to Austria. Furthermore, the prohibition against sinking ships without warning and the conduct of war under "cruiser" or "prize" rules were less problematic than they might have seemed, for U-boats could carry only a limited number of expensive torpedoes and U-boat commanders naturally preferred to sink ships by other means, notably gunfire or explosive charges. Moreover, the commodities essential for British life were usually carried on freighters, not large passenger liners. The danger to merchant seamen did not create as much of a public-relations problem as the threat to women and children.

The Introduction of the Q-ship

German U-boat commanders adjusted quickly to these changing legal definitions. However, as the war progressed the nature of submarine warfare made adherence to cruiser rules difficult. Though the British and French still had few really effective antisubmarine weapons, one that showed promise was the Q-ship, a ship disguised as a harmless steamer or sailing vessel that would attempt to lure a U-boat close enough to destroy it with concealed weapons. The Q-ship became a diminishing asset as the Germans grew wary after a few successes in the summer of 1915; nevertheless, it made plain the fact that if a submarine conducted *guerre de course* under cruiser rules it lost its greatest asset, its invisibility, and thereby greatly increased its vulnerability.

Completely by coincidence, on 19 August 1915 two episodes gave each side its own grounds to charge the enemy with a war atrocity. A number of German sailors from the U-boat *U.27* (which was itself sunk the same day, completely legitimately, by the Q-ship *Baralong*) were killed under questionable circumstances during the retaking of a steamer that their boat had previously stopped. Any potential German propaganda advantage from this incident was canceled, however, by the sinking without warning that day of the White Star liner *Arabic*, with the loss of two American lives, in an apparent violation of earlier German assurances to President Wilson.

German and American relations reached another crisis point, and so too did divisions within the German government. Naval leaders were embittered at the restrictions placed on U-boat operations, but Bethmann Hollweg was anxious to avoid a break with the

United States. He was supported by the chief of the German General Staff, Erich von Falkenhayn, though Falkenhayn was more concerned about the Netherlands being drawn into the war than the United States. At this time Bulgaria was on the verge of joining the Central Powers, and Falkenhayn anticipated a campaign in the Balkans, which would mean great difficulty in concentrating enough troops to counter any Dutch intervention.[23]

Once again the chancellor and the army prevailed. The kaiser ordered that even small passenger liners should not be sunk without warning, and he insisted on provision for the safety of passengers and crews. Shortly afterward, the kaiser also ordered that U-boats were not to be stationed in the western approaches to the British Isles, where the worst incidents had taken place. U-boats would instead concentrate in the North Sea, working under cruiser rules. His decision marked the effective end of the first submarine campaign. The acrimony in the navy against this decision was great enough to bring about the replacement of the chief of the Admiralty Staff by Adm. Henning von Holtzendorff, an opponent of Tirpitz. The latter resigned for a second time, and though his resignation was again not accepted, he was relegated to mere administrative duties rather than advisory status. Outside of the navy, however, Tirpitz's threat of resignation generated what has been described as "a wave of support" for him as a symbol of Germany's drive for victory.[24]

A Last-Ditch Return to Unrestricted Submarine Warfare

What had the first submarine campaign achieved? From August 1914 to September 1915 1,294,000 tons had been sunk, and probably only a state insurance scheme spreading the risks among individual companies and owners prevented a paralysis of British trade. However, new construction had added 1,233,000 tons of shipping, and the capture or detention of enemy ships had added a further 682,000 tons. The percentage of loss compared to the total volume of trade was small, less than that inflicted by privateers during the Napoleonic Wars.

Over a longer period of time, however, this situation might have changed. After the one-time gain represented by captures, shipping was likely to be a diminishing asset, with losses exceeding gains, as demonstrated by the final months of the campaign. In addition, the growing need for shipping caused by overseas campaigns in the eastern Mediterranean or Mesopotamia would heighten the problem. Furthermore, the British practice of devoting more resources to naval than to merchant-ship construction and repair exacerbated this problem, along with potential shortages of material and manpower in the yards. Therefore, the final verdict on the submarine *guerre de course* remains in doubt.[25]

Over the course of 1915 the submarine *guerre de course* opened a new area of operations in the Mediterranean.[26] Germany's ally Austria-Hungary had successfully employed a small U-boat force in the defense of the Habsburgs' Adriatic coast. However, the Austrians initially had little capacity to operate outside the Adriatic. The first German U-boats had arrived in the theater in response to a Turkish appeal for assistance at the time of the allied operations against the Dardanelles. On this occasion the Germans sent two small, UB.1-class boats, broken down into sections, overland by rail to the Austrian naval base at Pola, where they were assembled by German engineers. A large U-boat completed the trip directly from Germany, passing undetected through the Strait of Gibraltar.

These German U-boats soon made their presence felt at the Dardanelles, where they constituted a major—but not insurmountable—problem for the British and French. The Germans soon assembled an additional three UB boats at Pola, transferring two to the Austrian navy. The German Admiralty was well aware that an important portion of British and French trade passed through the Mediterranean and Suez Canal. There were also choke points through which traffic had to pass that facilitated U-boat operations, which would also be somewhat easier in the autumn and winter months of bad weather in the Atlantic. Austrian bases at Pola, and later in the Gulf of Cattaro, closer to the entrance to the Adriatic, could be used by the U-boats. The Mediterranean had another advantage as well—fewer American ships, or American citizens traveling on allied ships, than elsewhere meant fewer potential diplomatic complications.

By the end of October 1915 the Germans had half a dozen large U-boats operating in the Mediterranean. The Germans also sent materials and German workers to Pola for the assembly of six U-boats of the more capable UB.II class, considered ideal for Mediterranean operations. The Germans readily accepted the delay in completion of U-boats for northern waters that this diversion of labor and material implied. The results were encouraging. In November 1915, 152,882 tons were sunk, a major portion of the 167,043 tons sunk in all theaters; although the total fell the following month, the Mediterranean still represented more than half of all tonnage sunk.

There was a slight inconvenience, however. Italy was already at war with Austria-Hungary, more than a year before declaring war on Germany. Consequently, when operating on the surface against Italian ships, German U-boats had to fly the Austrian flag. This meant that when the Italian liner *Ancona* was sunk by a German U-boat in November 1915, with the loss of twenty American lives, it was Austria-Hungary that had to take the blame. To cover their tracks, certain German U-boats were retroactively added to the Austrian navy list as of the moment they passed Gibraltar.[27]

In 1916, the most successful cruise by any German U-boat during the war took place in the Mediterranean, when Kapitänleutnant Arnauld de la Perière in *U.35* sank fifty-four steamers and sailing craft (90,150-plus tons) from 26 July to 20 August. German U-boat strength in the Mediterranean continued to grow to the point where at the beginning of 1918 it was divided into two flotillas, one at Pola and the other at Cattaro. It was, however, increasingly difficult for the Germans to maintain U-boats far from home waters, largely because of a shortage of skilled labor. At the beginning of 1918 a substantial backlog of U-boats in the Adriatic awaited refit and repair.

In northern waters, the German naval command chipped away at the restrictions on the use of U-boats, although the rules seemed constantly to change, reflecting diplomatic considerations. The spreading installation of guns on hitherto defenseless merchant steamers caused problems. In mid-November 1915, the small U-boats of the Flanders Submarine Flotilla were authorized to sink without warning all enemy freighters between Dunkirk and Le Havre, and in January 1916 U-boats were allowed to sink without warning armed enemy freighters in the war zone. In March this was extended to all enemy freighters.

The adherents of unrestricted submarine warfare gained an important supporter when Holtzendorff, originally brought in as chief of the Admiralty Staff to oppose Tirpitz, converted to the idea. He in turn converted Falkenhayn, who now doubted that the British could be forced out of the war by operations on land. In February 1916 Reinhard Scheer assumed command of the High Seas Fleet. Scheer was equally offensive minded and a partisan of unrestricted submarine warfare. But the wavering kaiser and hesitant chancellor continued to avoid unrestricted submarine warfare while they sought an agreement with the United States. However, they did authorize what was termed "sharpened" submarine warfare—enemy merchant ships inside the war zone could be destroyed without warning, while those outside could be destroyed without warning only if armed. U-boats either inside or outside the war zone could not attack passenger liners while submerged, even if the ships were armed. This was far less than Tirpitz wanted, and this time his resignation was accepted.

The period of "sharpened" submarine warfare in March and April 1916 lasted less than two months before diplomatic pressure brought it to a halt. There were problems with the Netherlands, particularly after the sinking of the Royal Holland Lloyd liner *Tubantia* on 16 March, the largest neutral ship sunk during the war. The Germans did not admit responsibility, and the German envoy in The Hague tried to divert and frighten the Dutch by spreading a rumor the British might invade, thereby forcing the German army to take countermeasures.[28] Shortly afterward the French cross-channel steamer *Sussex* was torpedoed; it did not sink, but American lives were lost, and fragments of the torpedo belied German denials.

President Wilson took this final incident as a challenge. On 20 April 1916 he demanded that German U-boats cease operating in this manner. This demand amounted to an ultimatum, since Wilson threatened to sever diplomatic relations. The German navy had no choice but to yield to the civilian members of the government, and on 24 April U-boats in northern waters were ordered to operate under cruiser rules, provided ships did not offer resistance or attempt to escape.

This concession may have involved ulterior motives. Adm. Eduard von Capelle, Tirpitz's successor, apparently thinking that almost as much could be accomplished under cruiser rules, preferred deferring any unrestricted submarine campaign until more of the U-boats then under construction entered service. Scheer strongly disagreed with the view that prize rules could be successful, instead claiming they exposed U-boats to undue danger. He accordingly recalled his U-boats (except the UC-class minelayers) by wireless and announced the submarine *guerre de course* would cease. The large U-boats in the north would be used in support of operations by the High Seas Fleet. Scheer too may have had ulterior motives; his action might have been an attempt to stimulate on the part of German supporters of unrestricted submarine warfare a protest strong enough to force Bethmann Hollweg from office. His decision aroused considerable criticism in the German navy.[29]

Under Scheer's leadership the High Seas Fleet embarked on a period of more aggressive operations that culminated in the battle of Jutland on 31 May 1916. Regardless of German claims of a tactical victory in that action on the basis of larger British losses, the strategic position of the German navy did not change. Scheer recognized this in a memorandum to the kaiser on 4 July admitting that even the most successful sea battle could not compel the British to make peace and that victory could be achieved only by crushing Britain's economic life through U-boat action against British commerce.[30]

But in early October 1916, contrary to Scheer's wishes, the Admiralty Staff withdrew the U-boats from cooperation with his fleet and resumed a submarine campaign, restricted by prize rules. Although the Germans now reserved the right to sink any armed merchantman without warning, in fact during the closing months of 1916 only about 20 percent of ships destroyed were sunk without warning by torpedo. The majority were sunk by gunfire. Overall allied losses grew as more U-boats became operational, yet at the end of 1916 the British merchant marine remained at 94 percent of its size at the beginning of the war. No quick German victory was in sight.[31]

In a famous memorandum of 22 December 1916, Holtzendorff summarized the argument for unrestricted submarine warfare. Data for the memorandum had been compiled by the Admiralty Staff, including the reserve officers Dr. Richard Fuss, a banker, and Dr. Hermann Levy, a professor of economics at Heidelberg. These staff members estimated that unrestricted submarine warfare would force Britain to make peace in five months. They based these calculations on British requirements for imports, estimated shipping available after worldwide commitments (minus the estimated six hundred thousand tons per month that unrestricted U-boats would sink), and the terror this would inspire in neutrals. No possible American intervention would be effective during this period, nor would shipping be available to bring American troops to Europe.[32]

There were, as events would later show, a number of questionable assumptions in this argument.[33] But the preponderance of forces within the German government, including the all-powerful chief of the General Staff, Paul von Hindenburg, now sided with proponents of unrestricted submarine warfare. On 9 January 1917 the decision was taken to commence unrestricted submarine warfare on 1 February.[34]

What resources were available to the German navy for submarine warfare at this time? On 1 February 1917 the Germans had 105 U-boats—sixty-nine on the North Sea or Flanders coast, twenty-three in the Mediterranean, and the remainder in the Baltic or at Constantinople. New construction more than offset losses, and in the second half of the year U-boat strength did not fall below 120.[35] German attempts to increase the size of the U-boat fleet had been plagued by problems arising from the rapid expansion of German yards after the start of the war, the inexperience of subcontractors supplying engines, and a general shortage of trained labor after the mobilization of many experienced workers. Delays were frequent, and there was a tendency to favor smaller, albeit improved UB.III and UC boats, with their shorter construction time, over the larger and more complex craft, which might not be finished before the end of the war. The navy won concessions from the high command, which agreed to provide the names of skilled workers now eligible for potential release from the army and granted the navy priority in the transport of critical raw materials and components for U-boat construction. In February the navy contracted for fifty-one U-boats (mostly UB.IIIs) and, as hopes for a quick end to the war faded, another ninety-five at the end of June. In August the Reichsmarineamt suspended work on large warships and gave U-boat construction priority even over work on torpedo boats.[36]

The predictable incidents involving U.S. ships and American citizens occurred after the reintroduction of unrestricted submarine warfare, and these incidents led inexorably to the American declaration of war against Germany on 6 April 1917. Would U-boat operations force Britain out of the war before American participation had any effect? Failing that, would they prevent substantial American aid from crossing the ocean? At first the U-boats achieved spectacular success, and shipping losses jumped from 328,391 tons in January 1917 to 860,334 in April. The world's shipping tonnage was reduced by over two million tons, almost 1.25 million tons of which was British. New construction or transfer of ships from foreign flags could in no way compensate, and the predictions of Holtzendorff and the U-boat enthusiasts—Britain forced to make peace by November—seemed likely to be fulfilled.[37]

But this did not happen, of course. Faced with a new crisis, the British adopted a series of antisubmarine measures, including the gradual extension of a convoy system that blunted the German threat. Shipping losses soon began to decline. There may have been momentary spikes, but the long-term trend was downward. Losses fell from their April peak to 411,766 tons by December 1917. Losses were still painful, however, and British imports in 1917 fell by 20 percent from those of 1916, and strict rationing and control of nonessentials remained in force.[38] But the situation improved in 1918, with losses to U-boats down to 171,972 tons in September and 116,237 in October.[39]

Meanwhile, Entente countermeasures claimed more U-boats. The Germans began to feel the loss of experienced commanders, although the Admiralty Staff rejected a proposal to have wireless-equipped U-boats coordinate the attacks of other submarines, a forerunner of the World War II "wolf pack" system.[40] The Germans tried instead, at the end of 1917, to strengthen their effort with an expanded 1919 building program of 120 U-boats, but none were finished by the end of the war. In 1918 Admiral Scheer, now head of the navy, had even more grandiose plans for mass production, with 333 U-boats to be delivered in 1919 and a total of 405 to be ready in 1920.[41] It is questionable how realistic these plans were.

Conclusions

The worsening situation of the army, joined with domestic disturbances, compelled the Germans to sign an armistice on 11 November 1918. The submarine *guerre de course*, although close to achieving success for a few months in 1917, had ultimately failed to drive Britain out of the war or prevent massive American assistance from reaching European waters in 1918.

This experience raises several intriguing questions. Would the Germans have succeeded if they had begun and maintained without interruption their unrestricted warfare earlier in the war, coupled with the ambitious U-boat construction program they adopted only later, when it was already too late? Would adoption of an early form of coordinated U-boat attack have altered the situation? Perhaps.

On the other hand, would earlier adoption of the antisubmarine measures eventually used by the Entente have prevented the U-boat menace from ever reaching the proportions it did? These questions can never be answered. Regardless, in the long run *"Handelskrieg mit U-Booten"* was a failure.

Notes

The thoughts and opinions expressed in this publication are those of the author and are not necessarily those of the U.S. government, the U.S. Navy Department, or the Naval War College.

1. Michael Epkenhans, *Tirpitz: Architect of the German High Seas Fleet* (Washington, D.C.: Potomac Books, 2008), p. 23. A full analysis of Tirpitz and his policies is now available in Patrick J. Kelly, *Tirpitz and the Imperial German Navy* (Bloomington: Indiana Univ. Press, 2011).

2. Ivo Nikolai Lambi, *The Navy and German Power Politics, 1862–1914* (Boston: Allen & Unwin, 1984), pp. 140–47.

3. Epkenhans, *Tirpitz*, pp. 60–61. See also Michael Epkenhans, "Technology, Shipbuilding and Future Combat in Germany, 1880–1914," in *Technology and Naval Combat in the Twentieth Century and Beyond*, ed. Phillips Payson O'Brien (London: Frank Cass, 2001), pp. 63–67; and Lambi, *Navy and German Power Politics*, pp. 404–405, 423.

4. Holger H. Herwig, *"Luxury Fleet": The Imperial German Navy, 1888–1918* (London: Allen & Unwin, 1980), pp. 90–91.

5. Epkenhans, *Tirpitz*, pp. 61–62.

6. Tirpitz to Chief of the Naval Staff, 16 September 1914, in Adm. Alfred von Tirpitz, *My Memoirs* (New York: Dodd Mead, 1919), vol. 1, pp. 95–96.

7. Paul Schmalenbach, *German Raiders: A History of Auxiliary Cruisers of the German Navy 1895–1945* (Cambridge, U.K.: Patrick Stephens, 1979), pp. 13–14.

8. A short summary is in John Walter, *The Kaiser's Pirates: German Surface Raiders in World War One* (London: Arms and Armour, 1994), p. 35. Further detail is in the German official history: E. Raeder and Eberhard von Mantey, *Der Kreuzerkrieg in den ausländischen Gewässern* (Berlin: E. S. Mittler, 1922–37), vol. 1, pp. 20–23.

9. *Wolf* sank or damaged (by mines) thirty ships, representing over 138,000 tons. See Richard Guilliatt and Peter Hohnen, *The Wolf* (New York: Free Press, 2010), p. 293.

10. Four submarines were under refit at the outbreak of war, sixteen were under construction, and an additional eleven were immediately ordered but would not enter service for well over a year. V. E. Tarrant, *The U-boat Offensive, 1914–1945* (Annapolis, Md.: Naval Institute Press, 1989), p. 7.

11. See Paul G. Halpern, *A Naval History of World War I* (Annapolis, Md.: Naval Institute Press / UCL, 1994), pp. 29, 33.

12. Carl-Axel Gemzell, *Organisation, Conflict and Innovation: A Study of German Naval Strategic Planning, 1888–1940* (Stockholm: Esselte Studium, 1973), p. 142; Bernd Stegemann, *Die Deutsche Marinepolitik, 1916–1918* (Berlin: Duncker & Humblot, 1970), pp. 22–23.

13. Tirpitz, memorandum, 25 January 1915, in Walter Görlitz, ed., *The Kaiser and His Court: The Diaries, Notebooks and Letters of Admiral George Alexander von Müller, Chief of the Naval Cabinet, 1914–1918* [English translation] (London: Macdonald, 1961), pp. 58–60.

14. Raffael Scheck, *Alfred von Tirpitz and German Right-Wing Politics, 1914–1930* (Boston: Humanities, 1998), pp. 26–27.

15. R. H. Gibson and Maurice Prendergast, *The German Submarine War, 1914–1918* (London: Constable, 1931), pp. 27–28.

16. A. Gayer, "Summary of German Submarine Operations in the Various Theaters of War from 1914 to 1918," U.S. Naval Institute *Proceedings* 52, no. 4 (April 1926), p. 626; Tarrant, *U-boat Offensive*, pp. 15–18. The most detailed account of submarine operations may be found in the five-volume official history: Arno Spindler, *Der Handelskrieg mit U-Booten* (Berlin [vol. 5, Frankfurt, Ger.]: E. S. Mittler, 1932–66).

17. Eberhard Rössler, *The U-boat: The Evolution and Technical History of German Submarines* (London: Arms and Armour; Annapolis, Md.: Naval Institute Press, 1981), pp. 38–40, 44–47, 50; Harald Bendert, *Die UB-Boote der Kaiserlichen Marine, 1914–1918* (Berlin: E. S. Mittler, 2000), 11–14; Harald Bendert, *Die UC-Boote der Kaiserlichen Marine, 1914–1918* (Berlin: E. S. Mittler, 2001), pp. 11–15.

18. This formation is studied at length in Mark D. Karau, *"Wielding the Dagger": The MarineKorps Flandern and the German War Effort, 1914–1918* (Westport, Conn.: Praeger, 2003).

19. Epkenhans, *Tirpitz*, p. 65. See also Tirpitz, *My Memoirs*, vol. 2, pp. 139–47.

20. Ernest R. May, *The World War and American Isolation* (Chicago: Quadrangle Books, 1966), pp. 123–28; Tarrant, *U-boat Offensive*, p. 14.
21. Vice Adm. Sir Arthur Hezlet, *The Submarine and Sea Power* (London: Peter Davies, 1967), p. 48; Tarrant, *U-boat Offensive*, p. 18.
22. Epkenhans, *Tirpitz*, pp. 66–67.
23. Hubert P. van Tuyll van Serooskerken, *The Netherlands and World War I* (Leiden: Brill, 2001), pp. 156–59.
24. Scheck, *Alfred von Tirpitz and German Right-Wing Politics*, pp. 41–42; Epkenhans, *Tirpitz*, p. 67; Herwig, "Luxury Fleet," pp. 164–65.
25. Hezlet, *Submarine and Sea Power*, pp. 52–54; C. Ernest Fayle, *Seaborne Trade* (London: John Murray, 1920–24), vol. 2, pp. 127–35. On the navy's high consumption of industrial resources to the detriment of merchant-ship construction and maintenance, see Jon Tetsuro Sumida, "Forging the Trident: British Naval Industrial Logistics, 1914–1918," in *Feeding Mars: Logistics in Western Warfare from the Middle Ages to the Present*, ed. John A. Lynn (Boulder, Colo.: Westview, 1993), pp. 230–31.
26. The subject is examined in detail in Paul G. Halpern, *The Naval War in the Mediterranean, 1914–1918* (London: Allen & Unwin; Annapolis, Md.: Naval Institute Press, 1987).
27. Ibid., pp. 194–99, 204–205, 251–52.
28. Van Tuyll van Serooskerken, *Netherlands and World War I*, pp. 159–61.
29. May, *World War and American Isolation*, pp. 249–52; Gerhard Ritter, *The Sword and the Scepter: The Problem of Militarism in Germany* [English translation] (Coral Gables, Fla.: Univ. of Miami Press, 1969–73), vol. 3, pp. 171–77; Tarrant, *U-boat Offensive*, pp. 29–30; Gayer, "Summary of German Submarine Operations," pp. 640–41.
30. Adm. Reinhard Scheer, *Germany's High Sea Fleet in the World War* (London: Cassell, 1919), pp. 168–69.
31. Ibid., p. 245; Hezlet, *Submarine and Sea Power*, pp. 63–66.
32. Detailed analysis of the arguments in Avner Offer, *The First World War: An Agrarian Interpretation* (Oxford, U.K.: Clarendon, 1991), chap. 24; and Holger H. Herwig, "Total Rhetoric, Limited War: Germany's U-boat Campaign, 1917–1918," in *Great War, Total War: Combat and Mobilization on the Western Front, 1914–1918*, ed. Roger Chickering and Stig Förster (Washington, D.C.: German Historical Institute; New York: Cambridge Univ. Press, 2000), pp. 192–99. The memorandum is reproduced in Scheer, *Germany's High Sea Fleet in the World War*, pp. 248–52.
33. Herwig, "Total Rhetoric, Limited War," pp. 199–205.
34. Herwig, "Luxury Fleet," pp. 196–98; Ritter, *Sword and the Scepter*, chap. 8.
35. Spindler, *Handelskrieg mit U-Booten*, vol. 4, pp. 2–3.
36. Rössler, *U-boat*, pp. 47–49, 53–59, 63–65, 75–76, 78; Gayer, "Summary of German Submarine Operations," p. 653.
37. Hezlet, *Submarine and Sea Power*, pp. 88–89; Fayle, *Seaborne Trade*, vol. 3, pp. 92–93.
38. Spindler, *Handelskrieg mit U-Booten*, vol. 5, pp. 364–65; Hezlet, *Submarine and Sea Power*, pp. 95–97.
39. Spindler, *Handelskrieg mit U-Booten*, vol. 5, pp. 364–65.
40. Philip K. Lundeburg, "The German Naval Critique of the U-boat Campaign, 1915–1918," *Military Affairs* 27, no. 3 (Fall 1983), pp. 115–16; Spindler, *Handelskrieg mit U-Booten*, vol. 4, pp. 39–40.
41. Rössler, *U-boat*, pp. 78–87; Scheer, *Germany's High Sea Fleet in the World War*, pp. 333–37, 341–44; Gary E. Weir, *Building the Kaiser's Navy: The Imperial Navy Office and German Industry in the Tirpitz Era, 1890–1919* (Annapolis, Md.: Naval Institute Press, 1992), pp. 172–78. Professor Herwig condemns the Scheer program as "a national placebo" and "a propaganda effort"; Herwig, "Total Rhetoric, Limited War," p. 205.

The Anglo-American Naval Checkmate of Germany's *Guerre de Course,* 1917–1918
KENNETH J. HAGAN AND MICHAEL T. MCMASTER

By January 1917 the United States had spent almost two and a half years attempting to maintain a policy of qualified neutrality toward the two warring blocs in the stalemated war in Europe. One major obstacle to declaring neutrality was the growth of North America's highly lucrative transatlantic trade in foodstuffs and war materiel with Britain and France, leaders of the Triple Entente. Commercial and financial relations with Germany, the principal power in the opposing Triple Alliance, simultaneously withered, because the Royal Navy effectively closed hostile ports on the Continent. The German navy periodically struck out against this lack of economic neutrality in the only way it could—with unrestricted submarine attacks against allied and neutral shipping.

President Woodrow Wilson's diplomatic protests turned back the U-boats in 1915 and 1916, but in January 1917 overly confident German naval officers persuaded the government of Kaiser Wilhelm II that unrestricted U-boat attacks on heavily loaded transatlantic cargo ships, which were flooding Britain with American war munitions and food, could drive the United Kingdom out of the war within six months. According to this calculation, the United States would enter the war but mobilization would take so long that American military and naval participation could not alter the outcome.

On 31 January 1917, Berlin announced that starting the next day German U-boats would, without warning, attack and sink all enemy and neutral vessels found in or near British waters. Wilson was incensed. He ordered the severance of diplomatic relations with Germany and waited anxiously for the toll of sunken ships to mount. He also ordered U.S. Navy officers—first Rear Adm. William S. Sims and somewhat later Rear Adm. Hugh Rodman—to join their British counterparts in developing combined strategies to defeat the German U-boat offensive.

Birth of a Naval Coalition

Germany's declaration of unrestricted submarine warfare transformed Rear Admiral Sims from a man with a controversial career into an international warrior-diplomat. Within two months of assuming the position of President of the Naval War College at Newport, Rhode Island, he received a telephone call from Washington ordering him to report at once to the Navy Department. There he learned that in a cable dated 23 March 1917 the staunchly pro-British American ambassador in London, Walter Hines Page, had requested the immediate dispatch of "an Admiral of our own Navy who will bring our Navy's plans and inquiries." Page explained the benefits that would accrue to the United States: "The coming of such an officer of high rank would be regarded as a compliment and he would have all doors opened to him."[1] The ambassador further confided, "I know personally and informally that they hope for the establishment of full and frank naval interchange of information and cooperation."[2]

President Wilson had by this point decided to declare war on Germany. He informed Secretary of the Navy Josephus Daniels, "The main thing is no doubt to get into immediate communication with the Admiralty on the other side (through confidential channels until the Congress has acted) and work out the scheme of cooperation."[3] By 31 March 1917 Sims was under way on the fast passenger liner *New York,* traveling incognito in civilian clothes with a single aide, Cdr. John V. Babcock. While he was at sea, Congress on 6 April declared war against Germany. Sims now represented a belligerent power, a de facto ally of Great Britain.

The collective experiences of Admiral Sims served to prepare him well for the position he was now to fulfill. For almost two decades he had been closely acquainted with leading figures of the Royal Navy. In 1901, in the British crown colony of Hong Kong, Lieutenant Sims had met Percy Scott, the Royal Navy's leading specialist in gunnery and ordnance. In his memoir Scott would praise Sims as instrumental in the U.S. Navy's "wonderful strides in perfecting their shooting."[4] In 1906, as a lieutenant commander with navy-wide responsibility for improving the gunnery of American warships, he had been granted an informal preview inspection of the radically new *Dreadnought* as it was nearing completion. He was then so well-known and highly respected by senior officers of the Royal Navy that, on Christmas Day, First Sea Lord John Fisher and his family hosted Sims at an early-afternoon luncheon in their home. The hospitality was especially notable considering that Fisher was the preeminent First Sea Lord of the Edwardian era (1901–10). During the Christmas afternoon *en famille,* Fisher praised Sims for his technical vision and for boldly advocating the all-big-gun battleship, of which *Dreadnought* was the first iteration.[5]

Four years later, on 3 December 1910, Sims, now a commander, gave a speech in the London Guildhall that concluded, "If the time ever comes when the British Empire is seriously menaced by an external enemy, it is my opinion that you may count upon every man, every dollar, and every drop of blood of your kindred across the seas."[6] The assemblage roared its approval as Sims proposed three cheers: "For the King, the British people, and the integrity of the British Empire."[7]

President William Howard Taft had been less than enthusiastic, and in January 1911 he issued a public reprimand of Sims.[8] Neither was Josephus Daniels a particular friend of Great Britain. He recalled after World War I that Sims "had been selected in spite of and not because of the Guildhall speech."[9] Sims might dispute Daniels's interpretation of why he was chosen, but it is undeniable that the Guildhall speech of 1910 had endeared him to the British public as a popular figure and made him "the American Naval Officer most widely known to the British Navy."[10] The irony was not lost on Taft: "The ways of history are strange. When I was President I reprimanded an officer for saying exactly what he is doing now."[11]

Sims was the perfect choice to go to London. On 10 April 1917 he greeted an old comrade, Adm. John Jellicoe, now the First Sea Lord. Sims had first met Jellicoe in 1901 in China, where according to Sims their common interest in ordnance and gunnery "had brought us together and made us friends."[12] They had renewed their friendship in 1906, during Sims's visit to England to inspect *Dreadnought*, informally and unofficially. Their meeting on 10 April was a sobering experience for Sims, who had arrived in England confident that the British had the war at sea well in hand. On the contrary, Jellicoe somberly informed him, German U-boats were ravaging British and neutral shipping at such a rate that the allied powers unquestionably would lose the war for want of food and materiel, possibly as early as August 1917, certainly by October. Sims later remembered that he had asked Jellicoe, "Is there no solution for the problem?" The First Sea Lord had replied, "Absolutely none that we can see now."[13] Sims was dumbfounded: "I was fairly astounded; for I had never imagined anything so terrible." He realized, "The thing must be stopped."[14]

By the next morning Sims was well aware of the threat that the combined forces would be facing. He later recalled, "This morning at 10:30 I had another conference with Admiral Jellicoe, and Rear Admiral Sir Dudley R. S. de Chair, who is to be my 'opposite number' in Washington. . . . It was most satisfactory. We all agreed perfectly as to what should be done."[15] Never before had a senior officer of the Royal Navy sought the help of the U.S. Navy in a major war at sea; never before had the U.S. and British navies been formally and informally linked at the top levels of command and strategy making. Sims's friendship with Jellicoe was a key ingredient in this success. Throughout the remainder of the war he often met with him and occasionally also with Adm. Sir David Beatty, commander in chief of the Grand Fleet.[16]

By 30 April, with Sims's encouragement, the sea lords had agreed to experiment with destroyer escorts of convoys "as the general plan of campaign."[17] Washington proved more difficult to convince. For at least four months American naval strategy in World War I was the subject of an acrimonious transatlantic debate. From London, Sims and Page beseeched Washington to send every seaworthy destroyer to escort convoys of merchant vessels through U-boat-infested waters off the English coast. Wilson and Daniels agreed that the submarine constituted the chief threat to the allied cause, but they seriously doubted that the peril facing Great Britain was quite as extreme as portrayed by Ambassador Page and Admiral Sims, both of whom they regarded as too Anglophilic. What finally won over Wilson and Daniels was the accumulating documentation of the gradual but inexorable reduction in the monthly rate of U-boat sinkings of merchant ships as the Admiralty and Sims expanded the convoy system. In the summer of 1917 the administration finally agreed to curtail the capital-ship construction program authorized in 1916 and put the full weight of the American naval shipbuilding industry behind the construction of convoy escorts.

Sims divided most of his days and nights among his London residence in the Carlton Hotel, the Admiralty, and his own headquarters at the American embassy in Grosvenor Square. But because of his operational responsibilities he also spent time at the Royal Navy base in Queenstown (now Cobh), Ireland. Sims's operational authority had been spelled out by the U.S. Navy Department in a telegram of 29 April from the Secretary of the Navy. With the characteristic disdain for grammar and syntax (in favor of telegraphic brevity) typical of such messages, it stated: "Rear Admiral Sims detached all duty Newport assume command all United States destroyers operating from British bases including tenders and auxiliaries there to be sent later." Sims subsequently explained, "Putting me in command of the destroyers that are sent over here means not that I will handle them at sea but that I will have general control of their operations while carrying out my present duties in connection with the admiralty."[18] As a commander of operating ships, the admiral enjoyed the perquisite of a flagship, in this case USS *Melville*, a destroyer tender, which remained moored throughout the war at the base in Queenstown.

Given the competing demands on Sims's time—his planning and diplomatic duties with the Admiralty in London and the complexities of operating squadrons of destroyers in combat—he made the remarkable decision to put the U.S. ships under the operational control of Vice Adm. Sir Lewis Bayly, commanding in Queenstown. This subordination of operational units of the U.S. Navy to the Royal Navy was historically unprecedented, and it ensured that the forces fighting the German U-boat threat would be most efficiently used. At the time, Sims said his revolutionary action "will clear away many misunderstandings and considerable friction. I think it is now arranged so that there will be complete cooperation between headquarters and those in the 'field.'"[19]

Sims's Partnership with Bayly

Sims had been drawn to Queenstown because it was home for the American destroyers that escorted cargo vessels and troopships on the last legs of their journeys from New York or Hampton Roads to their destinations in southern England. Many of the destroyer skippers had served under Sims when he commanded the Torpedo Flotilla, Atlantic Fleet, from 1913 to 1915. But in a very short time the greatest source of Sims's personal attraction to Queenstown became the British commander, Admiral Bayly, whom the British strategist Sir Julian Corbett considered "the father of destroyer tactics and organization."[20] The less awestruck and more irreverent American sailors nicknamed the British sea dog "Old Frozen Face." Figuratively, they first fell under his gaze when Cdr. Joseph K. Taussig led the first American destroyer flotilla into Queenstown Harbor on 4 May 1917. He immediately called on Bayly, who asked when he would be ready for combat patrols. Taussig replied, "I shall be ready when fueled." He required neither repairs nor supplies. Bayly ordered, "You will take four days' rest. Good morning."[21] No wonder Bayly later said of Taussig's arrival—and of the joint patrols that immediately began—that they had made "all the difference" to the success of the convoys.[22]

The relationship between the two flag officers was as remarkable for its personal intimacy as it was for its strategic success. In May and June 1917 Bayly and Sims, who was promoted to vice admiral on 26 May, conspired to arrange for Sims to assume temporary operational command of the British and American destroyers operating out of Queenstown. The ostensible reason was Bayly's need for a week of rest and recuperation. The British admiral advised Sims to broach the idea with the First Sea Lord and if he approved, "we will arrange it between us without any frills, and if the Admiralty during my absence 'regret that you should have [done something other than what had been done],' I will take the blame. If they give you a DSO [Distinguished Service Order] keep it."[23] Sims rather disingenuously said the offer was the surprise of his life, but he was honest about feeling it to be a great "honour."[24] He leapt at the opportunity.

The tone of camaraderie was further exhibited when on 18 July 1917 Sims instantly agreed to the wisdom of a suggestion by Bayly that Capt. Joel Roberts Poinsett Pringle—the commanding officer of Sims's flagship, *Melville,* and also his chief of staff in Ireland—be assigned additional duty as a formal member of Bayly's staff. Many years later, Bayly reminisced about Pringle's superb performance in a billet without antecedent: "At my request, Captain Pringle's name was entered in the English Navy List—the first time that a foreign naval officer had ever appeared there as serving on a British Admiral's staff in time of war."[25]

The Sims-Bayly cooperation checked the U-boat attacks in the Atlantic. When Sims first arrived in London, as has been seen, Admiral Jellicoe predicted that the U-boats would strangle Britain through economic warfare at sea. This did not happen. By December 1917 the Entente was losing a more sustainable 350,000 tons of shipping per month, down from 900,000 in April. In October 1918, the month before the war ended, the U-boats could sink no more than 112,427 tons. This massive reduction in loss, with the corollary monthly increase in tonnage of materiel shipped from the United States, was accomplished between May 1917 and November 1918 by 1,500 convoys of eighteen thousand ships. The U.S. Navy provided almost 30 percent of the escorting destroyers in British waters, the Royal Navy just over 70 percent.

The American admiral repeatedly cited the Anglo-American disparity in combatant vessels to justify the subordination of his destroyers to the British operational commander. His was a perfectly defensible stance, but at the same time he undercut himself with official Washington by appearing unnecessarily sycophantic in his relations with the Admiralty. In early 1918 the British offered him honorary membership on the Board of Admiralty, an unprecedented distinction that would have made him privy to the innermost deliberations of the Royal Navy's central headquarters. Daniels forbade acceptance, saying later, "I regarded it as rather a love of glitter and foreign recognition and honor than anything else."[26]

Perhaps shown up by the generous British treatment of Sims, however, the U.S. Navy Department did refrain from taking any further action inimical to its London commander for the rest of the war. In December 1918 it even promoted him to the rank of full—that is, four-star—admiral. It was a bittersweet, ephemeral reward for Sims. Once he left the London command he had to revert to two-star rank, the highest permanent grade in the U.S. Navy at the time.[27]

American-British Cooperation in the North Sea

While Sims and Bayly were prosecuting the transatlantic naval war, two other British and American admirals joined forces to contain the German High Seas Fleet, not incidentally thwarting in the process the German U-boats and surface raiders that were preying on the vital maritime supply route from Norway to Britain and France. The Briton was Admiral Beatty, Jellicoe's successor as commander in chief of the Royal Navy's Grand Fleet, which had blockaded the Germans in their home ports since the battle of Jutland of May–June 1916. The American was Rear Adm. Hugh Rodman, commander of the U.S. Atlantic Fleet's Battleship Division 9. Four dreadnoughts composed the division: *New York, Florida, Delaware,* and *Wyoming.* They were the best coal burners in the U.S. battle fleet, and they had been chosen for that reason. Newer, oil-burning American dreadnoughts would have increased the overall thirst for scarce fuel oil once they reached their anchorages in Scapa Flow, but by contrast the British base had stored almost unlimited amounts of coal.

Rodman was a classmate of Sims from the small Naval Academy class of 1880, only four members of which reached flag rank. As cadet-midshipmen the two men had not been close friends. After graduation they had independently made the jolting transformation from sailors of the old navy into masters of the science and art of commanding the revolutionary all-steam warships of the new American navy, embodiments of Alfred T. Mahan's doctrine of "seapower." The personalities of Sims and Rodman were as different as dreadnoughts were from wooden men-of-war. Sims had made his reputation as an outspoken critic of the status quo and a proponent of new technologies, especially in gunnery. Rodman's reputation rested on his love of the sea, masterful ship handling, solicitous grooming of his junior officers, and—not least—his propensity to spin yarns for the bewitchment of all around him.[28]

Washington's decision to send American battleships to the Grand Fleet had been even more protracted than that to deploy substantial numbers of convoy escorts to Queenstown. The anguish over dispatching one of the four divisions of the U.S. battle fleet was particularly severe because the deployment directly challenged the prevailing strategic doctrine of the U.S. Navy, based on the lectures and writings of Mahan. Mahan's influence, first felt when he had been a professor at the Naval War College in the late 1880s, continued with the publication of his globally influential book *The Influence of Sea Power upon History, 1660–1783.* More than any other individual in or out of the service, Mahan was responsible for the navy's repudiation of its historical doctrine of *guerre de course* and coastal defense and its adoption of the neo-Nelsonian concept of commanding the seas with fleets of battleships prepared—even anxious—to engage similarly configured and deployed enemy battle fleets.

Rear Adm. William T. Sampson's victory over the Spanish at Santiago de Cuba in July 1898 was purposefully interpreted as an exemplar of Mahan's doctrine. In the next decade the United States added new and improved battleships and heavy cruisers to its battle fleet at the rate of as many as two per year. At the same time, the new navalists of Mahanian persuasion prepared strategic plans for fighting the two most likely enemies—imperial Japan ("Orange") and imperial Germany ("Black"). The core of the American operational concepts can be stated succinctly. First, be prepared at any time to defeat the enemy's approaching battle fleet in a massive and decisive sea battle, preferably annihilating the opponent's fleet, as happened to the Franco-Spanish fleet that had had the temerity to challenge Horatio Nelson at the battle of Trafalgar on

21 October 1805. Second, and accordingly, never under any circumstances dilute the power of the battle fleet by dividing it prior to war.

This thoroughly embedded doctrine thwarted all attempts to dispatch a number of U.S. battleships to the war zone during the first eight months of American belligerency. In July 1917, after first obtaining the endorsement of Admiral Sims, Admiral Jellicoe, the First Sea Lord, requested a detachment of four American coal-burning dreadnoughts. The American battlewagons would strengthen the Grand Fleet in its blockade of the German High Seas Fleet in the North Sea. They also would permit the British to retire some of their predreadnoughts, thereby freeing up crews to man 119 new antisubmarine destroyers under construction in British yards.[29]

Adm. Henry T. Mayo, commander in chief of the U.S. Atlantic Fleet, and the Chief of Naval Operations, William S. Benson, would have none of it. The two most senior officers in the U.S. Navy considered the Royal Navy to be insufficiently aggressive, and they regarded Sims as a supine instrument of the Admiralty. Of at least equal weight in the decision was their firm Mahanian conviction that they could not in good conscience fragment the battle fleet by detaching a quarter of its ships. To do so would emasculate the fleet should the United States find itself in a war with Japan, always a feared possibility given the perpetual tension over the American forward political and territorial presence in the Philippine Islands. The governing "Plan Orange" dictated that such a war was to be decided by an Armageddon fought in the Pacific by opposing battle fleets.

Four months passed before Benson changed his mind. Between July and November 1917 he was subjected to constant lobbying by Capt. William Veazie Pratt, Assistant Chief of Naval Operations and a cohort of Sims, and by the popular American novelist Winston Churchill. The skeptical Admiral Mayo attended an Inter-Allied Naval Conference in London on 4–5 September and was impressed with Sims and his arguments regarding the battleships, but Benson held firm in his opposition. Finally, President Wilson attached Benson as the naval representative to a delegation to Britain, a group with broad investigatory powers and headed by the president's most trusted intimate, Col. Edward M. House.

The Americans arrived in Britain on 7 November 1917. Within three days Benson had capitulated to the Sims-Jellicoe proposition that four American battleships be attached to the Grand Fleet. On 10 November the Chief of Naval Operations fired off a somewhat incoherent cable advising Daniels to modify the sacrosanct Mahanian doctrine: "The principle not to divide the fleet does not apply to this matter in my opinion. It would apply to the portion of the fleet necessarily kept in American waters by logistical considerations, rather than to a division to join the Grand Fleet."[30] For the first time, four principal American warships were to be integrated, as subordinate units, into a fleet commanded by an officer of the Royal Navy, the historical nemesis of the U.S. Navy.

Three days after Benson cabled Washington, Daniels appointed Rodman as commander of Battleship Division 9. Wasting no time, Rodman led his four coal burners to sea on 25 November. After an extremely rough passage across the gale-swept, wintry Atlantic, *New York, Florida, Delaware,* and *Wyoming* steamed into the massive base of the British Grand Fleet in the roadstead of Scapa Flow, in the Orkney Islands. The date was 7 December 1917, and that day's marriage of American and British capital fleets lasted to the end of World War I in November 1918. Following a series of postwar squabbles, the vows were to be renewed in 1939 and honored throughout World War II and the Cold War. The union persists in intimate amicability almost a century after Admiral Beatty first spied Hugh Rodman's flagship *New York* entering Scapa Flow.

Rodman Subordinates Himself to Beatty

As soon as Rodman arrived, he called on Beatty aboard his flagship, *Queen Elizabeth.* Rodman offered to place his four battleships under the operational command of the Royal Navy. This historic subordination was a voluntary act on the part of Rodman: "I realized that the British fleet had three years of actual warfare and knew the game from the ground floor up; that . . . there would be a great deal to learn practically." He could not conceive of harmonious cooperation if there were "two independent commands in one force . . . and [therefore] the only logical course was to amalgamate our ships and serve under the command of the British commander-in-chief."[31]

The American battleship division became the 6th Battle Squadron of the Grand Fleet, constituting about 12 percent of the fleet's capital ships.[32] In administrative matters Rodman at least nominally reported to Sims in London, but he enjoyed direct access to the U.S. Navy Department in the form of weekly reports he sent to Secretary of the Navy Daniels.[33] The multilinear network did not correspond to the navy's ideal of straight-line command responsibility; it had been cobbled together because of disagreements between strong personalities on how to conduct the naval war.

For fighting purposes the chain of command was clear and direct—Beatty was Rodman's immediate operational commander and would remain so until the war's end. This integration and subordination exceeded that of Sims's relationship with Bayly, where the arrangement was more one of partnership than abdication of independent command authority. Moreover, Rodman gave Beatty command of four of the major warships of the U.S. battle fleet, not simply destroyers and other escort vessels. The symbolism of Rodman's historic action was unmistakable.

The American admiral was adamant that this was in fact the chain of command: "Every movement order to my command that had any relation to hostilities that I received during the war after joining the Grand Fleet, emanated from Admiral Beatty and not from our liaison officer in London [Sims] or anybody else."[34] His admiration for Beatty was unbounded, and by the end of the war his Anglophilia at least matched that for which Sims had become notorious. Rodman readily stated, "I am free to admit that next to my own country and countrymen I admire the British more than any others on earth." It became his fervent hope and prophecy "that the feeling of comradeship and brotherhood that was engendered in the Grand Fleet will last for many years, and that our respective nations will stand together in the future as they did in the World War."[35]

Three days after steaming into Scapa Flow, Rodman led his newly rechristened 6th Battle Squadron on a battle patrol in company with the rest of the Grand Fleet. The immediate matters at hand were tactical: mastering the British codes of signaling, learning to take and keep station as one wing of the Grand Fleet, and measuring American accuracy in gunfire by British standards.[36] The alien codes were quickly learned, and station keeping came easily to Rodman and his experienced captains, but their ships' gunnery fell short of the expected standards of excellence. In early March 1918 Rodman bemoaned the poor scores in gunnery practice of one of his battleships. "In spite of her four years commission," he wrote somewhat awkwardly to the Secretary of the Navy, and the fact "that she has now the [American] gunnery trophy, and was flying the efficiency pennant, *she was not ready to fire under war conditions.*"[37] He criticized the U.S. Navy's system of emphasizing good scores compiled for salvos against fixed targets as inferior to the British system of firing against simulations of moving ships, although the fixed-target system had been refined in the prewar years by his Academy classmate, William S. Sims.

On the strategic level Rodman always kept in sight what he considered to be the crucial importance of the ability of the Grand Fleet to maintain its station in the North Sea until "the surrender of the whole German fleet."[38] By containing the High Seas Fleet, the Grand Fleet prevented it from marauding throughout the North Sea and the Atlantic approaches to England and France. In this generalized sense, therefore, Rodman and Beatty indirectly guarded shipments of men and cargo. In addition, Beatty regularly deployed some of his smaller ships expressly for antisubmarine actions.

When Rodman's squadron arrived, convoys of essential materiel from Scandinavia faced a new and more troubling menace—German surface raiders. The Admiralty's failure to halt the severe threat to shipping posed by German light cruisers and destroyers had contributed to the downfall of Jellicoe as First Sea Lord. Beatty cannily took the hint. In January 1918 he began to deploy battleships and other heavy warships to shield convoys originating in Norway from attacks by the German surface raiders. On

5 February, with elegantly understated British condescension, he confided to his wife, "I am sending old Rodman out on an operation of his own, which pleases him and gives them [the American sailors] an idea that they are really taking part in the war. I trust they will come to no harm."[39] They did not.

The 6th Battle Squadron stood out from Scapa Flow for its first independent operational patrol on 6 February. Rodman was accompanied by the British 3rd Light Cruiser Squadron and its protective destroyers. In another precedent-shattering subordination of one navy's capital ships to another's, an act that showed a keen sense for diplomacy, Beatty had placed both units under American command. Rodman's flotilla picked up an outbound convoy of cargo vessels, escorted it to Norwegian waters, loitered offshore, picked up an inbound convoy of about thirty ships, shepherded it to British waters, and reentered Scapa Flow on 10 February. No surface raiders were encountered, but jumpy lookouts spotted several submarine periscopes or conning towers and wakes, which were diligently attacked, with no confirmed hits. Rodman, Beatty, and the U.S. Navy Department accepted the sightings as genuine, but the shrewd and skeptical Capt. Henry A. Wiley, commanding officer of *Wyoming,* placed the blame on the lookouts' inexperience and on porpoises "bobbing up and down."[40] Seasoning in subsequent patrols sharpened the eyesight of the American spotters.

The independent and combined operations to protect the Norwegian convoys continued until late April, when Adm. Reinhard Scheer's entire High Seas Fleet finally ventured out to intercept a convoy and annihilate its protectors. The Germans miscalculated the date of the convoy's sailing, and the battle cruiser *Moltke* lost a propeller and had to be taken under tow. *Moltke*'s intercepted radio signals pleading for help alerted the Grand Fleet, but before Beatty's full force could confront the Germans they had fled to the safety of home waters.

This misadventure was the final excursion of the High Seas Fleet prior to its inglorious surrender, internment, and scuttling in Scapa Flow at war's end. Scheer's mission had been to destroy the capital-ship escorts of a convoy that earlier in the war would have been protected by destroyers and attacked by nothing heavier than light surface commerce raiders. However, the Rodman-Beatty convoy screens of capital ships had drawn out the German battle fleet. A Mahanian battle might have ensued; the United States and Britain conceivably could have lost control of the North Sea as a corridor to the Atlantic, and then the vital transatlantic convoys of men and materiel from North America would have been fatally imperiled.[41]

Rodman astutely perceived the potential danger and expressed his alarm in a general report to the Secretary of the Navy on 27 April: "I am of the opinion, which is shared by most, if not all of the flag officers of the Grand Fleet, that there are possibilities of a grave disaster to the supporting force."[42] Made privy to the alleged danger, Sims scathingly rebuked Rodman for his "lack of confidence in the ability of the Commander-in-Chief [Beatty] and the Admiralty to handle the fleet with safety to its detachment." He archly dismissed Rodman's concerns: "the danger which you have assumed has not at any time existed."[43] Sims forwarded his critical evaluation of the battle squadron's commander to the Chief of Naval Operations, Benson, noting sharply that "Rodman has been doing excellent work with the fleet but he is rather impulsive and liable to 'slop over' at times."[44]

From the comfort of London, Sims was disparaging a senior commanding officer who had been in combat at sea for five continuous months. In June the dispute between Rodman and Sims evaporated with the British Admiralty's decision to discontinue the practice of covering the Scandinavian convoys with battleships and heavy cruisers. Only more expendable light cruisers would be used for that purpose.[45] Thereafter, the allied naval strategy in the North Sea and beyond reverted to what it had been.

Throughout the rest of the war, the Grand Fleet would stand guard over the sheltered High Seas Fleet, while the transatlantic convoys covered by Sims and Bayly flooded Britain with materiel and France with American fighting men. There would never be an apocalyptic Mahanian battle in the Atlantic, only perpetual preparation for one. In its place there would be the painstaking, systematic defeat of the deadly modern practitioners of the ancient art of *guerre de course,* the U-boats of imperial Germany, a war-winning triumph that in the years between the two world wars would be dismissed as irrelevant to future planning.

Conclusions

The cooperation of Sims and Bayly and that between Rodman and Beatty protected the convoys of troopships carrying the balance-tipping force of two million American soldiers "without losing a single man."[46] But beyond the destroyers at Queenstown and the battleships at Scapa Flow, Admiral Sims, Commander, United States Naval Forces Operating in European Waters, directly or indirectly commanded naval detachments of varying sizes and compositions at Murmansk, in Russia, and in Brest and elsewhere on the coast of France; submarine chasers stationed at Plymouth, England; an American naval base at the British naval bastion in Gibraltar; more submarine chasers on the island of Corfu; the U.S. mine force in Scotland; all U.S. naval aviation bases; and six U.S. Navy port offices.[47] Ultimately a total of 196 officers staffed Sims's London headquarters.

There had not been anything remotely approaching this scale of overseas commands and operations in the entire history of the U.S. Navy, and the whole complex apparatus was improvised. There had been no prewar planning for cobelligerency with Great Britain, and as a result there had been no anticipation of this array of installations and operations. In a relatively brief period between April 1917 and November 1918, two British admirals and two American admirals had overcome their navies' historical distrust of one another in order to forge a victorious Anglo-American naval alliance.

Highly personal and born in reaction to a lethal sea war of unprecedented magnitude, the alliance would fragment in 1919. It would lie shattered throughout the two interwar decades. But as soon as Great Britain went to war with Nazi Germany in September 1939 it was reconstituted and reshaped, often under the guidance of officers who had served in World War I as disciples of Beatty, Rodman, Bayly, or Sims. Notable among the understudies was Cdr. Harold R. Stark, the personnel officer at Sims's London headquarters. He became Chief of Naval Operations in 1939, and the next year he wrote the comprehensive plan—known as Plan DOG (the traditional naval phonetic term for the letter D)—for fighting Germany and Japan. In April 1942, Stark was sent to London to establish a naval headquarters modeled on the "London Flagship" of 1917–18.[48]

Notes

The thoughts and opinions expressed in this publication are those of the authors and are not necessarily those of the U.S. government, the U.S. Navy Department, or the Naval War College.

1. Walter Hines Page, quoted in Josephus Daniels, *Our Navy at War* (New York: G. H. Doran, 1922), p. 36.

2. Walter Hines Page to Department of State, telegram, 23 March 1917, William S. Sims collection, cont. 76, folder "Special Correspondence [Walter Hines] Page 1917–26," Library of Congress, Washington, D.C. [hereafter LOC].

3. Woodrow Wilson to Josephus Daniels, 24 March 1917, quoted in Mary Klachko and David F. Trask, *Admiral William Shepherd Benson: First Chief of Naval Operations* (Annapolis, Md.: Naval Institute Press, 1987), p. 57.

4. Admiral Sir Percy Scott, *Fifty Years in the Royal Navy* (New York: George H. Doran, 1919), p. 156.

5. William S. Sims to Anne H. Sims, 25 December 1906, William S. Sims Collection, no. 168, box 6, folder 31, Naval War College, Newport, R.I. [hereafter NWC].

6. Sims, quoted in "Commander Sims's Guildhall Speech," *Times* (London), 11 January 1911. Sims later recalled a slightly different wording in William S. Sims to P. H. Kerr, Esq., 13 July 1918, William S. Sims collection, cont. 47, folder "Special Correspondence," LOC. This is Sims's version, sent upon request to Downing Street for the information of Winston Churchill.

7. Sims, quoted in "The American Naval Visit: Sailors at Guildhall," *Times* (London), 5 December 1910. See also "Memorable Scene at Guildhall," *Daily News,* 4 December 1910, William S. Sims collection, cont. 115, folder "Clippings 1910," LOC.

8. "Public Reprimand for Sims," *Sun,* [11?] January 1911, William S. Sims collection, cont. 115, folder "Clippings 1911," LOC.

9. U.S. Senate, *Hearings before the Subcommittee of the Committee on Naval Affairs,* 66th Cong., 2nd sess., vol. 1, p. 268.

10. John Langdon Leighton, *SIMSADUS London: The American Navy in Europe* (New York: Henry Holt, 1920), p. 11.

11. William Howard Taft, quoted in Elting E. Morison, *Admiral Sims and the Modern*

American Navy (Boston: Houghton Mifflin, 1942), p. 284.

12. William S. Sims, *The Victory at Sea* (Garden City, N.Y.: Doubleday, Page, 1920), p. 7.

13. Ibid., p. 9, quoting John Jellicoe.

14. Ibid., p. 10.

15. William S. Sims to Anne H. Sims, 11 April 1917, William S. Sims Collection, no. 168, box 9, folder 17, NWC. In his letter Sims misspelled the name as "De Chare"; the authors have inserted the correct name in the quotation.

16. William S. Sims to Lewis Bayly, 18 July 1917, William S. Sims Collection, cont. 47, folder "Special Correspondence, Bayly 1917," LOC.

17. John V. Babcock, quoted in Morison, *Admiral Sims and the Modern American Navy*, p. 351.

18. William S. Sims to Anne H. Sims, 29 April 1917, William S. Sims Collection, no. 168, box 9, folder 17, NWC.

19. William S. Sims to Anne H. Sims, 4 June 1917, William S. Sims Collection, no. 168, box 9, folder 18, NWC.

20. Julian Corbett, quoted in Morison, *Admiral Sims and the Modern American Navy*, p. 378.

21. William N. Still, Jr., ed., *The Queenstown Patrol, 1917: The Diary of Commander Joseph Knefler Taussig, U.S. Navy* (Newport, R.I.: Naval War College Press, 1996), p. 188 note 84.

22. Lewis Bayly, quoted in Arthur J. Marder, *From the Dreadnought to Scapa Flow*, vol. 4, *1917: Year of Crisis* (London: Oxford Univ. Press, 1969), p. 275.

23. Lewis Bayly to William S. Sims, 30 May 1917, William S. Sims collection, cont. 47, folder "Special Correspondence, Bayly 1917," LOC.

24. William S. Sims to Lewis Bayly, 1 June 1917, William S. Sims collection, cont. 47, folder "Special Correspondence, Bayly 1917," LOC.

25. Admiral Sir Lewis Bayly, *Pull Together: The Memoirs of the Late Admiral Sir Lewis Bayly* (London: G. G. Harrap, 1939), pp. 273–74.

26. Josephus Daniels in 1920 congressional hearings, quoted in Morison, *Admiral Sims and the Modern American Navy*, p. 391.

27. He was restored to the four-star rank on the retired list in 1930; Morison, *Admiral Sims and the Modern American Navy*, p. 518.

28. Hugh Rodman, *Yarns of a Kentucky Admiral* (Indianapolis, Ind.: Bobbs-Merrill, 1928), *passim*.

29. Jerry W. Jones, *U.S. Battleship Operations in World War I* (Annapolis, Md.: Naval Institute Press, 1998), p. 12.

30. Ibid., p. 17, quoting Benson.

31. Rodman, *Yarns of a Kentucky Admiral*, p. 268.

32. Jones, *U.S. Battleship Operations in World War I*, pp. 27–28.

33. Ibid., p. 33.

34. Rodman, *Yarns of a Kentucky Admiral*, p. 269.

35. Ibid., pp. 266–67, 280.

36. For signaling, see ibid., pp. 268–69; for wireless communications and gunfire, see Jones, *U.S. Battleship Operations in World War I*, pp. 32–34.

37. Jones, *U.S. Battleship Operations in World War I*, p. 43, quoting Rodman to Secretary of the Navy (Operations), *General Report*, 2 March 1918. Original source in italics.

38. Rodman, *Yarns of a Kentucky Admiral*, p. 266.

39. Jones, *U.S. Battleship Operations in World War I*, p. 34, quoting Beatty to his wife.

40. Ibid., p. 38, quoting Wiley.

41. Ibid., pp. 48–50.

42. Ibid., p. 50, quoting Rodman.

43. Ibid., pp. 50–51, quoting Sims to Rodman.

44. Ibid., p. 51, quoting Sims.

45. Ibid., pp. 51–52.

46. Sims, *Victory at Sea*, p. 357.

47. Ibid., insert between pp. 144 and 145. For U.S. minelaying, see Jones, *U.S. Battleship Operations in World War I*, p. 56; and Leighton, *SIMSADUS London*, pp. 66–75. For further details, see Sims, *Victory at Sea*.

48. For Stark, see Benjamin Mitchell Simpson III, *Admiral Harold R. Stark: Architect of Victory, 1939–1945* (Columbia: Univ. of South Carolina Press, 1989).

Logistic Supply and Commerce War in the Spanish Civil War, 1936–1939

WILLARD C. FRANK, JR.

The course and outcome of the Spanish Civil War of 1936–39 largely depended on the supply of arms, specialists, and troops from abroad. Airplanes, tanks, and personnel from Italy and Germany, with major support from Moroccan troops, provided the backbone for the forces in service to the rebel generals. Meanwhile, aircraft, tanks, and advisers for waging war on land, in the air, and at sea, mainly from the Soviet Union (the USSR), supported Republican militia and other forces of an emerging Popular Army, for which the International Brigades served as shock troops. The great majority of these forces came to the Iberian Peninsula by sea, a fact that gave rise to significant attempts to use the sea or deny its use by the naval and air forces of the Spanish belligerents and their international allies. The Spanish Republic, in particular, lacked the raw materials and workforce necessary for a war industry. In Catalonia, only forty of 246 industries did defense work; the rest were more engaged in the ongoing socioeconomic revolution. The Republic needed 150,000,000 bullets per month to wage war but produced only fifteen thousand. The rest had to come from abroad.[1] In addition, the economic viability of both sides was largely determined by trade, imports to provide for domestic needs and exports to pay for needed foreign military aid.

The struggle for control of sea-lanes in the Spanish Civil War was deeply intertwined throughout with major legal issues, starting with the legal status of the participants and their shipping.[2] The Spanish Republic in July 1936 declared the seas a war zone and adopted a blockade on the coasts and ports of the Nationalist enemy. Neither the Nationalists nor the intervening or neutral maritime nations recognized this blockade, which could not be made effective in the absence of a determined and strong Republican blockade by warship patrols. In time, however, Nationalist patrols mounted just such a determined and strong effort, producing a de facto, if never a de jure, blockade, particularly in the Strait of Gibraltar. Warships visited, searched, and frequently seized ships and cargoes, even when a seizure clearly violated the established rules of war.

Mine warfare, aerial bombardment, and interference within the territorial waters of the Spanish parties or neutral states, especially France, were special cases that recurred throughout the war. Attempts in May–June 1937 to give warships belonging to Britain, France, Italy, and Germany specific assurances of their neutrality and to provide them with safety zones in Spanish waters failed to gain agreement. In particular, submarine attacks were major issues to Britain and France, and they will be treated separately.

One of the most important legal issues was the international nonintervention system, with its deep roots in nineteenth-century international relations. On 15 August 1936 Britain and France inaugurated a Europe-wide effort to contain the Spanish conflict by an international ban on all intervention in Spain, direct or indirect, including all war material, even that already under contract. The Non-Intervention Agreement (NIA) would come into force whenever either inaugurating power—Britain or France—and Germany, Italy, the Soviet Union, and Portugal adhered. Language varied in many of the national declarations, but eventually twenty-seven states of Europe declared their adherence.[3] This agreement eventually led to the 1937 Nyon agreement outlawing the use of submarines in unrestricted warfare.

International Influence in the Spanish Civil War

Soon after the Spanish Civil War broke out on 17 July 1936, foreign countries took sides, supporting the Republicans or the Nationalists. For their part, Britain and France attempted to contain the Spanish conflict, but their efforts failed to be enforced by the interventionist states of Italy, Germany, and Portugal, on one side, or on the other by the non-European states of Mexico and—for a while—the United States, and eventually the Soviet Union and its Communist International (Comintern) affiliates operating throughout Europe.

When immediate self-control failed, the states party to the Non-Intervention Agreement formed on 9 September 1936 a Non-Intervention Committee (NIC) of European ambassadors in London and, on 14 September, a smaller Chairman's Sub-Committee comprising the major intervening states, which traded barbed accusations of gross violations of the NIA. Britain and France refrained from making accusations and rather tried to work as smoothly as possible toward some form of a practicable nonintervention scheme. Italy, Germany, and Portugal flagrantly violated the terms of their own declared prohibitions, even while accusing the Soviet Union of shipping military aid to the Spanish Republic at a stage when it was only sending nonmilitary supplies, such as clothing, medicine, and food.[4] On 7 October 1936, and particularly on 23 October, the Soviet representative on the NIC announced that the Soviet Union would abide by the NIA only so far as other states did so as well.[5] By late October, intervention with arms shipments to both sides was in full swing.

Further legal problems included foreign vessels, some of which carried arms or volunteers, sailing in the territorial waters of member states, where attacks by submarines, surface warships, mines, and aircraft persisted. So did the question of whether or when belligerent rights might be accorded to both sides. Britain and France were concerned lest granting belligerent rights lead to a wider war and refrained from doing so throughout the conflict. The League of Nations heard the Spanish Republican appeals against foreign intervention but referred all legal issues normally under its purview to the NIC for resolution.

The United States, as a non-European state, was not party to the nonintervention system but largely abided by its terms in separate and parallel measures. American ships did come under attack—the steamship *Exmouth* in August 1936, the destroyer *Kane* in August, and the gunboat *Erie* in December. The Nationalists seized the tanker *Nantucket Chief*, which carried petroleum from the Black Sea to the Republic, only to release it after an American protest. As for selling arms to the belligerents in Spain, the United States moved to follow a separate and parallel support for the policies of Britain and France in applying the Non-Intervention regime in Europe. Moral persuasion proving insufficient to prevent licenses to export arms, on 8 January 1937 the United States made it illegal to export arms to either side in Spain, an act that appeared to be strengthened by an embargo law of 1 May 1937 giving the president the authority to act if the export of arms "would threaten or endanger the peace of the United States."[6]

Germany and Italy sent their military aid to the Nationalists openly by sea to southern Spanish ports as special cargoes in national merchant ships protected by their warships.[7] Meanwhile, Stalin hesitated to get more deeply involved than to send cargoes of humanitarian aid to Republican ports. On 29 September, the Politburo, with the escalation of German and Italian military aid threatening to become decisive and the Non-Intervention Agreement a dead letter, changed course and started to send military aid in Soviet and Spanish merchant ships. The shipments went at first openly to Alicante, where German warships were watching and reporting the arrival of Soviet aid, and by mid-October to the more secure port of Cartagena, which, as the Republic's main naval base, was better defended by coastal and antiaircraft artillery and airfields.[8]

This prompted a Nationalist and Italo-Germanic response in the form of an Italian, German, and Nationalist commerce war against Soviet and Spanish Republican arms ships. Of the first twenty-nine voyages of such "Igrek" arms ships organized and dispatched by the Intelligence Department of the Politburo, twenty-four sailed from the Black Sea to Cartagena, two from Leningrad to the northern front on the Bay of Biscay, and three from third countries. Igreks at first sailed individually and alone, leading to the sinking or capture of six of the seven dispatched in December 1936. Yet by

6 December 1936 the Republic had received sufficient arms—including 136 aircraft, 106 tanks, 30 armored cars, and 174 pieces of artillery—to hold Madrid.[9]

To pay for the Soviet arms the Republican government decreed on 13 September 1936 that the finance minister, Juan Negrín, could dispatch Spain's national treasure of gold then held in the vaults of the Bank of Spain to "wherever he considers safest." Some gold had already been sent to Paris, the hub of European arms-traffic transactions.[10] A second possible alternative considered was the United States, which was in any event used as cover for the actual transfer to the USSR.

With Nationalist armies only twenty miles away from Madrid and pressing hard, Alexander Orlov, the senior operative of the People's Commissariat for Internal Affairs (NKVD) of the Soviet Union in Spain, received a telegram directly from Stalin ordering him secretly to send the Spanish gold to the Soviet Union, without signing any formal receipt. The USSR was by mid-October sending major arms shipments to the Spanish Republic and was in competition with Premier Francisco Largo Caballero on the guiding of the Republic politically. Several Soviet merchant ships were in Cartagena and ready after unloading arms to return home, with space for special cargoes. Stalin ordered Orlov to organize the transfer in Soviet merchant ships. Orlov had the gold brought to Cartagena, where it was temporarily stored in the Algameca munitions caves just outside the naval base. Sixty Republican submarine sailors were employed carrying the boxes of gold, from train to munitions depot to trucks, to be loaded into four Soviet merchant ships ready to depart from Cartagena for Odessa. From there the gold would be transported by special train to the State Bank in Moscow, where a receipt was promised. Commanding officers of Republican warships under sealed orders were stationed along the intended route to intercept any Nationalist search-and-seizure attempt.

By 6 November the four vessels, each with a portion of the treasure, had arrived in Odessa, where the gold was transferred to the train. Bank of Spain officials who accompanied the gold were put up in a hotel in Odessa but were not allowed to go home until the end of the war. Stalin clinched his intention by openly quoting an old Russian proverb: "The Spaniards will never see their gold again, as they don't see their own ears." The gold paid for arms shipments and the expenses of Soviet military specialists serving in Spain. An inventory of the gold, some 510 metric tons, was finally prepared and signed on 5 February 1937, but without any indication of its intended use, which Orlov later reported as dual in nature: to safeguard the gold from Nationalist capture and to serve as security for Soviet arms shipments to the Republican government.

In 1937, as Soviet arms shipments reached a peak, Stalin drew on the gold as payment; the gold treasure was soon exhausted in the purchase of military supplies in the Soviet Union and elsewhere in Europe. Spaniards learned to their surprise that further aid would only come on credit, a bill that the Spanish government would have to pay after victory had been attained. The Spaniards were in no position to dicker. After the war, Negrín and his sons tried to get the Soviet Union to return to the Spanish state the gold—which they understood had been sent for "safekeeping" to the USSR—only to be told that Spain actually owed the Soviet Union fifty million dollars of an eighty-five-million-dollar credit extended in 1937 and not repaid. There the matter remains to this day.[11]

The Maritime Conflict Deepens

The Soviet Union found itself deep in a maritime war it had not wanted. The Soviet ambassador in London, Ivan Maiskii, who was also the Soviet representative on the NIC, proposed sending a squadron of Soviet warships to Spanish waters to engage in surveillance of the Germans and Italians and perhaps to escort Soviet supply ships themselves. The top Soviet naval leader, Flagman of the Fleet 1st Rank (that is, Admiral) V. M. Orlov, however, wanted to avoid any possible complications and incidents in the Mediterranean, especially given a still-weak Soviet fleet and the absence of friendly bases in the western Mediterranean. He found reasons to delay and oppose the deployment of a Soviet squadron to Spanish waters. This came, however, just as Stalin was in the early stages of building a grand "sea and oceanic fleet" that could give the Soviet Union clout in foreign waters. Admiral Orlov was soon liquidated in Stalin's bloody purges of the Soviet military's top commands.[12]

Germany and Italy found it relatively easy to flout the terms of the NIA. Italy in late 1936 and early 1937 sent massive numbers of troops to Cadiz by ocean liner and escorted their munitions ships and troopships with its navy, which escort the NIC had officially sanctioned. It also reflagged Spanish Nationalist supply ships as Italian, a device that got around nonintervention rules. Italy soon abandoned the use of large Italian liners as troopships and relied instead on smaller vessels that flew the Italian flag for the first half of their voyages from Italian to Nationalist ports, especially Palma de Mallorca, and then the Spanish Nationalist flag when nearing Spanish territorial waters, where international naval patrols were in evidence. (Italy returned to the use of large troopships under escort later in the war for large-scale movements of forces and for repatriation of wounded and sick soldiers.) The deception worked smoothly; the Republicans understood its mechanisms but were unable to respond effectively against Italy, with its powerful navy on the loose.[13]

Italy was also first to engage in commerce war, deploying submarines, at first in pairs and later four at a time, from 8 November 1936, two days after Italy had signed the international Submarine Protocol outlawing attacks on unarmed merchant ships. Nationalist officers on board Italian submarines, if challenged, were to pose as the boats' commanding officers and pretend that the submarines were Spanish.[14] Italo-Nationalist instructions allowed these Italian craft to torpedo Spanish Republican warships whenever found but Soviet merchant ships only within the international three-mile limit, with positive identification of the target ship required in all cases.

Each submarine had a zone of operations on the Spanish east coast, usually off a major port. Since arms ships entered Cartagena only after dark, with no colors flying and names painted out, the Spanish-Italian rules of engagement could not be met, and Italian boats engaged in frustratingly fruitless patrols under orders that disallowed attack on unknown darkened shapes entering Republican ports at night. Success finally came on 22 November, when the chief Soviet naval adviser to the Spanish Republican navy, Capt. N. G. Kuznetsov, in hopes of reducing the threat of air attack to warships within the harbor of Cartagena, persuaded the Republican naval command to place valuable warships in the open roadstead outside the port—a supposedly safe anchorage, since the Nationalists had no submarines of their own. The Italian submarine *Torricelli*, with Spanish lieutenant commander Arturo Genova on board, slammed two torpedoes into the machinery spaces of the anchored Republican flagship, the cruiser *Miguel de Cervantes*, which was then laid up under repair for most of the rest of the war. "Authorities" in Cartagena, as well as Adm. J. F. Somerville of the Royal Navy, assumed that the Republican submarine *B-5*, missing in action, had defected to the enemy and was responsible. For the moment the Italian deception worked.[15]

Germany was second to attack. Out of fear of an international incident if German complicity were to come to light, and with Britain keeping a sharp eye on German fleet development since the 1935 Anglo-German naval treaty, Berlin deployed to the area two of its seven new oceangoing U-boats *(U-33* and *U-34)*, under complete secrecy and with all identifying marks removed. The Nationalists could only guess at the existence of this "Training Exercise URSULA," named for Adm. Karl Dönitz's daughter.[16] Rules of engagement allowed attacks on Republican warships and within territorial waters on any darkened warship or escorted merchant vessel. Secrecy was so effective that the existence of URSULA came to light only through the chance discovery in German archives during the 1960s of a coded handwritten note on one page in a folder that showed that the Germans secretly tried to make this operation appear to be a training exercise for merchant ships.[17]

Defective torpedoes frustrated the two skippers and their crews; the torpedoes ran foul and exploded, if at all, far from their intended targets. *U-34*, however, did attain one success, sinking the Spanish Republican submarine *C-3* patrolling on the surface off Málaga on 12 December. Yet a Republican investigation concluded that the loss of *C-3* had been due to an accidental internal explosion, whereupon the issue was dropped.[18] Three earlier German torpedo firings had achieved no success. In any case, *C-3* was a target of opportunity; Operation URSULA was officially over and not scheduled to be renewed. Identifying numbers and names were repainted off the northern Netherlands coast. Deception, the Oberkommando der Marine (OKM), or naval high command, reminded all, was "of the highest principle to avoid compromising Germany," especially in the eyes of the British.[19]

The OKM was far more cautious than was its Italian counterpart about risking charges of flagrant violations of the Non-Intervention terms. Hitler agreed with the war minister, Gen. Werner von Blomberg, that the principle of unity of command should prevail, without confusing movements, as one nation's submarines cleared the common operating area so that the other's submarines could move into vacated patrol stations. German and Italian admirals agreed on 14 December to divide their responsibilities, so that the Italians could continue clandestine submarine warfare in the Mediterranean, and the Germans to operate only with surface forces, to attain quality intelligence on ship movements.[20] The German navy thus felt relieved of any complications when an unexploded Italian torpedo ran up the beach near Barcelona on Christmas Day, 1936, and another near Tarragona the next day. All the world could see Italian responsibility in these actions, while Germany escaped blame. Fragments of exploded torpedoes in *Miguel de Cervantes* revealed the Italian origins of that attack. Italy had so far sent forty-two submarines into action in the Spanish War, had tracked 133 targets, and had launched twenty-seven torpedoes, which had damaged one cruiser and sunk two freighters, but it had stopped no Soviet arms ships. Germany had deployed two U-boats, tracked twelve targets, and launched four torpedoes but had sunk only one Spanish submarine.

Italy soon loosened its instructions to the extent of allowing cruiser and submarine bombardment of port facilities at night, but only if secrecy was maintained. Nevertheless, Italian shell fragments were found at Valencia and Barcelona after nocturnal bombardments; Italian admirals acted puzzled and surprised when British admirals raised the issue of responsibility.[21] In frustration, and with the war not going as well as hoped, Mussolini abandoned his ineffective clandestine submarine warfare in mid-February 1937. The large Italian destroyer *Giovanni da Verrazzano*, with its identifying marks painted out and a false third funnel installed, towed two motor torpedo boats to participate in the final assault on Málaga. In the nighttime confusion, the boats took *Da Verrazzano* for a Republican submarine and fired a torpedo that missed the destroyer by just a few meters. They escaped detection by returning to the Nationalist port of Ceuta before dawn.[22]

The Nationalists received reliable German and Italian intelligence on Soviet ship movements from the Black Sea, through the Turkish Straits and the Mediterranean to Spain. On 14 December, the cruiser *Canarias* caught the motorship *Komsomol* in a compromised position that made its cargo of "ore" from Poti to Ghent appear to mask a real cargo of arms. *Komsomol* could have been carrying both kinds of cargo. *Canarias* took the passengers and crew prisoner and sank *Komsomol* with gunfire.[23] Soviet naval advisers realized that all future arms ships would have to be better protected by the Republican fleet, especially during the last leg of the dash from the Algerian coast into Republican ports.

The NIC Control Scheme in Operation

Throughout the autumn and winter of 1936–37, the NIC wrestled with how to proceed. The Spanish belligerents refused to cooperate by allowing neutral observers on their territories and in their seaports, a problem that was supposedly settled in April 1937 by a land and sea Control Scheme, by which neutral observers would be stationed on land frontiers to observe and report to an NIC office in London any arms traffic or volunteers crossing frontiers, while neutral "observing officers" would be assigned to all merchant ships heading for Spanish ports, to ensure that no proscribed cargoes or volunteers were introduced into Spain. Further, the Spanish coast was divided among the powers with significant naval forces in Spanish waters. German and Italian warships were assigned to patrol the Republican coast and the British and French to patrol Nationalist coasts.[24]

The Control Scheme operated as planned in April and May 1937.[25] But on 29 May, as part of a planned Republican deception to facilitate the arrival of the arms ship *Magallanes,* the Republican fleet bombarded the harbor of Ibiza while Soviet SB-2 bombers attacked it from the air.[26] By mistake, the Soviet bombers hit the German pocket battleship *Deutschland,* lying at anchor off the island town, killing or mortally wounding thirty-one sailors. The German admiral, Hermann von Fischel, had assumed that Ibiza was a safe and secluded port in which to give his warships on NIC patrol some rest and recreation, while the Republicans assumed that it was an open port in enemy hands, against which a diversionary strike was acceptable. Neither assumption proved to be accurate.

That night's Republican squadron convoying *Magallanes* sailed through a German squadron that Hitler had ordered to gather. Hitler was determined to retaliate against the Cartagena naval base and its moored warships, which the Republic could have used to escalate its conflict with Germany, perhaps into a declared war between Germany and the Republic. The Republican defense minister, Indalecio Prieto, urged that the Republican navy seek out the German retaliatory squadron for combat, hoping thereby to provoke just such a German declaration of war against the Republic, which he hoped would lead in turn to a full European war, which he saw as the best circumstance for the survival of the beleaguered Republic.[27]

The Germans were especially sensitive to being labeled "baby killers," as they had been after the Scarborough and Hartlepool raids of 1914. A less volatile sentiment eventually prevailed in Berlin. Under Foreign Minister Konstantin von Neurath's moderating influence, Hitler abandoned his early demand for a strong retaliation against the well-defended naval port of Cartagena and instead ordered the bombardment by *Admiral Scheer* of the open, or at least poorly defended, city of Almería on the 31st, an attack that killed nineteen townspeople at first count and destroyed thirty-five buildings. Hitler had counted on the Republican battleship *Jaime I*, his target of choice, being present, but it had slipped out to return to Cartagena, which was thought to be safer. Germany immediately dropped out of the Control Scheme, while demanding further measures by the international naval forces in Spanish waters to counter what seemed to be a provocative act (i.e., the *Deutschland* bombing) by the Republic. As Nationalist Spain was primarily playing the grand-strategic role of masking the German rebuilding of its military during the "danger zone" of its early phases, Neurath diverted Hitler's demand for vigorous retaliatory action.[28]

Tension mounted further in June, however, when the German cruiser *Leipzig* reported that on four occasions between June 15 and 18 noises heard against the cruiser's hull had indicated attacks by "Spanish-Bolshevist submarine pirates." In an emergency four-party meeting, Germany demanded, first, the internment of all Republican submarines; second, "an immediate joint naval demonstration by the four powers off Valencia"; and third, a stern warning to the Valencia government that "any further attack would result in immediate military reprisals by the four powers." These were in addition to a proposed retaliation by three German U-boats then in the Atlantic: they would secretly enter the Mediterranean and, in a repeat of the 1936 Operation URSULA, attack Republican warships and escorted merchant ships as they approached Cartagena. Neurath refrained from going that far, but he instructed the other powers that if they doubted whether the attacks had actually taken place, such doubts must be "sharply rejected even to the point of walking out of the conference." Yet as Berlin reacted in bold certainty, Adm. Hermann von Fischel, on reflection, became increasingly doubtful that there had been any attacks at all. German tests in these same waters showed that hydrophone indications of torpedoes were often actually machinery noises from one's own ship. The supposed indications of attacks on *Leipzig* might have been caused by porpoises. Yet von Fischel asserted that even if these had been false alarms, "one has to expect the possibility of a submarine attack" sometime; he prepared an even more deniable retaliation plan, in which the same three U-boats would enter the Mediterranean, secretly, and use only electric torpedoes to sink just Republican submarines. Hitler remained "extremely wrought up" but refrained from retaliatory action on his own, while the Republican navy was especially careful to avoid any action that might trigger a German retaliation.[29]

The potential for a major incident remained high. On 30 June, a Republican convoy on the north coast encountered the German U-boat *U-35,* on the surface; the boat had black-white-red recognition stripes on its conning tower, but they were not seen by the two destroyers guarding the Republican convoy. The destroyers immediately went on the attack. *U-35* immediately submerged and eventually cleared the area. Both the Germans and Republicans were following established rules, each assuming hostile intent by the other, assumptions that were not always the case.[30]

The Creation of the "Rome–Berlin Axis"

Under constant Nationalist pressure to donate surface warships, Mussolini agreed at least to transfer in April 1937 two submarines, *Torricelli* and *Archimede,* with partly Spanish crews. In Italian waters they were Italian, but on war patrols or at their forward base at Sóller, in Mallorca, they were Spanish—with the designations *C-3* and *C-5,* respectively, as if these submarines, both previously lost, had actually defected to the Nationalist cause.[31] These two submarines sank or damaged seven Republican vessels; the world continued to assume direct Italian responsibility, thus indicating to Spaniards how others viewed Mussolini's outrages in the Mediterranean.[32]

In the case of Germany, locomotive carriers with extra-large hatches that could accommodate crated aircraft were disguised as Panamanian ships, since non-European states were not signatories to the Non-Intervention Agreement. These "Panama steamers" were directed to the northwestern port of Vigo whenever no foreign ships would be within sight. Even so, they were escorted during the last legs of their voyages by German or Nationalist warships, thus keeping the deception intact. French warships were assigned to patrol these waters for the Non-Intervention Committee, and Nationalist patrol boats from Vigo stood ready to rush out and "capture" any uncontrolled "Panama steamer."[33] The scheme became routine and remained effective. A total of 180 German and 290 Italian arms cargoes arrived in Nationalist Spain in disguise. All but the first two German arms vessels *(Usaramo* and *Kamerun)* arrived without the slightest incident. In such ways Nationalist Spain, Germany, and Italy—and to a lesser degree Portugal—became active coalition partners in their aid of Francisco Franco's Spain.

The Nationalist navy, making full use of good German intelligence reports, captured actual or suspected arms carriers, many of them under the Soviet flag, made possible by the fact that the Soviet merchant marine was relatively weak in the 1930s and only slowly recovering with new construction, and then put these ships and their cargoes in service to the Nationalist cause. All this was deeply frustrating to Soviet and Republican Spanish leaderships, which could neither enforce nor end the policy of nonintervention, to which the Soviet Union had signed on.

Since many merchant ships used in the European trade with the Spanish Republic were chartered from British firms or from companies located in the smaller European states, merchant ships from all over Europe found themselves at the mercy of Nationalist naval patrols and open to capture and confiscation. One exception comprised Dutch merchant ships, which as neutrals in compliance with the rules of nonintervention were escorted past Gibraltar toward the Dutch East Indies by Dutch warships stationed in the strait for that purpose. Dutch ships became relatively safe from Nationalist capture, a protection that ships of other, less powerful nations did not enjoy.[34]

Out of the Spanish Civil War was formed Mussolini's "Rome–Berlin Axis" and a concert of policy making in which Germany and Italy came ever closer politically, at least outwardly so, standing in opposition to Britain and France, the leaders of the Non-Intervention movement. Reports of violations noted by the "observing officers" stationed by the April 1937 NIC scheme on each merchant ship heading for Spanish ports went to a NIC office in London for record keeping. Thus the Spanish Civil War, particularly at sea, where escalation of the war was most likely, was a major impetus toward drawing together Italy and Germany on one side and Britain and France on the other.[35]

The Culmination of the Commerce War

The costs of this commerce war were significant. The Spanish Republic and the Soviet Union bore the greatest economic costs, including higher insurance and chartering fees. They paid the greatest military price as well, as Soviet and Comintern aid greatly slackened during the second half of the war, leaving the Spanish Republic without the means to achieve victory. The Germans attained the greatest economic benefits, especially in strategic raw materials, primarily ores, which they gained through special agreements with their Spanish Nationalist friends. Since Italy was mostly seeking prestige in Spain, Mussolini never asked for the practical strategic economic benefits that the Germans demanded and gained as a matter of course. Therefore, Fascist Italy came out of this war economically weakened, which made it more dependent on its Nazi German Axis partner.[36]

In early August 1937, prompted by dire warnings from excited but inaccurate Nationalist intelligence operatives, Italy was greatly to expand the maritime and geographical dimension of the Spanish conflict by vastly increasing Italian submarine, surface, and air attacks on merchant shipping throughout the Mediterranean. This expansion of the scope of the war would prove intolerable to France and Britain, which would organize Mediterranean and Black Sea states in September 1937 to take concerted action to contain the Spanish War, restore security to the trade routes of the Mediterranean, and contribute to the stabilization of Europe. This initiative would become an operational success, in that it ended indiscriminate attacks, but a strategic failure, in that it contributed to the onset of the Second World War under more favorable conditions for the Axis powers.

In response to Spanish pleas for decisive action at sea against alleged massive Soviet aid shipments coming to the Republic by convoy, Mussolini in August ordered his air force and surface and submarine navy to engage in a major clandestine naval and air campaign to cut the Republic's sea lines of communications from the Black Sea through the Mediterranean to Republican Spain. Soon reports came in from across the Mediterranean, from Spanish waters, the Sicilian Channel, and the Aegean Sea, of systematic Italian naval and air surveillance of merchant traffic in the Mediterranean, punctuated by attacks by aircraft, surface warships, and submarines. From 5 August to 5 September 1937, fifty-two submarines, forty-three surface warships, and squadrons of aircraft scouted over five hundred suspect merchant ships on Mediterranean sea-lanes, carried out fifty attacks, sank twelve ships, and captured two.[37] Italian responsibility was often very clear. On 11 August, the destroyer *Saetta*, "SA" painted on its side, shadowed the Spanish tanker *Campeador,* full of Romanian oil, in the Sicilian Channel, and after nightfall, while Spanish crewmen watched, it came abreast and sank the tanker with four torpedoes.[38]

On 15 August, a submarine with "C-3" painted on its conning tower surfaced near the Spanish arms ship *Ciudad de Cádiz* in the Aegean Sea, sank it with gunfire and torpedoes, and disappeared. The Spanish Republican submarine *C-3*, of course, accused of defecting to the Nationalists, had actually been sunk by the German *U-34* in 1936. The ruse was transparent, as the designation "C-3" was painted in red, in Italian style, not in white as in Spanish practice. The attacking submarine had two deck guns, while *C-3* had had only one. The sketch made by survivors matched the features of the actual assailant, the Italian *Ferraris*.[39]

On 31 August, the Italian submarine *Iride* launched a torpedo against the British destroyer *Havock,* having mistaken it for a Spanish Republican destroyer of similar silhouette. The torpedo missed, but *Iride* surfaced momentarily on launching the torpedo, advertising its Italian identity to British crewmen. For hours *Havock* depth-charged *Iride* as it made its escape. There were forty-seven similar incidents. Italian, German, and Spanish Nationalist authorities maintained that the assailants were Spanish, or alternatively Soviet. Yet Italian responsibility was obvious.[40]

The culmination of the commerce war was to come with the increased Italian-led attacks in August–September 1937 on Soviet and neutral shipping in the Mediterranean and the resulting international Nyon Arrangement of 14 September 1937, which was to affect deeply the way Europeans faced the crisis that ultimately led to the Second World War.[41] The Nyon Arrangement would be a regional security system that pitted two established powers, France and Britain, against Fascist Italy. This arrangement would arguably be triggered by Italian naval intervention in the Spanish Civil War, in which the international struggle for arms supply was decisive.

The League Takes Action: The Nyon Arrangement

Broad sentiment among Mediterranean states called for action. At first each state took its own measures. The French started escorting, the British ordered counterattacks, the Turks patrolled, and the Soviets readied destroyers for the Aegean. The French premier, Camille Chautemps, and the British prime minister, Neville Chamberlain, were cautious, with Chamberlain taking the lead in trying to woo Mussolini away from the Axis with Germany. The French foreign minister, Yvon Delbos, and the British foreign secretary, Anthony Eden, however, were ready for action to coerce Mussolini into changing his behavior. Eventually Chamberlain concurred. Officials considered and rejected several options: granting belligerent rights to both Spanish factions, retaliating against Nationalist Spain (such as sinking the Nationalist cruiser *Canarias*), negotiating a broad Mediterranean Pact, holding direct talks with Italy, and raising the issue in the League of Nations or the Non-Intervention Committee. None of the officials, however, would produce quick action to force Mussolini to back down.

With the League of Nations scheduled to meet in mid-September, Delbos proposed a conference of Mediterranean states in Geneva to provide international sanction for action. Eden quickly agreed and proposed a narrow agenda to gain rapid support and produce quick action. Delbos agreed. Mussolini saw the signs and withdrew his forces from the attack. The democracies did not know it yet, but they had won their victory even before the conference had met.[42]

Delbos and Eden built an effective coalition through compromise, and Maxim Litvinov, the Soviet foreign minister, helped make it effective. Delbos wished to invite the Spanish Republic, or if not, the Soviet Union. Eden would not invite the Spaniards and wished not to invite the Soviets, but he did accept the Soviet Union if, to keep bridges open for appeasement, Italy and Germany were also invited. To maintain momentum, Delbos and Eden compromised by inviting all Mediterranean and Black Sea states and Germany. To prevent Italy from sabotaging the conference, the Soviets sent a stinging rebuke to Italy for its aggression in the Mediterranean. The maneuver pricked Italian sensitivities, and the Italian foreign minister, Galeazzo Ciano, who had said he would attend, fell into Litvinov's trap and declined. So did Germany. The greatest troublemakers in the Mediterranean thus excused themselves. Now, with the right grouping of states for rapid practical action, Litvinov became uncharacteristically agreeable, allowing Delbos and Eden the leadership and scope to make this exercise in collective security a success.[43]

To encourage Italy to attend, Delbos moved the conference from Geneva, with its negative associations for Italy (over Ethiopia), twenty-five kilometers up the lake to the town of Nyon. Nine delegations, representing all Mediterranean and Black Sea powers but Spain and Italy, met in the town assembly hall on 10 September 1937. On Eden's nomination, Delbos assumed the presidency of the conference. Delbos charged the delegates to arrive at "a rapid agreement which will put an end to the state of piracy, and an immediate lessening of the intolerable tension which involves the risk of new and graver incidents." He hoped that "rapid success" would pave the way for wider conciliation and collaboration. All the delegations were ready for action.[44]

A second factor that favored success was that quick practical action came through an approach that was intentionally limited, focused, and technical. Delbos and Eden—working remarkably smoothly and rapidly, given past Anglo-French tension—developed a joint plan based on Eden's insight that only a limited technical, not a broad political, counter to submarine attacks would succeed. A political solution was unlikely, given Italian pride and temperament, and a broad agenda would introduce delays. British naval and diplomatic officials produced the basics—a focus solely on "unknown" submarines, assignment of navies to hunt down and destroy all submerged submarines found in the vicinity of an attack, and the legal basis for such action. The conference adopted the concept of submarine attacks as "piracy," which in international law is a matter of private, not state-sponsored, attacks at sea. The approach had the advantage of likely deterring further attacks by not accusing the Italians, while prosecuting the supposedly stateless submarines as outlaws, and perhaps even enlisting Italy in the effort.[45]

The conference also invoked article IV of the 1930 London Naval Treaty, which had been ratified—even by Italy—as the 1936 "Submarine Protocol," stipulating that submarines must follow the rules that applied to surface ships prohibiting the sinking of unresisting ships without first providing for the safety of passengers, crew, and ship's papers. Any submarine not following this rule could face destruction. Merchant shipping was to use prescribed routes under protection of British and French naval patrols.[46]

The French and British navies were to patrol routes in assigned zones in the western Mediterranean. In the Aegean—to obviate the presence of Soviet warships, which Greece and Turkey opposed—France made an additional contribution of a number of large four-stack destroyers *(contre-torpilleurs)*, greatly pleasing the British. Litvinov acquiesced. Thus the plan moved forward, with the Soviets remaining as partners but refraining from deploying naval forces in the Mediterranean. British and French warships, stretching available capacity, took up their patrols.

In two days the delegates had come to complete agreement. The nine powers created a restricted "arrangement"—rather than a political agreement—by means of which the

French and British accomplished their own goal of a rapid and tangible counter to the Italian ravages against shipping.[47] In doing so, they captured the initiative from the Axis powers for the only time in the prewar years. Public opinion across the political spectrum in both France and Britain was solidly for the Anglo-French collaboration and their unhesitating action against Mediterranean "piracy." Neither Mussolini nor Hitler was prepared for such rapid and resolute action. For a brief time the dictators were at a loss as to how to respond.

The success of the limited approach at Nyon allowed an extension to include the more difficult question of countering attacks by surface warships and aircraft. This Supplemental Agreement, signed in Geneva three days later, put the Nyon system on an even stronger basis.[48] Immediately, however, Britain urged Italy to join the Nyon regime. France went along so as not to jeopardize its British partnership. Ciano now joined, and Britain and France rewarded the Fascist regime's hunger for prestige with the empty designation of Italy as "a great Mediterranean power." The pirate of the Mediterranean was now one of its policemen, a state of affairs that elicited widespread anxiety and sarcasm. The naval staffs of the democracies diverted every Italian attempt to obtain patrol authority over the Dardanelles-to-Spain route except a hundred-mile stretch (to scare off Soviet aid), over Suez as a "gate" imprisoning Italy in the Mediterranean (to needle Britain), and over French strategic routes with Algeria. To meet Greek and Turkish objections, they precluded Italy from patrolling the Aegean. Although outwitted, Italy signed the Paris Agreement of 30 September that included Italy in the Mediterranean patrols, but only after it was granted its request that patrol zones not be made public for fear of adverse domestic reaction.[49]

Following the signing of these agreements, Soviet aid finally came to a ragged end. Three arms ships were marooned in Algiers, while Nationalist cruisers remained on the prowl. In October 1937 the large Igrek *Cabo Santo Tomé*, full of war material and defensively armed, came under Nationalist surface attack off the Algerian coast and was destroyed.[50] This left the eleven unarmed merchant ships of the Comintern-controlled firm France-Navigation to continue to ply a low-level clandestine arms trade between the Black Sea and Spain. They did so without major incident through 1938.[51] Yet major arms shipments neither from the Black Sea nor from Leningrad were any longer operating on the Soviets' behalf. Only in early 1939, with a special plea from the Republicans through their air force chief, Ignacio Hidalgo de Cisneros, were arms shipments renewed, from the Baltic Sea to French Atlantic ports for transshipment by land to Catalonia and Barcelona. The play of Stalin's mood was critical, especially in this period, when terror ruled at home.[52] Most of these supplies never made it to the fighting front, most being abandoned in France at the end of the Spanish Civil War.[53]

Shipments of Italian and German aid operated much more smoothly throughout the end of the civil conflict. Smaller Italian or reflagged Spanish vessels constantly plied the routes between Italy and Nationalist Spain, flying the Italian flag except when just off Nationalist ports, where Nationalist escorts took over escort duty. Large Italian ocean liners came back into use as troopships in 1938, for major troop moves or to return to Italy forces to be repatriated in the expectation of the implementation of the Anglo-Italian accords of April 1938. At the end of the war, large troopships repatriated, with great ceremony, the rest to Naples. Mussolini had hoped to gain prestige in Spain as a great military power, but Italy came out of the struggle weaker economically than it went in. Prestige did not come easily or cheaply.

German aid through "special steamers" containing military cargoes continued uninterrupted, thanks to the wide Atlantic, the Panamanian ruse, German escorts, and the isolated destination port of Vigo. Bulk cargoes of strategic minerals for the return voyage were common, as Germany obtained the maximum trade advantages it could gain from its Spanish friends. Pressure for trade concessions continued after peace was restored in Spain. German forces were repatriated in *Kraft durch Freude* (Strength through Joy) cruise ships, in which peacetime shipboard games substituted for military training. Back in Germany these forces, still in their Spanish uniforms, paraded in Berlin for Hitler before shifting back into their Wehrmacht uniforms for the much larger military contest just ahead. Soviet logistic support allowed the Spanish Civil War to continue until by late 1937, with the Nyon and Paris accords, the shifting of Moscow's strategic priorities to China, and a likely priority for a military buildup at home, active Soviet military support for the Spanish Republic became only a trickle. Such low-level support continued in 1938, as Stalin pondered his next moves in response to the aggressively ambitious Germany so close at hand.

By starting with a limited focus and expanding the Nyon system by degrees, Delbos and Eden had given direction and courage to the democracies and to the cluster of Mediterranean and Black Sea states that relied on them. Italy, although admitted into their company, had to be content with the trappings and not the reality of parity.[54] The Italians made a pretense of patrols, while the French and British, understanding the political and psychological value involved, maintained their own intensive patrols on assigned Nyon routes. The Nyon system remained in force until the end of the Spanish War eighteen months later.[55]

Conclusions

The respect the democracies gained at Nyon was later squandered when Franco declared peace on his own terms on 1 April 1939 after the last of the Republican forces had surrendered. Alert to the irony of pirates becoming policemen, Ciano called the agreement in Paris a great victory. Emboldened, he dispatched new "Legionary" submarines to Spain, but under stringent rules of engagement.[56] Soviet hopes for collective security, buoyed by Nyon, sank again with the Paris Agreement making Italy an equal partner. By October 1937 the Soviet Union had stopped sending military aid via the Mediterranean, a fact that marked the beginning of Stalin's abandonment of Spain and proved a major factor in the Republican defeat. Soviet aid now shifted to China. Significantly, the Spanish defeat came only five months before the onset of World War II, precluding an anti-Fascist Spain in the coming fight for Europe. The Soviet abandonment of Spain was a major step toward the USSR's giving up on the democracies and toward the Nazi-Soviet pact of 1939.[57]

The piracy and Nyon episodes also arguably prompted Hitler to begin his conquest of central Europe earlier than he had anticipated. Hitler realized that the prolongation of the internationalized Spanish crisis had allowed him to pass safely through the "danger zone" of early rearmament in 1936–37, while the eyes of Europe were fixed on Spain. Then in August 1937 the breadth of the Mediterranean erupted in flames, and warships of the opposing European camps had each other in their sights. A full-scale Mediterranean war seemed possible at any moment. To understand the effect of the Mediterranean crisis on Hitler, we should view it from the prevailing mind-set of Nazi Germany—that the Mediterranean crisis was a creation of Bolshevik Russia to advance nefarious schemes to wrap its tentacles around all of Europe. The German naval commander in Spanish waters, Rear Adm. Rolf Carls, was so certain that the rampages were Bolshevik sponsored that he spontaneously offered the cooperation of the German navy "in the strongest possible action" with the British navy to hunt down the pirates.[58]

This vision of a Mediterranean about to erupt into a major conflict prompted Hitler in his speech to his diplomatic and military chiefs on 5 November 1937—at the so-called Hossbach conference—to advance plans for the early conquest of Austria and Czechoslovakia.[59] With Fascist Italy on the rise and imperial Britain in decline, and with French-Italian hostility or Soviet intrigue providing the spark, the Mediterranean at any moment might explode into an Anglo-French-Italian war. It was Hitler's most likely scenario for the near future. With the West embroiled in the Mediterranean, Germany must make a lightning strike against the Czechs, earlier than previously anticipated. The piracy-Nyon crisis was the specific catalyst that prompted Hitler to formulate early moves in central Europe.

Yet even as Hitler laid out his plan, all of the Mediterranean outside Spanish waters was already calm again.[60] Nevertheless, Hitler mobilized his war machine for action, surprising his unprepared military, and he never let up. Sprung by the Mediterranean crisis, Hitler's momentum toward war continued without interruption straight into catastrophe. The piracy-Nyon crisis, therefore, was a major factor in the inauguration of the

Second World War earlier than otherwise would likely have been the case. In addition, the Soviet Union was at this very time responding to the Japanese assault on China with new aircraft and truck convoys carrying military aid (Operation Z). Major military aid to Spain ended in October 1937, just as it was picking up in support of Soviet interests in China.

The Nyon Arrangement was also the first concrete step in the Soviet Union's eventual abandonment of the West in favor of a pact with Hitler. Munich then continued the process of distancing the Soviet Union from the West, just as Stalin's terror was taking a firm grip on the Soviet psyche. Thus, in the end, the regional security scheme of Nyon brought but temporary calm to the Mediterranean. Although it produced short-term operational success, it was in a longer view a strategic failure. It sowed the seeds of World War II under conditions less favorable to the democracies than they would likely have been otherwise.

Notes

The thoughts and opinions expressed in this publication are those of the author and are not necessarily those of the U.S. government, the U.S. Navy Department, or the Naval War College.

1. For the failure of the Spanish Republic to provide an effective war industry and workforce, see Iurii Rybalkin, *Operatsiya "X": Sovetskaya Voennaya Pomoshch' Respublikanskoi Ispanii, 1936–1939* (Moscow: Aero-XX, 2000), pp. 25–28.

2. Legal issues are treated in detail in Norman J. Padelford, *International Law and Diplomacy in the Spanish Civil Strife* (New York: Macmillan, 1939).

3. Ibid., pp. 57–72.

4. Non-Intervention records are gathered in FO 899/1-39, Public Record Office / The National Archives, Kew, U.K. [hereafter PRO/TNA].

5. See warning statements of S. Kagan, 7 October 1936, and I. Maiskii, 23 October 1936, to the NIC, FO 899/1, PRO/TNA.

6. Padelford, *International Law and Diplomacy in the Spanish Civil Strife*, pp. 169–88. See also Richard P. Traina, *American Diplomacy and the Spanish Civil War* (Bloomington: Indiana Univ. Press, 1968); for the later temptation of Franklin D. Roosevelt to lift the embargo, or at least to deliver arms surreptitiously to the Republic via Canada, see Dominic Tierney, "Franklin D. Roosevelt and Covert Aid to the Loyalists in the Spanish Civil War, 1936–1937," *Journal of Contemporary History* 39, no. 3 (July 2004), pp. 299–313.

7. For German military aid and the support of the German navy in ensuring it, see "Tätigkeitsbericht der Schiffahrtsabteilung (OKM A VI) im Dienst des Sonderstabes W während des Spanienkrieges," Anlage 5, RM 20/1437, Bundesarchiv-Militärarchiv, Freiburg, Ger. [hereafter BA-MA]; Erwin Jaenecke, *Erinnerungen aus dem spanischen Bürgerkrieg*, 2 April 1956, German Air Force Monograph Project G/1/1b, U.S. Air Force Historical Research Center, Maxwell Air Force Base, Ala.; Dieter Jung, "Der Einsatz der deutschen Handelsschiffahrt während des Spanienkrieges 1936–1939," *Marine-Rundschau* 76, no. 5 (1979), pp. 322–29; and Axel Schimpf, "Der Einsatz von Kriegsmarineeinheiten im Rahmen der Verwichlungen des spanischen Bürgerkrieges 1936 bis 1939," in *Der Einsatz von Seestreitkräften im Dienst der auswärtigen Politik: Vorträge auf der Historisch-Taktischen Tagung der Flotte*, ed. Deutsches Marine Institut (Herford, Ger.: E. S. Mittler, 1983), pp. 76–103. For German efforts to overcome such logistical problems as refueling at sea, see Paul W. Zieb, "Logistische Probleme der Kriegsmarine," *Neckargemünd: Scharnhorst* (1961), pp. 83–88.

8. N. G. Kuznetsov, "Soobrazheniya po proveleniyu operatsii s igrekami," fond. 35082, op. 1, delo 18, listy 66–55; "Doklad morskogo sovnika glavnomu voennomu sovetniku, 30 Aug., 1937," f. 35082, op. 1a, d. 23, ll. 47–51; both Rossiiskii Gosudarstvennyi Voennyi Arkhiv, Moscow. See also *Soviet Shipping in the Spanish Civil War,* Memorandum 59, Research Program on the USSR (New York: East European Fund, 1954).

9. Rybalkin, *Operatsiya "X,"* pp. 42–43.

10. The banks of Paris, and especially the communist-controlled Banque Commerciale pour l'Europe du Nord, were a prime alternative to Moscow as the destination for the Spanish gold.

11. Alexander Orlov defected in 1937 and came to the United States in hiding. On the shipment of the Spanish gold to the Soviet Union, see Rybalkin, *Operatsiya "X,"* pp. 90–100; Frank Schauff, *Der verspielte Sieg: Sowjetunion, Kommunistische Internationale und Spanischer Bürgerkrieg, 1936–1939* (Frankfurt, Ger.: Campus, 2005), pp. 341–52; U.S. Senate, *The Legacy of Alexander Orlov,* prepared by the Subcommittee to Investigate the Administration of the Internal Security Act and Other Security Laws of the Committee on the Judiciary, 93rd Cong., 1st sess., August 1973 (Washington, D.C.: U.S. Government Printing Office [hereafter GPO], 1973), pp. 42–50; Alexander Orlov, "How Stalin Relieved Spain of $600,000,000," *Reader's Digest* 89, no. 535 (November 1966), pp. 37–50; and John Costello and Oleg Tsarev, *Deadly Illusions* (New York: Crown, 1993), pp. 258–63.

12. Adm. V. M. Orlov, Commissar of the Soviet Navy, correspondence and reports in f. r-1485, op. 3, d. 244, ll. 1–85, *passim,* Rossiiskii Gosudarstvennyi Arkhiv Voenno-Morskogo Flota, St. Petersburg [hereafter RGAVMF]; N. G. Kuznetsov, "Flagman flota i ranga V. M. Orlov," *Voenno-istoricheskii Zhurnal,* no. 6 (June 1965), pp. 66–71.

13. Italian military aid to the Nationalists is listed in archival documents under the cost of the shipments by the cost of each shipment in lire, not the total numbers of units sent. Large aircraft, such as bombers, readily flew to Mallorca. Transport by sea included ten large Spanish ships sailing under Italian auspices. From July 1936 until October 1937, Italian or reflagged Spanish ships made 128 voyages to Nationalist ports, deliveries that continued in 162 reflagged ships thereafter. With few exceptions, all made their voyages safely. See details in "Relazione finale sull'attività dell'Ufficio Spagna," buste 9–10, Archivio Storico-Diplomatico, Ministerio degli Affari Esteri, Rome [hereafter ASD-MAE].

14. Franco Bargoni, *L'Impegno navale italiano durante la Guerra Civile Spagnola (1936-1939)* (Rome: Ufficio Storico della Marina Militare [hereafter USMM], 1992), pp. 133–44, 195–210.

15. "Relazione finale sull'attività dell'Ufficio Spagna," buste 9–10, 129–31; see Willard C. Frank, Jr., "Naval Operations in the Spanish Civil War, 1936–1939," *Naval War College Review* 37, no. 1 (January–February 1984), pp. 32–34, 37–38.

16. German records disguised Operation URSULA as an exercise: "Übungsbefehle für Kriegspiel Ob.d.M," OKM to B.d.A., 6 November 1936, FT 1808, 20 November 1936; Italo-German naval agreement, Rome, 17 November 1936, RM 20/899; both BA-MA.

17. The secret file on "Unternehmen Ursula" is RM 20/899, BA-MA.

18. See Verdía report, 14 December 1936, AR 25-14 (XI), 9574, Archivo General de la Marina Álvaro [hereafter AGMAB].

19. See Willard C. Frank, Jr., "German Clandestine Submarine Warfare in the Spanish Civil War, 1936," in *New Interpretations in Naval History: Selected Papers from the Ninth Naval History Symposium Held at the United States Naval Academy, 18–20 October 1989,* ed. William R. Roberts and Jack Sweetman (Annapolis, Md.: Naval Institute Press, 1991), pp. 107–23. For responsibility for the sinking of *C-3,* see the letter of admission by the surviving watch officer of *U-34* at the time, Flottillenadmiral a.D. Gerd Schreiber, to Willard C. Frank, Jr., 6 January 1978. A copy is filed in the U-Bootsarchiv, Cuxhaven, Germany.

20. For the meetings of Italian, German, and Spanish admirals on 10 and 14 December 1936, see Bargoni, *L'Impegno navale italiano,* pp. 153–60.

21. "Missione Segrete," pp. 166–67, vols. 13–16, Archivio Centrale dello Stato, Rome [hereafter ACS].

22. Ibid., messages 3–8 February 1937, p. 167, vols. 15–16, ACS; Bargoni, *L'Impegno navale italiano,* pp. 181–84.

23. See Moreno's war diary in N.C. 25-13, vol. 7, SHA, AGMAB, Viso del Marques; the account

of the captain of *Komsomol* is G. A. Mezentsev, "V fashistskom plenu," in *Pod znamenem ispanskoi respubliki, 1936–1939* (Moscow: Nauka, 1965), pp. 541–75. Mezentsev maintained that *Komsomol* carried no war equipment on this voyage, only bulk ore for Ghent. The actual cargo could readily have been both. He was released from prison in late 1937.

24. The NIC formed several administrative agencies: the NIC itself; the Chairman's Sub-Committee; the NIC International Board, under Vice Admiral van Dulm of the Netherlands, to record the results of observation; a naval patrol composed of warships of Britain, France, Italy, and Germany, each nation's warships to patrol a section of the Spanish coast, the Italians and Germans opposite the Republican coast and the British and French along the Nationalist coast; and administrators and accounting agencies. For NIA-NIC records, see FO 899/1–39, PRO/TNA.

25. Bargoni, *L'Impegno navale italiano*, pp. 243–54.

26. The Italian auxiliary cruiser *Barletta* had been bombed and damaged by Soviet aircraft in the Palma roadstead on 26 May 1937, a factor that led Admiral von Fischel to move *Deutschland* from Palma to Ibiza; see ibid., pp. 255–60. For German experience in the *Deutschland* incident, see Kriegstagebuch des Befehlshaber der Panzerschiffe (KTB des BdP), RM 50/10, BA-MA; for Soviet and Spanish experience see V. L. Bogdenko, "Stranitsy starykh bloknotov," in *Leningradtsy v Ispanii: Sbornik vospominami*, 3rd ed. (Leningrad: Lenizdat, 1989), pp. 157–98. See also Willard C. Frank, Jr., "Misperception and Incidents at Sea: The *Deutschland* and *Leipzig* Crises, 1937," *Naval War College Review* 43, no. 2 (Spring 1990), pp. 31–46.

27. Indalecio Prieto, "El bombardeo alemán de Almería," in *Convulsiones de España: Pequeños detalles de grandes sucesos* (Mexico: Oasis, 1968), vol. 2, pp. 95–100.

28. *Documents on German Foreign Policy, 1918–1945*, ser. D, vol. 3, *Germany and the Spanish Civil War* (Washington, D.C.: GPO, 1950) [hereafter DGFP], pp. 297, 299–300, 326–27, 343–44, 348, 349; *Documents on British Foreign Policy, 1919–1939*, 2nd ser. (London: HMSO, 1980) [hereafter DBFP], vol. 18, p. 836; "Denkschrift," 14 June 1937, RM 20/1222, BA-MA.

29. KTB *Leipzig*, RM 92/5070, BA-MA. See Frank, "Misperception and Incidents at Sea,"

pp. 40–42. Spanish archival records show all Republican submarines in port.

30. KTB des Befehlshaber der Aufkläungsschiffe, RM 50/7, BA-MA.

31. Spagna, Fondo di Guerra, b. 44; "Eventuale cessione di nave da guerra al governo spagnolo," ASD-MAE; Bargoni, *L'Impegno navale italiano*, pp. 266–72.

32. *Times* (London), 30, 31 July 1937.

33. German military aid, mostly from Hamburg to Seville and later to Vigo as a safe port for 180 special steamers, brought 16,524 personnel, 978 noncrated vehicles, and 117,882 tons of war equipment in crates. These ships returned with 16,846 troops who had been replaced and holds filled with strategic minerals. See "Zusammenstallung der insgesamt während des Spanienkrieges durch geführten Sondertransporte," Anlage 5 zu OKM AVI, 4385/31, M1388/80769, BA-MA; and tables of aid dispatched from Germany as cited in documents in the Archivo del Ministerio de Asuntos Exteriores, Madrid, and reproduced in Lucas Molina Franco and José María Manrique, *Legion Condor: La historia olvidada* (Valladolid, Sp.: Quirón, 2000), pp. 45–77.

34. For the Dutch escort system in the Strait of Gibraltar, March 1937–September 1938, see Netherlands, Departement van Defensie, *Jaarboek van de Koninklijke Marine, 1936–37* (The Hague: Algemeene Landsdrukkerij, 1938), pp. 125–27, 174–75, 187–88, 212–13, 234–35, 240–41, 244–49, 252–54, 258–60, 264, 267, 274–75, 301–303, 310–13, 321–23, 328, 344; Ph. M. Bosscher, *De Koninklijke Marine in de Tweede Wereldoorlog* (Franeker, Neth.: Wever, 1984), vol. 1, pp. 43, 108, 121, 160–61; and L. L. von Münching, "Een herinnering aan veertig jaar geleden: De Nederlandse zeemacht tijdens de Spaanse Burgeroorlog (1936–1938)," *Marineblad* (1976), pp. 342–55.

35. For an elaboration of this theme, see Willard C. Frank, Jr., "The Spanish Civil War and the Coming of the Second World War," *International History Review* 9, no. 3 (August 1987), pp. 400–408.

36. One may compare the prestige demanded by Mussolini and the economic gains in raw materials demanded by Germany as compensation for participation in the Spanish War. For main treatments, see John F. Coverdale, *Italian Intervention in the Spanish Civil War* (Princeton, N.J.: Princeton Univ. Press, 1975), translated as *I Fascisti italiani alla Guerra di*

Spagna (Rome: Laterza, 1977); and Robert H. Whealey, *Hitler and Spain: The Nazi Role in the Spanish Civil War, 1936–1939* (Lexington: Univ. of Kentucky Press, 1989).

37. For the operational level of Italian action, see Bargoni, *L'Impegno navale italiano*, pp. 280–316, 444–56; and for the strategic level, see Patrizio Rapalino, "Il ruolo della strategia marittima nella Guerra di Spagna (1936–1939)," *Rivista Marittima* 134, no. 7 (July 2001), pp. 13–31, elaborated in his *La Regia Marina in Spagna, 1936–1939* (Milan: Mursia, 2007).

38. For intelligence reports on *Campeador* and *Saetta*, see 1 BB9 270, Service Historique de la Défense–Marine, Vincennes, France.

39. For intelligence reports on *Ciudad de Cádiz* and *Ferraris*, see ADM 116/3917, PRO/TNA.

40. Vice Adm. Sir Norman Denning and Capt. Sir Patrick Beesley of the Royal Navy's Operational Intelligence Centre (OIC) confirmed to the author in interviews in 1977 that the OIC was just starting to establish a tracking system for foreign warships when the August–September 1937 Italian antishipping campaign began. With decoding only partial and still corrupt, the OIC was able to corroborate Italian responsibility but not to generate specific evidence of the names of the submarines responsible. The OIC quickly inaugurated a submarine-tracking system that would become invaluable in the Second World War. For a summary of these efforts, see Denning Report, 25 January 1938, "Movement Section of N.I.D.," ADM 223/286, PRO/TNA.

41. See Willard C. Frank, Jr., "The Nyon Arrangement 1937: Mediterranean Security and the Coming of the Second World War," in *Regions, Regional Organizations and Military Power: XXXIII International Congress of Military History, 2007*, ed. Thean Potgieter (Stellenbosch, S.A.: South African Military History Commission / African Sun Media, 2008), pp. 563–89.

42. The Delbos-Eden collaboration, with its frequent twists and turns but also clear sense of purpose and direction, is detailed in French and British archival documents and in John E. Dreifort, *Yvon Delbos at the Quai d'Orsay: French Foreign Policy during the Popular Front, 1936–1938* (Lawrence: Univ. Press of Kansas, 1973), esp. pp. 55–78; and A. R. Peters, *Anthony Eden at the Foreign Office, 1931–1938* (New York: St. Martin's, 1986), pp. 289–95. See also Schauff, *Der verspielte Sieg*, pp. 298–304.

43. Admiralty to CinC [Commander in Chief] Med, 2130 and 2215, 7 September 1937, ADM 116/3522; record of meeting on "the proposed Mediterranean Conference," 8 September 1937, CP 211 (37), CAB 24/271; First Sea Lord to CinC Med, 1530, 9 September 1937, ADM 116/3523; all PRO/TNA.

44. Ministère des Affaires Étrangères, Commission de Publication des Documents Relatifs aux Origines de la Guerre 1939–1945, *Documents diplomatiques français, 1932–1939*, 2nd ser. (Paris: Imprimerie Nationale, 1970), vol. 6, pp. 553–72; Note du Cabinet de l'état-major général de la Marine, no. 127, 7 September 1937, in ibid., pp. 705–706.

45. For the twisted logic used to make the "unknown" submarines seem legally to be stateless "pirates," see S. G. Fitzmaurice note, "Possible Legal Justification for Application of Proposed Mediterranean Submarine Scheme to Non-participating Countries," 7 September 1937, W16879/23/41, FO 371/21359, PRO/TNA.

46. "Arrangement de Nyon/The Nyon Arrangement," 14 September 1937; "Agreement between the French and British Naval Staffs," 12 September 1937, ADM 116/3522, PRO/TNA, reproduced in DBFP, vol. 19, pp. 271–73, 299–301.

47. "Agreement between the French and British Naval Staffs"; "Arrangement de Nyon/The Nyon Arrangement" Eden postconference report in British Delegation to Foreign Office [hereafter FO], no. 32, 15 September 1937, ADM 116/3522, PRO/TNA.

48. Supplementary agreement, in British Delegation (Geneva) to FO, 17 September 1937, W17556/16618/41, FO 371/21406, PRO/TNA.

49. See documents on the British courting of Italy in file "Anglo-Italian relations" in R5523/1/22 and R6244/1/22, FO 371/21161, PRO/TNA; messages 15–21 September 1937, FO 371/21406, PRO/TNA; and Bargoni, *L'Impegno navale italiano*, pp. 318–30.

50. For the destruction of *Cabo Santo Tomé* on 10 October 1937, see "Partes de campaña of *Cánovas del Castillo* and *Dato*," 12 October 1937, AN 25-13 (21 & 23), 9732 & 9740, AGMAB.

51. There is no full study of the activities of the communist firm France-Navigation. Its ships, however, figure prominently in Soviet lists of "Igreks" of 1938. See Dominique Grisoni and Gilles Hertzog, *Les Brigades de la mer* (Paris: Grasset, 1979).

52. For the impact of Stalin's reign of terror at home, especially in the armed forces, on operations in Spain, see Rybalkin, *Operatsiya "X,"* pp. 82–89; and Schauff, *Der verspielte Sieg, passim.*

53. The pleas of Hidalgo de Cisneros for arms seemed to catch Stalin in one of his more generous moods. Despite a new Soviet loan for military equipment, however, there was little opportunity for its transport to or employment in Spain before the end of the war.

54. Galeazzo Ciano, *Diary, 1937–1943* (New York: Enigma, 2002), pp. 8–10.

55. French, British, and Italian archives maintain extensive records of the early stages of the Nyon patrols but less for later months. For the main records of the French "Dispositif spécial en Méditerranée," see Service Historique de la Défense, Marine, Toulon. The main British records are in "Nyon Arrangements," vols. 3–12, ADM 116/3524–33, PRO/TNA. The main Italian files are in c. 3124 and 3128, USMM.

56. For operations of the Legionary submarines, see Bargoni, *L'Impegno navale italiano,* pp. 330–45.

57. Litvinov to Maiskii, 29 October 1937, no. 390, conveying Stalin's desires, in Ministerstvo Inostrannykh Del SSSR, *Dokumenty Vneshnei Politiki SSSR* (1937; repr. Moscow: Izdatel'stvo Politicheskoi Literatury, 1976), vol. 20, pp. 578–79. The Soviet abandonment of Spain as a prime destination for military assistance may also be traced in Rybalkin, *Operatsiya "X,"* pp. 45–47.

58. The assumption by Admiral Carls that the piracy was Bolshevik inspired is reported in Rear Admiral Gibraltar to Admiralty, no. 713, 20 August 1937, W15831/23/41, FO 371/21358, PRO/TNA.

59. The "Hossbach Memorandum" is in DGFP, ser. D, vol. 1, pp. 29–39. See also Jonathan Wright and Paul Stafford, "Hitler, Britain and the Hossbach Memorandum," *Militärgeschichtliche Mitteilungen,* no. 42 (2/1987), pp. 77–123, which argues a projected Mediterranean war for 1938. For a contrary argument that Hitler was not serious but used the prospect of a Mediterranean war only to frighten his generals, see Hans-Hening Abendroth, *Hitler in der spanischen Arena: Die deutschspanischen Beziehungen im Spannungsfeld der europäischen Interessenpolitik vom Ausbruch des Bürgerkrieges bis zum Ausbruch des Weltkrieges, 1936–1939* (Paderborn, W. Ger.: Schöningh, 1973), pp. 176–79. Hitler, however, might have employed such fear in his drive toward the "triumph of the will" in crises manufactured to be implemented in 1938.

60. After Nyon the Republic imposed such stringent rules of engagement on its submarine operations off Nationalist ports and in the Strait of Gibraltar that Soviet commanders of Spanish submarines complained that they were allowed no opportunity for success. See reports of V. Yegorev and S. Sapozhnikov, November–December 1938, f. r-1529, op. 1, d. 118, ll. 2–24, RGAVMF. Italian skippers had made similar complaints in their submarine campaign of November 1936–February 1937.

The German U-boat Campaign in World War II
WERNER RAHN

The peace treaty of Versailles reduced Germany to the status of a third-rate naval power. Submarines and military aircraft were forbidden altogether.[1] As a result, the German navy lacked weapons that most other modern navies acquired as a matter of course. All British attempts to abolish the submarine altogether for all nations were thwarted by France's opposition, however, and the Anglo-German Naval Agreement on 18 June 1935 allowed Germany to have a surface fleet with a tonnage up to 35 percent of that of the British Empire. The 35 percent ceiling applied not just to the total tonnage but also to the individual categories of warships. Only in the case of U-boats was Germany allowed to achieve first 45 percent and later even 100 percent of British submarine strength.

One week after the Naval Agreement was announced, the German navy commissioned the first small, 250-ton U-boat, uncovering its secret activity in this matter. Since 1918, though the U-boat had not been basically improved, it had been given better torpedoes with trackless and bubble-less ejection and noncontact pistols; minelaying capacity for all U-boats; the ability to transmit and receive signals both surfaced and submerged; greater diving depths; and increased power of resistance, through welded pressure hulls.[2] Nevertheless, widespread opinion prevailed in all navies, including the German navy, that the U-boat had lost the crucial role it had achieved in World War I as one of the most effective naval weapons. Contrary to this opinion, the U-boat Staff, centered on Capt. Karl Dönitz, was convinced that antisubmarine-warfare weapons were greatly overrated and had not made decisive progress since 1918.[3]

The German experiences in World War I acted as the starting point for the development of "wolf pack" tactics against an enemy's sea routes. In 1917–18, a number of U-boats had successfully attacked on the surface under cover of darkness. In his book *Die U-Bootswaffe*, published in 1939, Dönitz drew attention to the advantages of night attacks. Despite this, World War II night attacks would take the British by surprise. Their escort forces would be unable to cope with the tactic, particularly as the early form of sonar known as ASDIC (Anti-Submarine Detection Investigation Committee) had an effective range of no more than about 1,400 meters, which left it ineffective against U-boats operating on the surface.[4] Captain Dönitz also recognized that the concentration of merchant shipping in convoys could be countered by a similar concentration of U-boats. During the initial stages of World War II, U-boats would as a rule be on their own when conducting reconnaissance west of the British Isles, although supported by effective radio intelligence. But before long, U-boats concentrated in large numbers to attack convoys.

Strategic Setting and Operational Concept

At the outbreak of war in September 1939, the German navy confronted an enemy that was ten times stronger and that enjoyed the additional benefit of an excellent strategic position.[5] Therefore, the German Naval Command, forgoing a struggle for naval supremacy, concentrated on an offensive concept of naval warfare aimed solely at destroying the maritime transport capacity of the English-speaking powers. In Directive No. 9 of 29 November 1939, entitled "Principles for the Conduct of the War against the Enemy's Economy," Hitler considered paralyzing "Britain's economy through interrupting it" the "most effective means" to defeat that nation.[6] However, Hitler expected a short war, limited to Europe, and did not want to jeopardize the hope of better relations with Britain with a radical war on its economy. From the start, in contrast, the Naval Staff was convinced that the conflict with Britain would be a long one.

In the summer of 1940, after the defeat of France, "Germany was the dominant power in Europe and had a military position and sufficient freedom of action to make Britain's defeat inevitable, if not quick and easy. Considering the enormously greater resources of a German-controlled Europe, Britain's position, without outside help, was hopeless."[7] After the failure of the Luftwaffe in the Battle of Britain, however, invasion of the British Isles no longer presented the possibility of a quick victory. The only strategic choice was to strangle the Atlantic supply line by naval and air attacks before the United States could mobilize its strength.

The surface force of the German navy was insufficient for a successful war against Britain. To supplement it, the navy concentrated on a weapon that had proved its worth during the First World War—the U-boat. From the experience of the Great War, the German navy knew that employment of the U-boat against the enemy's merchant fleet could be successful only if as many U-boats as possible were continuously deployed along the enemy's sea-lanes in the Atlantic. Time was the most important factor in each of the four major strategic considerations:[8]

1. In an economic war waged against a country that depended on supplies by sea, success could only be achieved in the long run. It was therefore a question of continuously weakening the enemy's maritime transport capacity to a degree that exceeded the rate at which new merchantmen could be built.
2. From the summer of 1940 onward, it became apparent that the British war effort was being increasingly supported by the resources of the United States. Therefore, the Naval Staff was intent on "putting Britain out of action soon, before the effects of even greater American aid make themselves felt."[9]
3. Since it took around two years to build U-boats and to make them operational in the quantities envisaged by the navy, plans had to be made at a very early stage in order to have the necessary concentration of forces.
4. While a numerically increasing U-boat fleet held out the prospect of success, the Naval Staff had to take into account that the enemy would do everything to strengthen his antisubmarine-warfare effort, in view of the threat that was looming.

From their analyses of the U-boat campaign in the World War I, Dönitz and his staff saw the introduction of convoys in 1917 as the decisive turning point and as the main cause for the U-boats' eventual failure. Dönitz's intention, therefore, was to succeed "technically and tactically in meeting the concentration of ships in convoys with a concentration of U-boats."[10] The main problems were, first, to find the convoys and then, second, to concentrate the available boats for attacks. This required efficient reconnaissance support by the Luftwaffe, a sufficient number of U-boats, and freedom of communication. Sea-air cooperation would fail in the long run; the number of operational U-boats would rise only slowly, from twenty-seven in June 1940 to fifty-three in July 1941; and free communications would always make an individual boat vulnerable.

Dönitz planned deployments to maximize the number of available U-boats and their real success rate. The result was "the effective U-boat quotient," by which U-boat Command meant the average sinking per U-boat per day for all boats at sea.[11] Operating out of French bases from July 1940, U-boat Command increased the number of boats at sea west of the British Isles. The monthly sinking rate climbed steadily from July to October, in which month the effective U-boat quotient reached 920 tons. However, owing to the limited number of boats and their need for replenishment, U-boat Command could not maintain this attrition rate on British transport capacity. Furthermore, because of the Admiralty's successful rerouting policy, the long search for convoys now began and ended mostly in failure.[12]

Although the average number of U-boats in the Atlantic in 1941 increased from twelve in February to thirty-six in August, the number of ships they sank declined drastically. Noticing this serious discrepancy, staff officers at U-boat Command began to suspect that the U-boat positions must be known to the British, especially since their own radio intelligence sometimes found that convoys at risk suddenly got orders to change course, thereby evading the U-boat lines.[13]

Late in June 1941, fifteen U-boats were on patrol, spread over a large area in the center of the North Atlantic. Since no ship sightings were being reported, Dönitz decided in mid-July to concentrate these boats in a scouting line farther to the east. On 17 July, a convoy was detected by air reconnaissance, but the boats could not attack—the convoy was rerouted away from the outpost patrol when the British deciphered German radio signals.

In a report to the Naval Staff on 22 August, Dönitz stated that in view of the enemy's reinforced defenses and air surveillance, it was now "necessary to employ approximately three times as many U-boats as before in order to achieve any decisive successes in attacking convoys." Consequently, he demanded a greater concentration of forces.[14] The Naval Staff perhaps agreed with Dönitz on this fundamental issue, but given the critical situation in the Mediterranean, they already expected that additional U-boats would eventually have to be employed there.[15]

At the beginning of September 1941, the number of operational U-boats had increased to over seventy, of which thirty-eight were at sea, operating in two groups: one southwest of Ireland and the other between Iceland and Greenland. At that time, deciphered ULTRA intelligence from U-boats' radio traffic was available to the British after approximately forty to fifty hours. From 11 September on, however, British deciphering was impeded when the U-boat Command Staff, ever suspicious and concerned about security, ordered an additional superencryption for position information. For several weeks thereafter this presented British radio intelligence with a considerable new problem in locating and identifying U-boat positions.[16]

Under these circumstances, the Admiralty was unable to reroute every convoy. On 9 September, despite following an evasive course, convoy SC.42 was detected east of Greenland and within three days had lost sixteen of its sixty-two ships in the war's largest convoy battle to that time. The escorts sank only two U-boats. This success did not blind Dönitz to the fact that the vast majority of the convoys were sighted more or less accidentally. He believed that the reason his forces were failing to detect convoys must be that the enemy had learned about the close formation of U-boat groups "by means of sources or methods which we have not yet grasped."[17]

On 28 September 1941, Dönitz became highly suspicious when two U-boats that had met at a remote bay in the Cape Verde Islands in a rendezvous arranged via radio message only narrowly escaped a torpedo attack by a British submarine. He asked the Naval Staff for an immediate investigation into cryptographic security. The Naval Intelligence Division reached the conclusion that "without any contradiction from any of the experts who have been involved in extensive work, . . . and the most important experts from the OKW [Supreme High Command of the Armed Forces], the procedures based on Key-M are regarded by far as the best of all known methods for ensuring the secrecy of military intelligence in time of war."[18] Remarkably, it seems that no attempt was made to have the security of the ciphers investigated by independent scientists outside the military. The result was a dangerous underestimation of the human and technical resources that the enemy had concentrated on code breaking.[19]

The average number of thirty-six U-boats at sea did not vary from August to October, but the number of ships they succeeded in sinking decreased noticeably in October. By late autumn 1941 U-boat Command was clearly losing the race against time. One reason for this failure lay "in the absence of independent thinking" within the small staff of U-boat Command: "Although Dönitz is credited with encouraging the open expression of opinion within his command, the U-boat professionals who surrounded him were built in his image and shared his convictions."[20] The evasive action taken by the convoys on the basis of ULTRA information led to a reduction in losses of around 65 percent in the second half of 1941. Without ULTRA, the Admiralty would have faced a much greater number of U-boat attacks on convoys, leading to probable losses of another 1.5 million gross register tons (GRT).[21]

Long-Range Operations in the Western and Southern Atlantic 1942

Members of the Naval Staff had anticipated the eventual entry of the United States into the war, but they had not expected it as early as December 1941. They were caught by surprise, therefore, and were largely unprepared for a fast push by U-boats as far as the coastal waters of North America.[22] The fight against Allied merchant tonnage in the Mediterranean and off Gibraltar during the autumn had resulted in heavy U-boat losses—from September to December 1941 a total of twenty-three U-boats were sunk—with only small successes. Now, the rapid extension of operations to the whole Atlantic gave Dönitz the opportunity to concentrate his forces in areas where a significant rise in sinkings could be expected.

This applied initially to the sea-lanes along the Eastern Seaboard of America, which—although three thousand nautical miles away—promised to be a rewarding area of operations as long as the shipping there was uncontrolled and largely unprotected. Dönitz intended to take advantage of these favorable conditions as quickly as possible.

Consequently, on 9 December 1941 he applied to the Naval Staff for the immediate deployment of twelve large Type IX B and C boats. With their great sea endurance (thirteen thousand nautical miles at ten knots) and their large stocks of torpedoes, they seemed particularly suited for this task. Dönitz meant to "roll the drums," as he said in his war diary. The Naval Staff, however, still mindful of the critical situation in the Mediterranean, released only six U-boats for missions off the American east coast, a decision that Dönitz regretted.[23]

Between 7 and 9 January 1942, seven medium-sized Type VII U-boats arrived at the Newfoundland Grand Banks, and a few days later the first five Type IX boats reached the U.S. east coast. Within a fortnight both groups had sunk fifteen ships, totaling 97,242 GRT. More boats were to follow, so that between six and eight boats were operating at the U.S. East Coast and Newfoundland at any given time. By February 1942, another group of five large boats was available, which meant that the long-range operations could be extended. The successes of German and Italian submarines in the Atlantic alone rose from 295,776 GRT in January to 500,788 GRT in March 1942.[24] These figures underlined again the principle of "economic use of U-boats," particularly since the losses in these areas of operations had remained extremely low.

Even after two and a half months U-boats continued to score impressive successes off the U.S. east coast. Bearing in mind that the U.S. Navy had been involved in escorting convoys since September 1941, it seems to modern eyes especially remarkable that this weakness in American coastal waters should have prevailed for so long. Several months passed before the U.S. Navy could protect its own sea-lanes. Dönitz was gratified to state in March 1942 that the enemy defenses were thin, badly organized, and untrained. Dönitz expected convoys to be formed eventually, but for the time being the sea-lanes were too numerous and there was chronic shortage of escort vessels.[25]

Dönitz's hopes revolved around the new U-tankers that were expected to become operational during spring 1942, greatly extending U-boat endurance in distant waters.[26] The first U-tanker, *U-459*, reached its area of operations about five hundred nautical miles to the northeast of Bermuda on 23 April 1942. By 5 May it had replenished no fewer than fifteen boats, most of them of the medium-sized Type VII C. In the spring of 1942 boats of this type operating off Halifax, Nova Scotia, could remain at sea for an average forty-one days without replenishment but achieved an average sea endurance of sixty-two days after one refueling, and up to eighty-one days if resupplied twice.[27]

From mid-June 1942, two to three U-tankers constantly deployed in the Atlantic. They were stationed beyond Allied air surveillance. This opened up entirely new operational perspectives for U-boat Command, both in the hitherto untouched busy waters of the Caribbean and the Gulf of Mexico, and in the North Atlantic, where a resumption of convoy battles was envisaged. Dönitz decided to launch U-boat attacks into even more distant areas, and by mid-July 1942 from four to six boats, supported by a U-tanker, were operating in the central Atlantic, off Freetown, Sierra Leone, to pressure Allied supplies to the Middle East and their links with India. The southward move was to be concealed for as long as possible. Beginning on 19 August, four large Type IX U-boats and their tanker *U-459* left their bases in western France.[28] The plan provided for the boats just mentioned to arrive off South Africa without prior warning. From 7 October, the boats, operating independently, succeeded in sinking fifteen ships, of 108,070 GRT, within six days. By the end of the month the score had risen to 156,235 GRT, or some 28 percent of all German U-boat successes in October.[29] These U-boats were followed by a second group, of three Type IX D_2 boats (1,616 tons). They advanced into the Indian Ocean along the African coast as far as Laurenço Marques, and by the middle of December 1942 they had sunk twenty-five ships, 134,780 GRT.

From the point of view of overall strategy, these wide-ranging U-boat operations actually came at least six months too late. The successes achieved off Cape Town and in the Indian Ocean from October 1942 hardly impaired the buildup of Allied military potential in Egypt, which was already well advanced and was to be the vital precondition for the successful British defense of El Alamein and the offensive in North Africa that followed.

Differing Views of Resources and Priorities

With a total of 249 boats in service on 1 January 1942, the German navy had achieved—in purely numerical terms—impressive strength. However, this figure can easily give a false impression, for only ninety-one were actually frontline boats. On that date one hundred new boats were in training or undergoing trials, and a further fifty-eight were training boats only. Of the frontline boats, twenty-six were already tied to the Mediterranean, six to Gibraltar, and four to the Norwegian region. For the decisive struggle against the Allies in the North Atlantic there remained fifty-five boats, of which only twenty-two were at sea—approximately half on station and half in transit. If just from ten to twelve boats were actually engaged in operations against Allied shipping—that is, 12 percent of the boats available, or, speaking purely quantitatively, 4.8 percent of the total potential of the German submarine weapon—one can hardly speak of a strategic concentration, let alone a "war-winning" fresh start.[30] Nevertheless, the U-boats had shown considerable capability, and their losses before the end of December 1941 had remained at a very low level, an average of 2.5 boats per month.

Despite all the efforts of the preceding years, at the end of 1941 the numbers of U-boats had not been brought to a level that could have better exploited this favorable position in the Atlantic through increased concentration of force. The subsidiary status of the

Atlantic in the framework of German strategy became very clear when Germany declared war against the United States on 11 December 1941. Of the ninety-one frontline boats, only six were immediately assigned to the American east coast. Given the number of boats tied up in the Mediterranean and the Norwegian Sea, Dönitz at the beginning of January 1942 was seriously worried "that we will finally arrive too late for the Battle of the Atlantic." He therefore called for a radical concentration of U-boats to cope with this task: "We are in a tight spot, but one which clearly shows that everything, absolutely everything should be invested into the U-boat arm, and that the fiction that we are still a naval power with surface forces collapses as soon as any kind of demand of the naval war is made of us anywhere."[31]

Beginning in July 1942, following the introduction of the convoy system in American coastal waters, U-boat Command shifted the main field of U-boat activity back to the North Atlantic. Dönitz aimed to bring his U-boats into action for several days outside the range of enemy air cover. This necessitated the earliest possible detection of convoys. On any given day the boats could cover 320 to 370 nautical miles, whereas the convoys could manage a maximum of only 240. Dönitz used this speed advantage to subject a given convoy, once detected, to repeated attacks by the same U-boat group, a mobile operation that could last several days.

U-boats sailing from Germany or from western France initially attacked convoys heading from Great Britain to the Newfoundland Grand Banks. Afterward, they were resupplied by U-tankers south of the North Atlantic routes, before attacking eastbound convoys in a second operation. During some operations, the U-tankers followed the convoy at a distance of only fifty to a hundred nautical miles, so as to refuel U-boats immediately after their attacks. Until the end of 1942, Allied airpower in the North Atlantic could provide close cover only for the convoys known to be at risk. At selected rendezvous points, the Germans were therefore conducting their supply operations undisturbed and without any problems. However, these prolonged operations, alternating between combat and replenishment, stretched the German crews to their limits.

After 1 February 1942, British radio intelligence was badly affected by a "blackout" in the deciphering of U-boat Command's Atlantic communications. The Admiralty's Operational Intelligence Centre was reduced to relying on conventional sources, but since March 1941 the Admiralty had known that the German navy was developing supply U-boats. Although there had been factors hinting that German U-boats were being replenished at sea as early as in June 1942, it was only in August that prisoner-of-war interrogation definitely established the existence of U-tankers.[32] Until the end of 1942, however, the British apparently underestimated their operational significance. Given the achievements of British radio reconnaissance, it is scarcely credible that they failed to pick up the radio traffic in the vicinity of the supply sites, particularly as directional signals were frequently sent in the autumn of 1942 to guide U-boats to the U-tankers.

Tonnage Race: Facts, Figures, Fallacies

Within the German navy there were two differing views of the U-boat war. On the one hand, the Naval Staff wanted to concentrate on attacking sea lines of communication and individual transports vital to the war effort. On the other hand, Dönitz right from the start upheld a conception of tonnage warfare. To him, the U-boat war had always been a struggle against the Allies' merchant tonnage, not ships performing particular missions.

On 15 April 1942, Dönitz briefly summarized this conception in his war diary:[33]

1. The enemy merchant navies are a collective factor. It is therefore immaterial where any one ship is sunk, for it must ultimately be replaced by new construction.
2. What counts in the long run is the preponderance of sinkings over new construction. Shipbuilding and arms production are centered in the United States, while England is the European outpost and sally port.

In Dönitz's estimate of the enemy's potential, every ship sunk counted double—not only in terms of mere tonnage but also through elimination of its cargo, which was lost to the enemy's arms industry. He believed that the war at sea would be decided by the "race" between the number of ships sunk and the rate of new construction: if for a lengthy period of time Germany were to sink more ships than its enemies could replace, the British economy and defense effectiveness were bound to slacken and eventually collapse.

In Dönitz's opinion, the tonnage war also constituted an effective contribution to the prevention of a second front and at the same time to the direct defense of France and Norway. Since the Allies were constantly increasing their shipbuilding capacity, the time factor played a crucial role. It was imperative that the maximum tonnage be sunk as fast as possible, before the enemy could make good the number of ships sunk. In a briefing to Hitler on 14 May 1942, Dönitz emphatically advocated the sinking of enemy tonnage "where as much as possible can be sunk as cheaply as possible, i.e., incurring only minor losses." He also stressed to Hitler the importance of sinking ships as soon as possible: "What is sunk today is more effective than what is not sunk until around 1943."[34] Optimistically, he pointed out that the race between the Allied building of new merchantmen and the sinking of them by U-boats was "in no way hopeless." In this context, he presented the following quantitative analysis of Allied shipbuilding capacities. In 1942, the United States would build new ships totaling 6.5

million GRT, and the British Empire would build new ships totaling approximately 1.6 million GRT, giving a combined total of 8.1 million; in 1943 the Allies would reach approximately 10.3 million GRT. From this calculation, Dönitz drew the conclusion that it was necessary to sink 700,000 GRT per month in 1942 to offset the newly built ships. Anything sunk in addition to 700,000 GRT would decrease the Allied maritime transport capacity in absolute terms.

Dönitz was of the opinion that taking into account the successes achieved by other means of naval warfare, including mines, surface forces, and the Luftwaffe, as well as by the Japanese and Italian naval campaigns, this result had already been achieved. Going by the German reports on sinkings, his assessment was surely correct from a subjective point of view. For March 1942 alone the Naval Staff had calculated a total loss of 362 enemy ships of 1,095,493 GRT. However, Allied losses were in fact considerably lower; in the first quarter of 1942 they really only averaged 644,568 GRT per month. Thus, on the basis of Dönitz's own capacity calculations for the year 1943, the Axis powers would eventually have had to increase their monthly rate of ship sinking to over 860,000 GRT on average to thwart the Allied strategy directed against Europe.

On 27 August 1942, Section 3 of the Naval Staff—responsible for the evaluation of intelligence—concluded that if the number of sinkings remained the same and Allied plans for new ships were maintained, the difference between new ships and sinkings would decline considerably and "in purely numerical terms the tonnage of the enemy powers would have reached its lowest point at the turn of 1942–43."[35] Section 1 of the Naval Staff (i.e., the Operations Section) had to reconsider the strategic objectives of the economic war. Should priority be given to hitting the supply traffic heading for Great Britain, or should the enemy's overall tonnage potential be the prime target?

In a lengthy study published on 9 September 1942, Section 3 reduced this question to its crucial point: "In an economic war, can a decisive influence be gained by the sinking of ships alone, regardless of where and whether laden or unladen, or must specific tonnage in specific areas be sunk in order to achieve this aim?" The study also estimated the enemy's monthly output of new ships at 750,000 GRT currently, rising to over 800,000 GRT by the end of 1942, and expected to reach a monthly average level of some 900,000 GRT in 1943. During the first eight months of 1942 the enemy had suffered a net loss (i.e., the difference between the number of newly built ships and ships sunk) of 3,380,000 GRT, which had forced the delay, at least for the time being, of a second front in Europe. The U-boats had accounted for 78 percent of this achievement, making the performance of this arm the absolute "yardstick for the success of the merchant war."[36]

The study by Section 3 pointed out that whereas in 1942 the U-boats had achieved their greatest results by launching surprise attacks into areas where the enemy's defenses were weak, it would hardly be possible to repeat such successes. The number of U-boats would have to be significantly increased, and the U-boats' fighting power against defenses would have to be enhanced "in order to keep previous monthly sinkings achieved at the same level." For the vital strategic question of the race between launchings and sinkings, Section 3 produced a pessimistic prognosis: "However, in view of increased output of new ships from the end of 1942, a permanent increase in the monthly sinkings to about 1,300,000 GRT will be necessary in order to achieve a reduction in tonnage equivalent to the current one. Given the situation, it is questionable whether such a high rate of sinkings can be achieved on a permanent basis."[37]

The obvious discrepancy between what was desired and what was feasible was soon manifest in the Atlantic, where the U-boats were considered to be "the main pillar in the supply war." By falling back on the argument that "in history no war has been won by employing just one means of warfare," the Naval Staff was unconsciously moving to a view that was close to actual conditions.[38] In light of enormous American capacities and of U.S. production methods in shipbuilding, Germany could never win a race between the building of new ships and sinkings. The Naval Staff was forced to recognize "that the enemy is about to match the figures for sinkings with its figures for building, or at least to reduce the previous losses of tonnage."[39]

A comparison of Allied shipbuilding capacity and German estimates for 1942 and 1943 (see the figure) reveals that both Dönitz and the Naval Staff were accurate in their predictions and regarded a further increase in 1943 as possible. Actual new Allied tonnage in 1943, however, exceeded the German estimates by nearly four million gross registered tons.[40]

Within the German Naval Command hopes were expressed that in the long term the enemy would reach material limits in his output of new ships and face difficulties in providing crews for the new vessels. These hopes, however, proved unrealistic. In the pure tonnage war, the race became hopeless in the last quarter of 1942. Undoubtedly the Allies continued to have a strategic Achilles' heel in the problem of sea transport, as the Naval Staff explained once again: "They have or manufacture enough, but they cannot transport enough for waging war, the economy, and their food supply. Against this weakness the import war must continue to be waged with all means."[41] Whether Germany would succeed in warding off an Allied offensive by conducting a "supply war" remained for the moment an open question.

The answer was not long in coming. On 8 November 1942, only three weeks after the Naval Staff was still arguing that the shortage of shipping capacity was frustrating the enemy's strategic plans and depriving him of his operational freedom of action, the Allies took the initiative in North Africa and in the Mediterranean. Now, Grand Adm.

The Tonnage Race: The Battle for Sea Transport Capacity 1939–1943

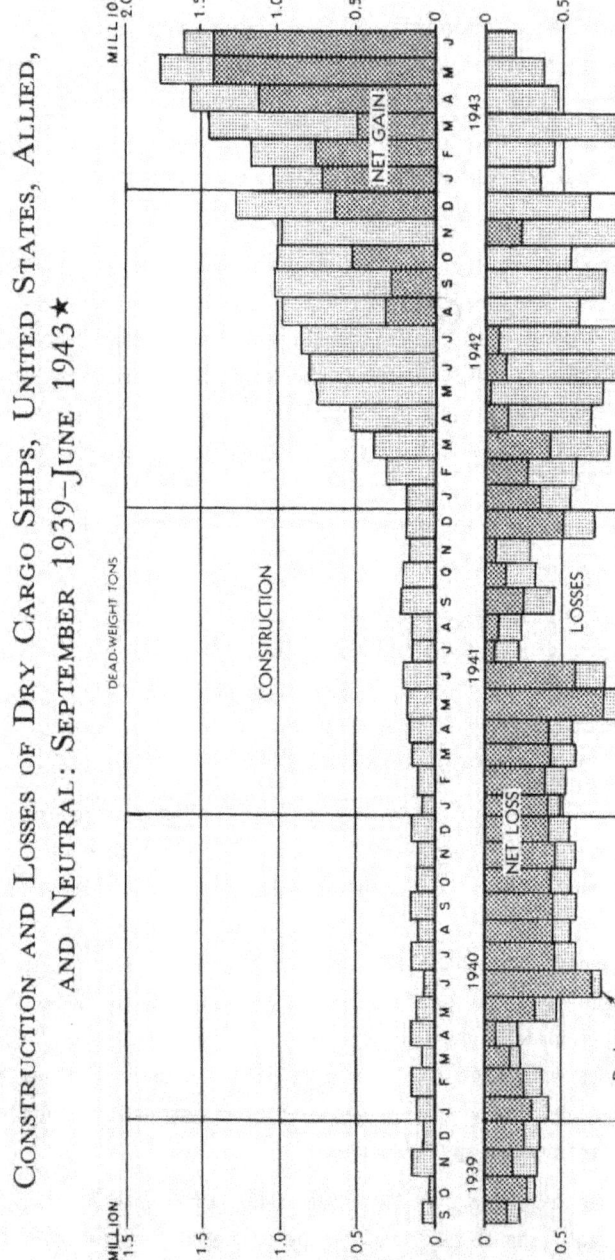

CONSTRUCTION AND LOSSES OF DRY CARGO SHIPS, UNITED STATES, ALLIED, AND NEUTRAL: SEPTEMBER 1939–JUNE 1943 ★

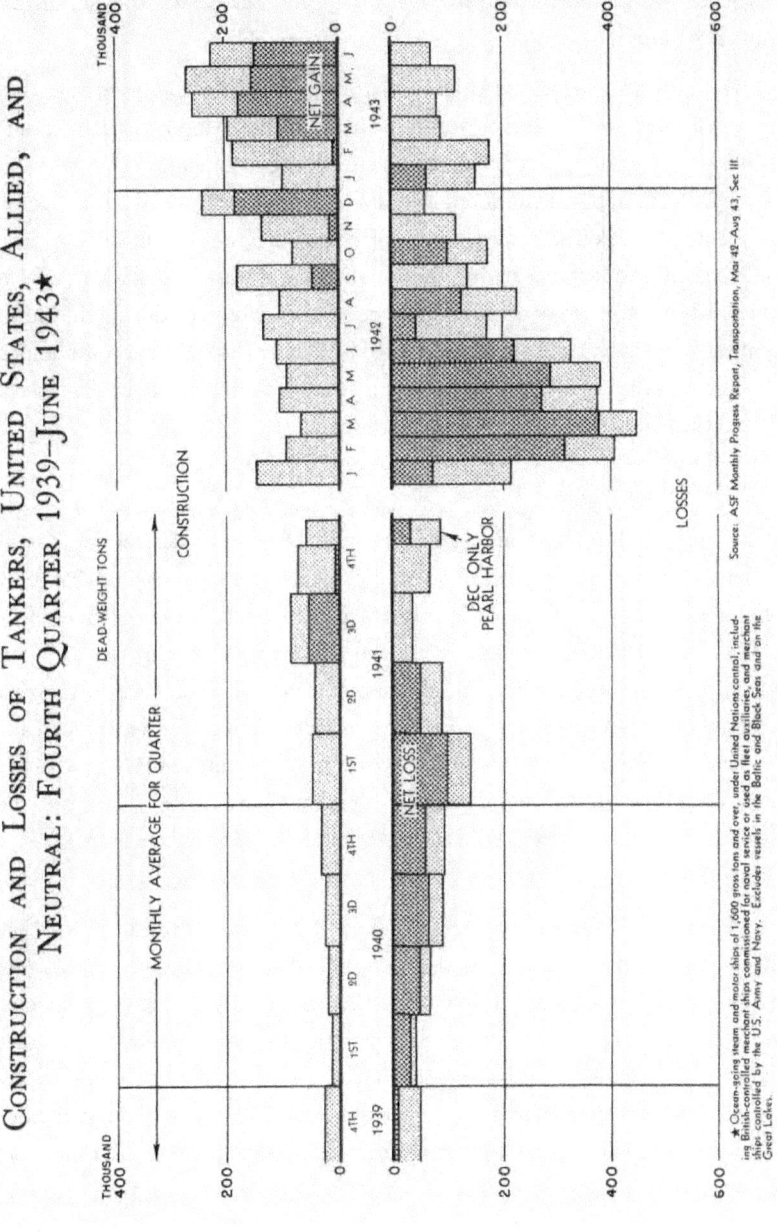

Erich Raeder was forced to admit to Hitler that the enemy was in a position to use his available sea transport capacity for strategic purposes and could launch another major operation in December.[42] This assessment was based on new calculations concluding that in September the enemy had already offset his monthly losses from sinkings by building new ships.[43]

The Naval Staff was well aware that in the battle against enemy shipping capacity, the manpower factor played an important role. Therefore, the manpower bottleneck facing the Allies was to be intensified by increased effectiveness of weapons—for example, with a new torpedo ignition, which should have a tremendous destructive power. However, the Naval Staff shrank from endorsing the principles of a war of annihilation, since it would be "the first instance within the annals of naval warfare that an order to use force against shipwrecked personnel would be given."[44] However, during the Nuremberg trials after the war the "*Laconia* order," issued by Dönitz to all U-boat commanders on 16 September 1942, was interpreted by the prosecution as a demand for the destruction of shipwrecked persons, in violation of international law.[45]

The Defeat of the U-boats in 1943

Between 1 and 20 March 1943 alone, U-boats in the Atlantic sank no fewer than seventy ships, sixty of them in the context of operations against convoys. During that month the shipping capacity lost to attack by Axis U-boats throughout the world reached 643,337 GRT, 512,303 GRT in the Atlantic. But Allied antisubmarine warfare operations grew more effective month by month. The size of naval and naval air forces involved rose continuously, and there was a steady improvement in weapons technology and in tactical and operational command and control. On 13 December 1942 the British had achieved their great breakthrough in deciphering the U-boat code.[46] Thereafter the entire radio traffic of U-boat Command's Atlantic operations was once again read, with only brief delay.

At the beginning of January 1943 Dönitz had at his disposal 212 frontline boats, of which 164 were earmarked for the North Atlantic. Of these Atlantic U-boats, more than ninety were at sea on any given day. Further increases were foreseen. In view of these figures, the operational advantages provided by ULTRA gradually declined, for the simple reason that the convoys now had very little chance of evading U-boat formations.[47] But the Admiralty's convoy routing proved so successful notwithstanding that Dönitz presumed the enemy countermeasures were based on actual knowledge of U-boat positions. Doubts about the security of the German cipher system reemerged, because some of the British reports of U-boat positions that were regularly deciphered by B-Dienst (the naval code-breaking office) coincided closely with actual conditions and could not be explained by the results of enemy reconnaissance or radio direction finding alone.⁴⁸ However, the German experts went down a false trail, wrongly blaming Germany's technological inferiority in the area of radar.

At the beginning of March 1943 there were fifty boats in three patrol lines in the North Atlantic. Four Allied convoys ran into these and experienced serious losses. But the volume of Allied naval transport in the Atlantic increased significantly, because more and more troops and war materiel had to be brought across to Britain. Between 6 and 11 March some forty U-boats engaged in action against two eastbound convoys, which suffered the loss of sixteen ships, 79,836 GRT overall. Only one U-boat was lost. Several days later, forty boats were operating against the eastbound convoys SC.122 (sixty ships) and HX.229 (forty ships). The battle lasted four days and was the most bitterly fought of the entire war. The Allies lost twenty-one ships, 140,842 GRT.⁴⁹

Dönitz described this fighting as "the greatest success achieved in a convoy battle to date;"⁵⁰ the tactical concept of a purposeful and massive employment of U-boats in different groups appeared at last to be producing vital results. One important reason for these successes—in addition to the large number of boats in the area of operations—was that B-Dienst had once again broken Allied radio codes for convoy routing (Naval Cypher No. 3). With the support of the B-Dienst, between February and March 1943, 54 percent of all North Atlantic convoys were sighted and reported by U-boats, and 24 percent were actually attacked by U-boats groups. This was the highest level of location of and attack against convoys in the entire war. Yet a glance at the actual situation in the North Atlantic revealed that in these critical days, aside from the convoys HX.229 and SC.122, another fourteen convoys were also under way, either eastward or westward. All except one of them escaped unscathed.

During this critical phase, there was some doubt within the Admiralty as to whether the convoy system could be maintained at all in the future, particularly as the large numbers of U-boats operating in the Atlantic made it very difficult to take evasive action. Yet there was no alternative. Some months later, in December 1943, by then secure in the knowledge of the coming Allied victory, the Admiralty made an analysis of the bitter and fluctuating convoy battles of March, concluding that "the Germans never came so near to disrupting communications between the New World and the Old as in the first twenty days of March 1943."⁵¹

After the war both British and American naval histories quoted this sentence from the "Monthly A/S [antisubmarine] Report" of December 1943 without qualifying it, thereby encouraging the erroneous belief that in March 1943 the German U-boat campaign in the Atlantic was within reach of imminent, decisive victory and Britain on the verge of defeat. There is no doubt that the high losses of that month brought supply problems and caused damage to the British war economy, but on the whole Britain's supplies and the preparations for the strategic offensive in Europe were no longer directly threatened—the massive American shipbuilding program was already taking effect. From the autumn of 1942 the monthly rate of new building exceeded losses by so clear a margin that the U-boats could only delay, not prevent altogether, the Allied seizure of the strategic initiative.

In April 1943, faced with the heavy fighting and high submarine losses of the spring of 1943 (nineteen boats in February and seventeen in March), Dönitz—since 30 January 1943 the new commander in chief of the navy—told Hitler that he had serious doubts whether the U-boat war in the Atlantic would be successful in the long run. His hopes and demands concentrated on increasing the monthly output of the U-boat building program to thirty boats, "in order to prevent the ratio of losses to newly built boats from becoming too unfavorable."[52] However, by 24 May 1943 Dönitz had no choice but to call a halt to attacks on convoys in the North Atlantic and withdraw the boats to the south. In May 1943, because of heavy Allied air cover, forty-one boats had been lost, a figure out of all proportion to their successes.

At the same time the mobilization of the American shipbuilding industry reached its first peak. From January to May 1943 Allied merchant-ship building exceeded sinkings by more than three million gross registered tons. The increasing buildup of Allied maritime transport capacity sustained the British people and their war production, securing the decisive strategic base for the Allied offensive in Europe. Paying a U-boat for the sinking of each Allied merchant ship as of the spring of 1943, Dönitz had clearly lost the "tonnage race," and his staff knew the facts and figures. Neither more nor better U-boats could make a difference.[53]

Conclusions

In the spring of 1943, Dönitz and many officers in the Naval Command still believed that only the deployment of more boats of the improved Type VII C would force a strategic decision in the Atlantic in the face of Allied antisubmarine warfare, which Dönitz himself noticed and assessed realistically. First, enemy reconnaissance aircraft and escort vessels had been fitted with precision radar equipment, resulting in frequent surprise attacks on surfaced U-boats at night and during bad visibility. These attacks often damaged or sank the boats—increasingly so in the Bay of Biscay, where the British had finally gained such air superiority that the passage of the boats through this area quickly became the most dangerous phase of an operation.[54]

Second, and no less important, the Allied development of long-range air reconnaissance in the Atlantic became more and more intensive. On 3 September 1942, with deepest concern, Dönitz had predicted "that the day was coming when in almost all areas of the

North Atlantic—the U-boats' principal battle ground—the situation in the air would be just as bad around the convoys. This would reduce the chances of success of the U-boats to an unacceptable degree unless adequate countermeasures were taken."[55]

A third development was the Allies' own increasing experience. Once a submerged U-boat had been discovered by an experienced antisubmarine warfare group with ASDIC, its slow submerged speed did not allow drastic evasive maneuvers. After being fixed and suffering the effects of ever-improving weapons, there was little chance of escape.

There were clear alarms that the established concept of the U-boat no longer promised success. The U-boat in service until then had, in fact, been only a submersible—a mobile torpedo boat with long seagoing endurance and the ability to vanish from the surface for relatively short periods of time. Underwater the boat was very slow (though at the lowest creeping speed it could remain submerged for nearly fifty hours). The exploitation of ULTRA information, the growing surveillance of the Atlantic, and the advent of powerful submarine-hunting groups and, especially, offensive airborne antisubmarine operations soon showed that the previous concept of U-boat operations was doomed to failure.

In the face of these changes, from the summer of 1943 the German U-boat authorities underestimated the situation and planned badly. On the one hand, they began building modern and powerful submarines. On the other hand, they continued building and employing the old submersibles well into 1944, although their performance capabilities no longer met the requirements of the changed warfare conditions.

The few missions of the Types XXI and XXIII boats put the superior capabilities of the new U-boat designs to the test.[56] No boat of these types was detected or destroyed by antisubmarine warfare forces. The only losses occurred during Allied air attacks on naval yards and bases or during surface cruising in mine-infested German coastal waters. The Allied antisubmarine warfare forces and the "technology developed during the war were designed to exploit the old U-boat's reliance on the surface. How the Allies would have dealt with a deep diving, very fast U-boat which did not need to surface to fire its torpedoes remains a mystery—one of the great what ifs of the War."[57]

After 1945, the U.S. Navy faced the possibility of a future Soviet submarine threat based on high-speed Type XXI U-boats. However, in the meantime the navy used two U-boats of this type in a series of tests to determine the effectiveness of its own antisubmarine warfare techniques. On 4 June 1946 the Chief of Naval Operations, Fleet Adm. Chester W. Nimitz, wrote to President Harry Truman that these tests "demonstrated that this particular submarine can with relative immunity attack a convoy or task group screened by the usual means and in deep water is virtually immune from destruction by any ship or aircraft or combination of both yet developed."[58]

Notes

This chapter is mainly based on Dr. Rahn's contribution "The War at Sea in the Atlantic and in the Arctic Ocean" to Horst Boog, Werner Rahn, Reinhard Stumpf, and Bernd Wegner, eds., *Germany and the Second World War,* vol. 6, *The Global War: Widening of the Conflict into a World War and the Shift of the Initiative 1941–1943* (Oxford, U.K.: Clarendon, 2001), pp. 301–466, and on his contribution "Die deutsche Seekriegführung 1943 bis 1945 [German Naval Warfare 1943–1945]," to *Das Deutsche Reich und der Zweite Weltkrieg,* vol. 10, *Der Zusammenbruch des Deutschen Reiches 1945,* part I, *Die militärische Niederwerfung der Wehrmacht* (Munich: Deutsche Verlags-Anstalt, 2008), pp. 3–271; cf. also his articles "The Atlantic in the Strategic Perspective of Hitler and Roosevelt, 1940–1941," in *To Die Gallantly: The Battle of the Atlantic,* ed. Timothy J. Runyan and Jan M. Copes (San Francisco: Westview, 1994), pp. 3–21, and "The Campaign: The German Perspective," in *The Battle of the Atlantic 1939–1945: The 50th Anniversary International Conference,* ed. Stephen Howarth and Derek Law (London: Greenhill Books, 1994), pp. 538–53.

The thoughts and opinions expressed in this publication are those of the author and are not necessarily those of the U.S. government, the U.S. Navy Department, or the Naval War College.

1. See Werner Rahn, "German Naval Strategy and Armament, 1919–39," in *Technology and Naval Combat in the Twentieth Century and Beyond,* ed. Phillips Payson O'Brien (London: Frank Cass, 2001), pp. 109–27.

2. Rear Adm. Eberhard Godt, "The War at Sea," unpublished essay, November 1945, quoted by Rahn, "German Naval Strategy and Armament," p. 119; Clay Blair, Jr., *Hitler's U-boat War: The Hunters 1939–1942* (New York: Random House, 1996), pp. 30–39.

3. Karl Dönitz (1891–1980) joined the navy in 1910, was chief of U-boat Command from 1935 until January 1943, and was promoted to rear admiral in October 1939 and to grand admiral on 30 January 1943. From 1943 to 1945 he was commander in chief of the navy and in October 1946 was sentenced by the International Military Tribunal at Nuremberg to imprisonment for ten years. In 1958 he published his memoirs, *Zehn Jahre und zwanzig Tage. Erinnerungen 1935 1945,* translated as *Memoirs: Ten Years and Twenty Days,* with an introduction and afterword by Jürgen Rohwer (Annapolis, Md.: Naval Institute Press, 1990). See Blair, *Hitler's U-boat War,* pp. 35–49; Peter Padfield, *Dönitz: The Last Führer—Portrait of a Nazi War Leader* (New York: Harper and Row, 1984); and Dieter Hartwig, *Großadmiral Karl Dönitz: Legende und Wirklichkeit* (Munich: Paderborn, 2010).

4. S. W. Roskill, *The War at Sea 1939–1945,* vol. 1, *The Defensive* (London: HMSO, 1954), pp. 130, 355.

5. Gerhard Wagner, ed., *Lagevorträge des Oberbefehlshabers der Kriegsmarine vor Hitler 1939–1945* (Munich: J. F. Lehmann, 1972). For an unaltered reprint see *Fuehrer Conferences on Naval Affairs 1939–1945* (Annapolis, Md.: Naval Institute Press, 1990). For another translation see the U.S. Navy's seven-volume typewritten edition: *Fuehrer Conferences in Matters Dealing with the German Navy* (Washington, D.C.: Secretary of the Navy, 1947); Michael Salewski, *Die deutsche Seekriegsleitung 1935–1945* (Frankfurt, Ger.: Bernard & Graefe, 1970–75); Ministry of Defence (Navy), ed., *The U-boat War in the Atlantic, 1939–1945* (London: HMSO, 1989); Timothy P. Mulligan, *Neither Sharks nor Wolves: The Men of Nazi Germany's U-boat Arm, 1939–1945* (Annapolis, Md.: Naval Institute Press, 1999); Marc Milner, *Battle of the Atlantic* (St. Catharines, Ont.: Vanwell, 2003).

6. Walther Hubatsch, ed., *Hitlers Weisungen für die Kriegführung, 1939–1945* (Frankfurt, Ger.: Bernard & Graefe, 1962 [repr. 1983]), pp. 40–42; *Fuehrer Directives and Other Top-Level Directives of the German Armed Forces 1939–1941,* pp. 73–76, typescript translation [Office of Naval Intelligence], Washington, D.C., 1948, Henry E. Eccles Library, Naval War College, Newport, R.I.

7. Alan J. Levine, "Was World War II a Near-Run Thing?," *Journal of Strategic Studies* 8, no. 1 (March 1985), p. 41.

8. Based on Vice Adm. Kurt Assmann, "Why U-boat Warfare Failed," *Foreign Affairs* 28 (1950), pp. 659–70, esp. 665–66.

9. Werner Rahn and Gerhard Schreiber, eds., *Kriegstagebuch der Seekriegsleitung 1939–1945,* Teil [part] A (Bonn: Herford, 1988–97) [hereafter KTB/Skl], vol. 16, p. 238 (20 December 1940).

10. Godt, "War at Sea," p. 6.

11. John Terraine, *The U-boat Wars 1916–1945* (New York: Putnam, 1989), p. 270.

12. Graham Rhys-Jones, "The Riddle of the Convoys: Admiral Dönitz and the U-boat Campaign 1941," unpublished paper, April 1992, Newport, R.I.

13. Patrick Beesly, *Very Special Intelligence: The Story of the Admiralty's Operational Intelligence Centre, 1939–1945* (London: Greenhill Books, 2000); F. H. Hinsley, *British Intelligence in the Second World War: Its Influence on Strategy and Operations* (Cambridge, U.K.: Cambridge Univ. Press, 1979–90).

14. U-boat Command, memorandum 2235, 22 August 1941, quoted by Rahn, "The War at Sea in the Atlantic and in the Arctic Ocean," p. 356.

15. KTB/Skl, vol. 24, p. 306 (19 August 1941).

16. Hinsley, *British Intelligence in the Second World War*, vol. 2, p. 173.

17. U-boat Command, War Diary, 16 September 1941, RM 87/18, fol. 53–54, Bundesarchiv-Militärarchiv, Freiburg, Ger. [hereafter BA-MA]; Rhys-Jones, "Riddle of the Convoys," pp. 42–46. See Rahn, "German Naval Strategy and Armament," p. 357.

18. See Rahn, "The War at Sea in the Atlantic and in the Arctic Ocean," p. 349ff.

19. David Kahn, "Codebreaking in World War I and II: The Major Successes and Failures," in *The Missing Dimension: Governments and Intelligence Communities in the Twentieth Century*, ed. Christopher Andrew and David Dilks (London: Macmillan, 1984), pp. 138–58, see esp. 157ff.; David Kahn, *Seizing the Enigma: The Race to Break the German U-boat Codes, 1939–1943* (Boston: Houghton Mifflin, 1991), pp. 203–11; R. Erskine, "Codebreaking," in Beesly, *Very Special Intelligence*, pp. 265–70.

20. Rhys-Jones, "Riddle of the Convoys," p. 64; Mulligan, *Neither Sharks nor Wolves*, pp. 42–55.

21. See Rahn, "The War at Sea in the Atlantic and in the Arctic Ocean," pp. 367–68.

22. See Werner Rahn, "Long-Range German U-boat Operations in 1942 and Their Logistical Support by U-Tankers" (Eighth Naval History Symposium, Annapolis, Md., September 1987). Published in German as "Weiträumige deutsche U-Boot-Operationen 1942/43 und ihre logistische Unterstützung durch U-Tanker," in *Die operative Idee und ihre Grundlagen. Ausgewählte Operationen des Zweiten Weltkrieges*, ed. Militärgeschichtliches Forschungsamt (Bonn: Herford, 1989), pp. 79–97.

23. U-boat Command, War Diary, 9 and 10 December 1941, RM 87/19, fol. 95–100; Dönitz, *Memoirs*, pp. 197–98; Salewski, *Die deutsche Seekriegsleitung*, vol. 1, pp. 508–509.

24. Jürgen Rohwer, *Axis Submarine Successes 1939–1945* (Annapolis, Md.: Naval Institute Press, 1999).

25. U-boat Command, War Diary, 13 March 1942, "Lage amerikanischer Raum," RM 87/20, fol. 197–98; Samuel Eliot Morison, *History of U.S. Naval Operations in World War II*, vol. 1, *The Battle of the Atlantic, September 1939–May 1943* (Boston: Little, Brown, 1947 [repr. 1954]), passim.

26. The Type XIV U-tankers (1,688 tons displacement) carried, besides 203 tons for their own needs, 439 tons of diesel fuel, as well as about fifty tons of other supplies (provisions, water, engine oil, spares, etc.).

27. According to war journals of U-boats engaged off Halifax. Eberhard Rössler, *The U-boat: The Evolution and Technical History of German Submarines* (London: Arms and Armour; Annapolis, Md.: Naval Institute Press, 1981), pp. 166–67.

28. See U-boat Command, War Diary, 11 August 1942, with a long Skl comment, RM 87/22, fol. 134–36; Rohwer, *Axis Submarine Successes*, pp. 275–76; Dönitz, *Memoirs*, pp. 256–63.

29. Figures according to Rohwer, *Axis Submarine Successes*; S. W. Roskill, *The War at Sea 1939–1945*, vol. 2, *The Period of Balance* (London: HMSO, 1957), p. 269; Beesly, *Very Special Intelligence*, p. 150.

30. See Rahn, "The War at Sea in the Atlantic and in the Arctic Ocean," p. 348, table; Dönitz, *Memoirs*, p. 197.

31. Conversation by teletype between Dönitz and German U-boat Command in Italy, 4 January 1942, in Rahn, "The War at Sea in the Atlantic and in the Arctic Ocean," p. 364.

32. Beesly, *Very Special Intelligence*, pp. 144–45; Hinsley, *British Intelligence in the Second World War*, vol. 2, pp. 230, 683.

33. U-boat Command, War Diary, quoted by Dönitz, *Memoirs*, pp. 228–29; Ministry of Defence (Navy), ed., *U-boat War in the Atlantic*, vol. 2, p. 15.

34. Commander of U-boats, "Notes for a Conference with the Fuehrer, 14 May 1942,"

published in Wagner, ed., *Lagevorträge des Oberbefehlshabers*, pp. 393–96. In *Fuehrer Conferences* (Naval Institute Press, 1990), pp. 280–83, the translation of the document is incomplete.

35. KTB/Skl, vol. 36, pp. 549–50 (28 August 1942). See Rahn, "The War at Sea in the Atlantic and in the Arctic Ocean," p. 331.

36. See Rahn, "The War at Sea in the Atlantic and in the Arctic Ocean," p. 331

37. Ibid.

38. Naval Staff, memorandum, 20 October 1942, annex 1, "Stand und Aussichten des U-Bootkrieges," in Salewski, *Die deutsche Seekriegsleitung*, vol. 3, pp. 293–303, quotation on p. 303. See Rahn, "The War at Sea in the Atlantic and in the Arctic Ocean," p. 337.

39. "Stand und Aussichten des U-Bootkrieges," 20 October 1942, in Salewski, *Die deutsche Seekriegsleitung*, vol. 3, p. 300.

40. Compare the diagram "The Tonnage Race," in Rahn, "The War at Sea in the Atlantic and in the Arctic Ocean," p. 336.

41. German Naval Staff, situation assessment, 20 October 1942, in Salewski, *Die deutsche Seekriegsleitung*, vol. 3, pp. 275–312, quotation on p. 291. See Rahn, "The War at Sea in the Atlantic and in the Arctic Ocean," p. 337.

42. Wagner, ed., *Lagevorträge des Oberbefehlshabers*, p. 428; *Fuehrer Conferences* (Naval Institute Press, 1990), pp. 299–300.

43. KTB/Skl, vol. 39/II, pp. 640–41 (24 November 1942).

44. Ibid., vol. 37, p. 224 (11 September 1942). Cf. also translation: "War Diary German Naval Staff, Operations Division," part A, vol. 37, p. 127. For translation, cf. John B. Hattendorf, "The War Diary of the German Naval Staff, 1939–1945," *Documentary Editing* 18, no. 3 (September 1996), pp. 58–62.

45. On 12 September 1942, *U-156* had sunk the British troopship *Laconia* and started rescue operations. On 16 September 1942, a U.S. bomber attacked the boat with lifeboats in tow, killing many survivors. The bombing led to the so-called *Laconia* order, which instructed all U-boat commanders to forgo any attempts to rescue the crews of ships they had sunk, on the grounds that rescue contradicted "the elementary necessity of war for the destruction of enemy ships and their crews." In 1946, the court at Nuremberg did not consider it proven beyond doubt "that Dönitz had deliberately ordered the deliberate killing of shipwreck survivors." For details see Rahn, "The War at Sea in the Atlantic and in the Arctic Ocean," pp. 387–89.

46. Erskine, "Codebreaking," pp. 269–71.

47. Hinsley, *British Intelligence in the Second World War*, vol. 2, p. 563.

48. U-boat Command, War Diary, 12 and 13 January 1943, RM 87/25; Ministry of Defence (Navy), ed., *U-boat War in the Atlantic*, vol. 2, p. 88; Hinsley, *British Intelligence in the Second World War*, vol. 2, p. 557. See Rahn, "The War at Sea in the Atlantic and in the Arctic Ocean," pp. 392–94, esp. note 156.

49. Jürgen Rohwer, *The Critical Convoy Battles of March 1943* (Annapolis, Md.: Naval Institute Press, 1977).

50. U-boat Command, War Diary, 20 March 1943, RM 87/26, fol. 105.

51. Roskill, *Period of Balance*, p. 367; Morison, *Battle of the Atlantic*, p. 344. The original is "Monthly A/S Report December 1943," CB 04050/43, ADM 199/2060, Public Record Office/The National Archives, Kew, U.K.

52. Wagner, ed., *Lagevorträge des Oberbefehlshabers*, pp. 475–77; *Fuehrer Conferences* (Naval Institute Press, 1990), pp. 316–18; David Syrett, *The Defeat of the German U-boats: The Battle of the Atlantic* (Columbia: Univ. of South Carolina Press, 1994), passim.

53. Levine, "Was World War II a Near-Run Thing?," pp. 49–50; Philip Pugh, "Military Need and Civil Necessity," in *Battle of the Atlantic*, ed. Howarth and Law, pp. 30–44. For the ratio of successes to losses achieved by German U-boats see Rahn, "The War at Sea in the Atlantic and in the Arctic Ocean," p. 404, table; and the German series *Das Deutsche Reich und der Zweite Weltkrieg*, vol. 10, *Der Zusammenbruch des Deutschen Reiches 1945*, part 1, *Die militärische Niederwerfung der Wehrmacht* (Munich: Deutsche Verlags-Anstalt, 2008), p. 168, table.

54. Roskill, *Period of Balance*, passim; S. W. Roskill, *The War at Sea 1939–1945*, vol. 3, *The Offensive* (London: HMSO, 1961), passim; Hinsley, *British Intelligence in the Second World War*, vol. 2, pp. 163–79, 525–72.

55. U-boat Command to OKM/Skl, memorandum 3642-A1, 3 September 1942, RM 7/2869, BA-MA.

56. Type XXI: 1,621 tons; maximum speed underwater, 17.2 knots (for one hour); range

11,500 nautical miles (nm) at twelve knots or, underwater, 340 nm at five knots. Type XXIII: 234 tons; maximum speed underwater, 12.5 knots (for one hour); range 3,100 nm at six knots or, underwater, 194 nm at four knots. For details see Werner Rahn, "The Development of New Types of U-boats in Germany during World War II: Construction, Trials and First Operational Experience of the Type XXI, XXIII and Walter U-boats," in *Les marines de guerre du dreadnought au nucléaire* (Paris: Service historique de la Marine, 1990), pp. 357–72.

57. Marc Milner, "The Battle of the Atlantic," *Journal of Strategic Studies* 13, no. 1 (1990), p. 63; Milner, "The Dawn of Modern ASW: Allied Responses to the U-boat 1943–1945," *RUSI Journal* 135 (Spring 1989), pp. 61–68.

58. Jeffrey G. Barlow, *From Hot War to Cold: The U.S. Navy and National Security Affairs, 1945–1955* (Stanford, Calif.: Stanford Univ. Press, 2009), p. 163ff.

The Shipping of Southeast Asian Resources Back to Japan
National Logistics and War Strategy
KEN-ICHI ARAKAWA

A major goal of the Greater East Asia War, according to the imperial decree declaring war, was to secure the so-called Southern Resources Area and to deliver raw materials to Japan, in order to counter what Tokyo perceived to be the economic aggression of the United States, Great Britain, and the Netherlands in the form of a complete embargo. In other words, Japan tried to replace its supply of strategic materials, which it had largely imported from the embargoing powers, by taking Southeast Asian resources by force.

Previous research on this subject has focused on the problem of Japanese shipping. Shipping was indeed a bottleneck for Japan's wartime economy, and Japan's unexpectedly high shipping losses precluded the importation of sufficient resources from Southeast Asia, resulting in the disintegration of the home economy. However, if the war is viewed in its entirety, Japan's shipping losses were less than predicted during the first year of the war—that is, through the fall of 1942—because of faulty American torpedoes, among other reasons. It was not until 1943 that Japan's shipping losses increased rapidly, owing to increasingly effective U.S. submarine and air attacks. After 1943, the military momentum too lay on the Allies' side, and the possibility of an ultimate Axis victory evaporated rapidly. Because of the increasing and enormous shipping losses suffered from this time onward, Japan's resource transportation system functioned less and less efficiently.

This chapter examines the origins of the strategic concept for shipping southern resources back to Japan, the plans developed to realize the concept, the organizations implementing the plans, and the actual results. Regarding the ships used, the "C" ships (merchant ships for civilian use) and "A" and "B" ships (merchant ships requisitioned by the army and navy, respectively) played key roles. It is important to emphasize that the origins of the Southern Resources Shipment System predated Japan's war with the United States by more than a year.

Origins of the Southern Resources Shipment System

Japan's incursion into central China in 1937 sparked international concern, including in the United States. Tokyo first began seriously to consider the shipment of resources from Southeast Asia in the summer of 1940, when the U.S. government began turning its threats of economic sanctions into action. Specifically, Japanese authorities were spurred by the rapid development of events in the European war and had drafted an Emergency Plan for Materials Mobilization, a study concerning the impact of a war with the United States and Great Britain on the supply of and demand for major strategic materials. The findings were extremely discouraging.

According to this study, if Japan joined the war on the side of the Axis powers and cut itself off economically from the United States and Great Britain, its supply of strategic materials would dwindle to a third of 1940 levels. Faced with such dire predictions, the Japanese military responded not by avoiding war but by frantically rushing to stockpile strategic materials before the outbreak of such a war, through Special Imports and Early Imports Programs and, after the outbreak of war, by developing plans to seize the Southern Resources Area by force and thereby to establish a long-term, self-supplying structure for national self-preservation.

On 6 June 1941, the Japanese Supreme Command adopted a document, entitled "An Outline of Southern Policy," which stated:

> The purpose of the Southern Policy is to insure Japan's survival and self-defense, and to expand Japan's overall defense capabilities. In order to realize this, relations (especially economic relations) with the countries in the south, i.e. French Indochina, Thailand, the Dutch East Indies, and others (in order of priority) will be maintained and strengthened. The main method by which this objective will be attained will be diplomacy. If, however, the United States, Great Britain and the Netherlands should implement an embargo against Japan, or if China should join these countries and cooperatively establish an anti-Japanese encirclement and pressure Japan, Japan will have no recourse but to resort to military force in order to insure her survival and self-defense. If the fall of the British home islands in the European War should become certain, the policies included herein, in particular the diplomatic measures to be carried out with respect to the Dutch East Indies, shall be strengthened and our objectives attained.[1]

By early June 1941 Germany had already abandoned its plans to invade Britain and had completed its preparations to invade the Soviet Union. Operation BARBAROSSA began on 22 June 1941. On 25 June, Japan's "Plan to Promote the Southern Policy" was approved by the Imperial Supreme Command, Government Liaison Conference (hereafter "Liaison Conference"), and the cabinet and was sent to the emperor for his approval.[2] The emperor questioned its necessity, to which Army Chief of Staff Sugiyama Hajime replied that it was necessary for the establishment of the Greater East Asia Co-prosperity Sphere and that it included plans for the occupation of southern French Indochina, so as to establish a stable East Asian defensive order. Such steps had to be taken before an embargo and other economic sanctions were put into place.

A Liaison Conference session of 2 July ratified the "Outline of Imperial Policy to Deal with the Change of Events."[3] Its main points concerned parallel efforts to advance both northward and southward. The measures for advancing to the south that had been agreed on in the earlier "Plan to Promote the Southern Policy" would be strengthened, regardless of the risk of war with Britain and the United States. In response to this decision, the United States, Great Britain, and the Netherlands froze all Japanese assets in their respective countries in early August and established a total embargo against Japan.

Development and Implementation of the Complete Embargo on Japan

Before the complete embargo, Japan had envisioned the use of trade to secure southern resources—that is, such resources would be peacefully imported to meet Japan's needs. The changes from peaceful importation to the "returning" by force of raw materials from occupied areas and from trade to "compulsory trade" were closely related to the freeze of Japan's overseas assets and the complete embargo in August. The severing of trade with the United States, Britain, and the Netherlands included trade with those nations' economic spheres of influence, such as India, Burma, Malaya, the Straits Settlements (with Singapore as the capital), Hong Kong, the Dutch East Indies, Australia, New Zealand, and the Philippines. As of 1935, Japan relied on these areas for more than 90 percent (in terms of import value) of its crude oil needs. Of that total, fully 76 percent was imported from the United States and 16 percent from the Dutch East Indies and British Borneo.

Japan's leaders, in particular its military leaders, feared that if the embargo continued the empire would collapse. They therefore decided on war against the United States, Britain, and the Netherlands to preserve Japan's very existence, and they turned to securing the Southern Resources Area. If Japan was to maintain its 1935 levels of imports of the main strategic materials, however, the Southern Resources Area would have to supply almost six times the amount it had in 1935.

After the outbreak of the Sino-Japanese War in 1937, many consumer products in Japan were in short supply. Sugar, for example, became a rationed item and more difficult to obtain. Japan was not the only country that in anticipation of war stockpiled strategic materials not available domestically. From 1939 to 1940, or soon after the outbreak of war in Europe, such imports as processed sugar to the Netherlands decreased, but these same imports to Britain increased rapidly by approximately 450 percent. Soon after the start of the Pacific War, Prime Minister Tōjō Hideki likewise ordered the army's requisitioned A (civilian) ships to carry sugar from Taiwan and other areas on their return journeys to Japan. Owing to the higher rate of imports, Japan's foreign currency reserves were drained more rapidly after the outbreak of the Sino-Japanese War. This in turn resulted in strict foreign-currency controls. Imports of consumer products that had to be purchased with foreign currency were given low priorities and were reduced. Whenever possible, such products were purchased from yen-bloc countries, such as Taiwan.

Because of these factors, Japanese imports from the Dutch East Indies decreased rapidly after 1937, until they reached almost zero in 1940. Meanwhile, Malaya exported more than it imported—retaining a particularly large trade surplus with the United States, and importing goods mainly from the rest of Asia. In 1940, more than 56 percent of its exports, by value, were rubber and related products, and more than 25 percent were accounted for by tin. In the same year, 26 percent of its imports were food, and more than 35 percent were manufactured goods. Broadly speaking, when Japan began the Greater East Asia War in 1941, the Dutch East Indies lost more than 60 percent of its peacetime trade, while British Malaya lost more than 80 percent of its exports and just under 40 percent of its imports.

The Decision for War and the Outlook for Southern Resource Shipments

The "Outline of Imperial Policy," adopted by the Imperial Conference on 6 September 1941, set a deadline for diplomacy, stating, "If by early October there is no reasonable hope of having our demands agreed to in the diplomatic negotiations mentioned above, we will immediately make up our minds to prepare for war against America (and Britain and the Netherlands)." On 16 October, the cabinet of Prime Minister Konoe Fumimaro resigned, to be replaced by a Tōjō cabinet. When Tōjō assumed the prime minister's position, the emperor directed him to "return to the state of a blank piece of paper and reconsider the problem of peace or war, regardless of the decision of September 6." In accordance with these instructions, the Tōjō cabinet in late October debated whether "to make war or to sustain perseverance." At this time, a systematic evaluation was conducted regarding the resources that could be obtained in the south if war broke out and the problems entailed in bringing them back to Japan. As a result of these deliberations, the Liaison Conference of 1 November 1941 "decided upon war with the United States, Britain and the Netherlands, with the outbreak of hostilities scheduled for early December."[4]

When discussing whether "to make war or to sustain perseverance," Japan's leaders were most interested in the impact on Japan's national power, in particular on the outlook for importing critical strategic raw materials. A key problem concerned the availability of shipping capacity to transport such raw materials to Japan in the event of war. During this debate, forecasts were made of shipping losses and maritime transport capabilities. A comparison of figure 1, "Projected and Actual Shipping Losses," and figure 2, "Projected and Actual Civilian Shipping Capacity," shows that actual shipping capacity was less than predicted from the second quarter of 1942 (April through June) onward, or half a year before actual shipping losses exceeded predictions, which was not until the fourth quarter of that year (October through December).

FIGURE 1
Projected and Actual Shipping Losses (thousands of gross tons)

YEAR	PROJECTED	ACTUAL	LOSSES TO SUBMARINES	PERCENTAGE LOST TO SUBMARINES
1942 1st qtr.	250	178	80	45
1942 2nd qtr.	250	167	70	42
1942 3rd qtr.	250	209	155	74
1942 4th qtr.	250	403	202	50
1943 1st qtr.	250	366	243	66
1943 2nd qtr.	250	371	329	89
1943 3rd qtr.	250	386	320	83
1943 4th qtr.	250	667	472	71

Source: Oi Atsushi, *Kaijō goeisen* (Tokyo: Asahi Sonorama, 1983), pp. 374–85.

FIGURE 2
Projected and Actual Civilian Shipping Capacity (thousands of gross tons)

YEAR	PROJECTED	ACTUAL	DIFFERENCE
1942 1st qtr.	7,719	8,611	892
1942 2nd qtr.	10,561	9,291	−1,270
1942 3rd qtr.	13,590	10,726	−2,864
1942 4th qtr.	13,461	10,853	−2,608
1943 1st qtr.	13,215	9,148	−4,067

Sources: Projections from Sanbo Honbu [Japan, Army General Staff], ed., *Sugiyama Memo (Jyo)* (Tokyo: Hara Shobo, 1967), vol. 1, p. 467. Actual capabilities from Senpaku Uneikai, ed., *Senpaku Uneikai Kaishi (Zenpen) Jyo*, 1947, pp. 589–90, National Institute for Defense Studies Archives, Ministry of Defense, Tokyo, Japan.

The earliest recorded plan for the transfer of southern resources to Japan, including the issue of shipping, was that known as "Details of Ship Allocations" (hereafter referred to as the "October 1941 Plan"). This was part of the "Calculations of the Supply Capabilities for Critical Materials," dated 22 October 1941, which was one of the so-called Sugiyama Memos. This plan determined that an annual volume of 960,000 tons of critical materials had to be transported from the Dutch East Indies, the Philippines, and Malaya (the "A" regions, Indochina and Thailand being the "B" regions), and assigned the necessary C (civilian) ships to this task.

A plan entitled "Charts Regarding Freighters" (hereafter the "December 1941 Plan") provided more detail; it was included in the "General Plan on Measures for Managing the Southeast Asian Economies," dated 18 December 1941. According to this detailed plan, 2,814,000 tons of materials were scheduled to be shipped back from the

A regions in 1942. Of this total, the C ships would carry 960,000 tons (or 34 percent of the total), and the army's A ships and the navy's B ships would carry 1,854,000 tons (or 66 percent). Exports to Japan were prioritized into four classifications, with the C ships assigned the two top classifications, and with the lower classifications transported, if possible, by the A and B ships.

A third plan was the top secret "Allocation of Ships according to the Materials Mobilization Plan" (hereafter the "June 1942 Plan"). The drafter of this plan is unknown. It was originally dated 11 March 1942 and revised on 11 June of that year. It called for the transport in 1942 of 2,300,000 tons of materials from the A region (roughly five hundred thousand tons less than called for in the December 1941 Plan). C ships would carry 1,203,000 tons (or 52 percent of the total, which was 25 percent greater than that assigned under the December 1941 Plan). For their parts, the A and B ships would transport 1,095,000 tons (or 48 percent of the total, an approximately 40 percent reduction from the December 1941 Plan).

As can be seen in figure 2, the total projected volume for C ships during 1942 was 45,331,000 tons, but the actual total was 39,481,000 tons, or 87 percent of the planned volume. Comparison of the June 1942 Plan and the Shipping Administrative Association's "Chart of Maritime Transport, Planned and Actual" (hereafter the "SAA's Chart") reveals that of the 3,112,000 tons of materials scheduled for transport in 1942 from the A and B regions on C ships, a total of 1,601,000 tons, or 51.4 percent of the total, was actually transported. The June 1942 Plan also shows that while A ships were assigned the transport of 832,000 tons of materials, B ships were assigned 1,553,000 tons, just under twice as much as the A ships.

How did the plans for transporting the southern resources fare in 1942? Accurate records have not yet been found. However, figure 3, "Actual Imports of Southern Resources in 1942," is assembled from the following three sources: the Tanaka Documents (believed to have been compiled around 1943); statistics compiled by the U.S. Strategic Bombing Survey, compiled around 1950; and the Iwatake Study, prepared in 1980, based on the Ministry of Finance's *Japanese Foreign Trade Annual*.

When compared, these three sets of data show that the figures given by the Strategic Bombing Survey are the highest (excluding shipments of raw rubber). These data were compiled after the war, by a surveying team sent from the United States with the cooperation of the Allied General Headquarters to interview Japanese politicians, industrialists, high-level bureaucrats, and military officers. Of the three sources given above, the Strategic Bombing Survey's data are thus believed to be the most reliable. The Iwatake Study, based on Ministry of Finance reporting, excludes materials that did not pass through customs. The Tanaka Documents are based on the reports sent by the army,

FIGURE 3
Actual Imports of Southern Resources in 1942 (thousands of gross tons)

RESOURCE	TANAKA DATA	USSBS* DATA	IWATAKE DATA	TARGET VOLUME
Nickel	21.1			100
Manganese	17.4			100
Iron ore	135.2	215	131.7	800
Tin	2.8	3.8		20
Bauxite	238.1	305	2	400
Copper ore	0.6	6.8	0	100
Rubber	20.8	29.7	44.2	200
Hemp	14.4	—	10.5	80
Copra	21	—	0	350
Crude oil	330.7	1,295.2	—	1,100
Rice	—	1,527.7	1,341.8	286
Scrap iron	—	9	2	164

* U.S. Strategic Bombing Survey

Sources: Tanaka data from Tanaka Shinichi, *Nihon Senso Keizai Hishi* (Tokyo: Computer Age, 1975), pp. 228–29; U.S. Strategic Bombing Survey, *Nihon Senso Keizai no Hōkai* (Tokyo: Nihon Hyoronsha, 1950); Iwatake data from Iwatake Teruhiko, *Nanpō gunseika no keizai shisaku* [Economic Policies under the Southern Army's Administration] (Tokyo: Ryukei Shosha, 1995), vol. 2, charts 2–15; target volume from "Besshi Daiichi Kamotsusen Kankei," in Committee No. 6, "Nanpo Keizai Taisaku Yoko," 12 December 1941, National Institute for Defense Studies Archives, Ministry of Defense, Tokyo, Japan, and "Showa 17-nendo Nanpo Chiiki yorino Busshi Shutokuryo narabini korega Haisen Kubun Keikakuhyo," 11 June 1942, National Institute for Defense Studies Archives, Ministry of Defense, Tokyo, Japan.

navy, and Shipping Administrative Association (SAA) to the Planning Board. This set of data is probably the least reliable of the three, because it is questionable whether the army and navy accurately informed the Planning Board of the tonnage of materials carried by the A and B ships under their respective jurisdictions.

The data on crude oil and rice shipments included in figure 3 merit special attention. The figures for crude oil shipments from the Strategic Bombing Survey are approximately four times greater than those from the Tanaka Documents, even greatly exceeding the import target for that year. The transport of oil to Japan came under the sole jurisdiction of the armed forces, and the amounts transported were highly classified. It is believed that the army and navy were wary of each other and thus did not send the Planning Board accurate reports of the amounts their respective ships transported back to Japan; however, it is also likely that actual oil imports did indeed exceed predictions in 1942. Rice imports in 1942 were also five times the target amount. If these figures are accurate, it means that in the first year of the war Japan allocated more than 70 percent of its maritime shipping capacity (excluding tankers) assigned to the Southern Resources Area to shipments of rice, whereas the original plans called for an allocation of only about 10 percent of shipping capacity for this basic staple.

Organizations and Procedures Involved in the Shipments of Southern Resources

On 20 November 1941, the Liaison Conference approved the "Outline of Administrative Control of Occupied Areas of Southeast Asia." The document specified that the main objectives of Japan's administration of the occupied areas were the reestablishment of order, the rapid securing of raw materials critical for national defense, and the establishment of a self-supply system for the Japanese armies operating in the area.

To attain these objectives, it was decided that the army and navy would administer the occupied areas for the time being. The Liaison Conference would be responsible for approving matters critical for such administration, and the army and navy chain of command would be responsible for giving the orders necessary to implement such decisions in the occupied areas. As for raw materials, a central organization, in which the Planning Board would play a key role, would make all necessary decisions. Important issues required the Liaison Conference's approval, after which the army and navy would be responsible for orders to units in the occupied areas.

On the basis of this "Outline of Administrative Control," an "Army-Navy Agreement Regarding the Military Administration of Occupied Areas" was signed on the same day, 20 November. This agreement specified the respective administrative areas for which the army and navy would assume responsibility. The army was assigned primary responsibility, with the navy having secondary responsibility, over Hong Kong, the Philippines, British Malaya, Sumatra, Java, British Borneo, and Burma. Meanwhile, the navy had primary responsibility over Dutch Borneo, the Celebes, the Molucca Islands, Lesser Sunda Islands, New Guinea, the Bismarck Islands, and Guam.

The military services dealt with the problem of administering occupied areas in various ways. The navy established a Political Affairs Office for Southeast Asia within the Navy Ministry to handle in a unified manner issues involving politics, economic development, and raw-material extraction. In the occupied areas high-ranking officers, typically fleet commanders, would be directly responsible for administration, acting on the orders of the navy minister. The army, on the other hand, did not set up a new organization but rather used the same system it had during the Sino-Japanese War.[5] In other words, the basic structure for the administration of occupied areas came under the jurisdiction of the Army General Staff, with administrative problems handled by the army minister. In occupied areas, "military administrations" were established directly under army commands at the start of the war in the Pacific.

Since existing records do not detail the procedures by which the army actually arranged and carried out the sending of southern resources back to Japan, one must make an educated guess, but the outlines probably were as follows. First, the Planning Board informed the army "how much" of "what" items was to be sent "from where," and "by when." This guidance was probably ordinarily accompanied by a ship assignment plan. Upon receiving this request, the Army Section of the Supreme Headquarters—more specifically, the Operations Department and the Third Department of the General Staff and the Military Preparedness Bureau of the Army Ministry, among others—coordinated with the sections and departments involved and drafted a transport plan; the Military Preparedness Bureau was responsible for shipping plans.[6] The Army Supreme Headquarters then sent the necessary directives to the Shipping Transportation Headquarters, located in Ujina, Japan, which had overall responsibility for the requisitioned ships under the army's control.

The next plan was the "Outline of Southern Economic Measures," which was drafted by the Sixth Committee of the Planning Board and approved by a meeting of cabinet ministers on 12 December 1941. It stated clearly that the primary objective was the securing of resources in the south. This outline divided the southern area by importance to Japan. British Borneo was added to the Dutch East Indies, British Malaya, and the Philippines as an "A" region; French Indochina and Thailand continued to be designated "B" regions. Most of the A regions were areas assigned to the army. The outline further stated that force would be used to secure the resources in the A regions and to prevent their export elsewhere.

The Special Military Compulsory Trade method was a form of forced trade under the jurisdiction of the armed forces. According to the "Outline of Administrative Control of Occupied Areas of Southeast Asia" mentioned above, ships, railroads, and port facilities were to be controlled by the armed forces, along with all matters pertaining to trade and currency exchange. As a rule, local currencies remained in use, except for certain cases in which special military coupons were issued. In other words, all matters concerning compulsory trade fell under the jurisdiction of army and navy units or the military administration offices. This was because "it came to be believed that since such compulsory trade would take place in areas of military operations, which involve special dangers, the form of military control which would enable the strongest direct intervention by the military, i.e., purchases, transport and sales by the military itself, would be most appropriate."[7]

The "Outline of Southern Economic Measures" specified the collection of goods, allocation, and compulsory trade as follows: "(1) Purchases and imports in connection with the supply to Japan of collected goods and materials shall for the time being be paid for by public accounts; exports of goods to the areas in question shall be handled likewise. (2) Exchanges of goods within the areas in question shall be handled in accordance with (1) above."[8]

These measures show that civilian goods handled by the military administration were to be treated like military goods—that is, goods to be used by the military units in the field. The "Public Accounts" mentioned here were the "Special Accounts of Extraordinary War Expenditures." This compulsory trade method should therefore be classified as under military rather than national administration. This system had a number of merits. First, it enabled the compulsory trade to be fully controlled, which in turn enabled the systematic and organized regulation of the supply and demand of various materials. Second, transportation costs were borne not by the parties actually involved in the trade but rather by the military. Finally, problems related to currency exchange rates could be avoided, since payments were not made with differing currencies. On the other hand, it can be imagined that the procedures that had to be taken to engage in such trade were overly complicated and inefficient, because the military was responsible for the accounting and other necessary activities, in which it had little experience.

Under this method, the military's accounting departments and the sea transport units acted simultaneously as foreign-currency banks, maritime transport companies, and insurance companies. This method was ultimately used for trade between Manchuria / North China and the A regions to the south, as well as between the occupied A regions and the B regions in French Indochina and Thailand. In addition, since it was decided to use the Special Military Compulsory Trade method for the occupied areas in the south, all trade involving at least one party in an occupied area fell under Special Military Compulsory Trade.

Military Developments and Shipments of Southern Resources during 1942

From the Japanese perspective, for half a year after Pearl Harbor the war progressed much better than anticipated. The British surrendered in Singapore in February 1942. By the time the Dutch forces on Java surrendered on 9 March 1942, the entire Southern Resources Area was in Japanese hands, with the exception of the Philippines, where the last American forces fighting on Corregidor capitulated on 7 May.

For that first half of 1942, unofficial records exist discussing the volume of materials the army shipped back to the home islands, including one entitled "Mainly an Outline of Maritime Transport during the First Half of the Greater East Asia War," by the former commander of the army's ship transportation forces, Lt. Gen. Saeki Bunrō (hereafter the "Saeki Memo"). The issue of army shipping during this period will be considered using the figures given in the Saeki Memo.

According to the June 1942 Plan, 5,500,000 tons were to be shipped back from the southern area (A and B regions combined) in 1942, of which the army's A ships were to carry 832,000 tons. According to the Saeki Memo, the Southern Army transported

654,000 tons on A ships from the outbreak of war in December 1941 through May 1942. This meant a yearly total of 1,308,000 tons, which would be 157 percent of what was planned. Of the army shipments to Japan, 69.1 percent, or 452,000 tons, was rice, which constituted 37 percent of all rice shipments to Japan.

The Saeki Memo also gives data for shipments on A ships of the Southern Army in April 1942. According to the "Details of Ship Allocations," approximately 82 percent, or 98,000 tons, of the 120,000 total tons to be shipped back to Japan in one month from the A regions came from the Dutch East Indies. According to the June 1942 Plan, however, no A, B, or C ships were allocated to the Dutch East Indies in the second quarter (April through June) of 1942. Nevertheless, the Saeki Memo states that 28,500 tons were shipped from the Dutch East Indies on the Southern Army's A ships. If this is true, it means that the target amount given in the June 1942 Plan for the first quarter of 1943, which was twenty-four thousand tons, had been met about a year earlier.

It should be remembered that the materials shipped back on A and B ships enjoyed extraterritoriality and were not required to pass through customs. It is therefore unlikely that accurate import-export records of such shipments ever existed. The volume of such shipments was also supposed to be reported to the Planning Board, but the army and navy both had a tendency to deflate their reports. The Saeki Memo also states that 240,000 tons of rice were shipped back from the southern area (excluding Taiwan) in April 1942. According to the shipment plans of individual items for 1942 given in the SAA's Chart, however, 259,000 tons of rice shipments were scheduled for April 1942, with 156,300 tons actually shipped, which would mean that the Saeki Memo's data are in excess for April by about eighty-four thousand tons.

Estimates of Actual Amounts of Southern Resources Shipped during 1942

Given the scarcity of documents, details of the actual amounts shipped back from the southern area can only be estimated. One set of data that is useful in this respect is the "Transport Plans and Records" compiled by the SAA immediately after the war, as part of its efforts to prepare an association history. However, data revealing how the total amount transported in ships came from the southern area are available only from July 1942, perhaps because the SAA itself was established only in April 1942. Another factor to remember is that the data are mainly about shipments on C ships, since it was those ships that the SAA was responsible for operating; the SAA data are therefore incomplete for shipments made on A and B ships. Furthermore, bauxite, tin, and other metals are lumped together as "non-ferrous" materials, meaning that data on shipments of individual metals are also incomplete.[9]

The amounts of iron ore, nonferrous metals, rice, and crude oil shipped by A and B ships combined and C ships have been estimated. The results are given in figure 4.

FIGURE 4
Estimates of Southern Resources Shipments of A, B, and C Ships, 1942 (thousands of gross tons)

	PERIOD	IRON ORE	OTHER METALS	RICE	CRUDE OIL
C ships	April–June 1942	15.9	1.2	450	180.1
	July–Sept. 1942	19.2	1.3	352	295.9
	Oct.–Dec. 1942	53.3	37.4	186	491.8
	Jan.–March 1943	—	106.2	117	327.4
	C ship total	88.4	146.1	1,105	
USSBS data		215	354.1	1,527.7	1,295.2
A/B ships (percentage)		126.6 (58.9)	208 (58.7)	422.7 (27.7)	

Source: Senpaku Uneikai, ed., *Senpaku Uneikai Kaishi (Zenpen) Jyo*, pp. 615–76. Under "other metals," nickel and manganese shipments are from the Tanaka data, while tin, bauxite, and copper ore are from the USSBS data.

According to these estimates, just under 60 percent of the total volume of iron ore and nonferrous metals shipped and just under 30 percent of the total amount of rice went to Japan on ships controlled by the military—the A and B ships.

In other words, an average of 48 percent of the total volume of these three items shipped back to Japan came on the military's A and B ships. This figure exceeds the total that the A and B ships together were scheduled to carry from the southern area in the plan for 1942, which was 43 percent of all such shipments. From this, it may be concluded that the scheduled amounts of crude oil, rice, and nonferrous metals (especially bauxite) were more or less actually shipped in 1942, through the effective use of the A and B ships. In particular, the target volume for crude oil, which was 0.6 to one million tons in the "Outline of Southern Economic Measures," was greatly exceeded, with 1.295 million tons actually shipped back to Japan.

If Japan had not suddenly lost a large portion of its merchant shipping, this shipping system, while perhaps not the most efficient possible, could have provided Japan the necessary raw materials during the war. But from the second half of 1942 onward, Japan's war fortunes reversed. Following the disastrous defeat at Midway, especially during the battle for Guadalcanal, which took place from August 1942 to the end of the year, Japan tried to meet its growing shipping needs by re-requisitioning ships that were due to be returned from government service, but it ultimately lost most of them to submarine attacks.

While the actual volume shipped declined from 1943 onward, the ratio of shipments from the southern area to all shipments combined increased from mid-1943. This means that Japan prioritized shipments from the southern area. For example, the total amounts of nonferrous metals shipped back were approximately three times the total for 1942, with respect to both the C ship totals and the Strategic Bombing Survey. In 1944, American air attacks on Japanese shipping became more frequent and more effective, reflecting the deterioration of the overall situation, and shipping losses were four times the losses incurred in 1942. Not until late 1943, however, did the United States fix its defective submarine-launched torpedoes, dramatically improving their accuracy.[10]

The rapidly increasing efficiency of the U.S. submarine campaign is reflected in the shipping figures. For example, shipments of nonferrous metals, which had increased greatly in 1943, were only 42 percent of that total in 1944, according to the Strategic Bombing Survey's figures, while similar figures for crude oil shipments show a 30 percent decline. It is interesting to note that much of the remaining transport capacity was used to ship rice and sugar. Just as the Saeki Memo indicates, in the period immediately after Pearl Harbor much effort had focused on shipments of rice and sugar. Japan's shipments of resources from the southern area seemingly came full circle and by 1944 were back to basic staples.

The Unification of Shipping Operations

All Japanese merchant shipping, in theory at least, came under the control of the army, the navy, or the SAA, but in practice it was as if the army and navy each controlled all of the ships, since either could, when necessary, ask the SAA to operate its ships for the army's or navy's purposes.[11] The United States, by comparison, established the federal government's War Shipping Administration on 17 February 1942, which exercised sole control over all merchant shipping. Some Japanese realized that shipping coordination was getting worse. From the fall of 1942, voices from both the army and navy called for the unified operation of all of these ships, to increase Japan's maritime shipping capacity. But these voices were largely ignored.

It was not until 8 March 1945 that this idea was realized in a meeting of department and bureau chiefs of the Army and Navy Ministries. The decision reached was as follows: "'A,' 'B' and 'C' ships and related port facilities will be put under unified operation. Specifically, this will be done as follows: The sections within the Army and Navy that are responsible for operating the ships will be combined, and a unified transportation headquarters will rapidly be organized (so that it may begin working by the start of April), and will work to realize the unified control of shipping."[12]

The Headquarters of War Shipping Administration was thus set up within the Supreme Headquarters to control the merchant marine in a unified manner. Its first commander, Admiral Nomura Naokuni, was appointed on 1 May 1945. By that time Japan's merchant marine totaled approximately 1.8 million tons, or 35 percent of the total at the start of the war, while the Americans had 32.8 million tons, or 240 percent of their December 1941 total. The Americans thus had eighteen times the shipping available to the Japanese. By this late date Japan had lost too much merchant capacity for unification to affect the outcome of the war, even had the use of the A-bomb and the Soviet declaration of war, both occurring in August 1945, not pushed Japan into a more rapid surrender.

Conclusions

The Pacific War is often characterized as a war of supply, especially maritime supply. Whichever side could send to the front a force of superior combat capability and keep that force supplied eventually won the attrition campaigns. In this context, it is often overlooked that for six months after Pearl Harbor the Japanese ships requisitioned by the military—the A and B ships—transported more supplies than was originally planned. The main problem concerned the C ships, which did not keep up with demand. When the civilian shipping organization was reformed in April 1942 into the SAA, this new organization had to overcome more problems than expected. Because of these challenges, it was not until mid-1943 that Japan was able to place top priority on shipment of southern resources back to Japan. But by then the needs of the military and the civilian sector clashed, and it was too late to make compromises.

By 1943 the low standards of the Japanese escort forces—especially their inadequate antisubmarine capabilities, including detection technology inferior to that used by the British and Americans—became the main reason for the gap between plans and reality. By 1944 the U.S. redesign of its flawed torpedoes, in combination with its use of cryptography to locate Japanese convoys, greatly increased the U.S. submarine kill rate. Thereafter, Japanese shipping losses grew alarmingly, a trend that was accelerated by the decline in the general war situation and insufficient (albeit ever-increasing) efforts by Japan to build new merchant ships. Thus did the vaunted Southern Resources Shipment System crumble away, and by early 1945 shipments from the south to Japan had ceased entirely. The seaports of Southeast Asia, occupied by Japan but bypassed by the advancing Allied forces, had huge stockpiles of raw materials that could not be sent to Japan but simply sat.

In retrospect, Japan's only chance for winning the Pacific War was probably during the spring to fall of 1942, if Tokyo had placed more importance on destroying Allied maritime supply lines. By severing the maritime transport capabilities of the United States, which were the arteries of its war machine, it might have been possible to retain a relative superiority in supply capability. Given the severe limitations of Japan's maritime escort capabilities at the time, however, even this strategy would have been difficult to execute. From 1943 onward the difference in industrial production capability of the two sides became even more pronounced, resulting in an ever-increasing superiority of the Allies' combat power. After 1943, therefore, it was no longer possible for Japan to regain the momentum by the mere introduction of a few new merchant ships.

Notes

The thoughts and opinions expressed in this publication are those of the author and are not necessarily those of the U.S. government, the U.S. Navy Department, or the Naval War College.

1. Sanbō Honbu [Japan, Army General Staff], ed., *Sugiyama Memo (Jyo)* (Tokyo: Hara Shobo, 1967), vol. 1, pp. 217–18.
2. Ibid., pp. 227–31.
3. Ibid., p. 260.
4. Ibid., pp. 378–79.
5. In an interview with this author in November 1998, Iwatake Teruhiko stated that the army had temporarily established a "Southern Policy Department" within the Military Affairs Bureau of the Army Ministry, before the navy did so, using a departmental directive, which meant that this organizational change could be made without a formal revision of the laws governing the Army Ministry.
6. Bōei Kenshūjo Senshishitsu, ed., *Rikugun Gunju Dōin* [The Army's Economic Mobilization] (Tokyo: Asagumo Shinbunsha, 1970), vol. 2, p. 507.
7. Iwatake Teruhiko, *Nanpō gunseika no keizai shisaku* [Economic Policies under the Southern Army's Administration] (Tokyo: Ryukei Shosha, 1995), vol. 1, p. 93.
8. Dai toa syo [Department of Great East Asia], ed., *Southern Economic Measures* (n.p.: NIDS Archives, 31 July 1943).
9. Another useful set of data is the aforementioned Strategic Bombing Survey's records. Data on shipments of individual metals may be found here, but only the totals shipped back on A, B, and C ships are given for each metal, so it is not possible to identify from these records how much of a certain metal was shipped by, for example, the army.
10. George W. Baer, *One Hundred Years of Sea Power* (Stanford, Calif.: Stanford Univ. Press, 1993), pp. 233–34.
11. Iwatake Teruhiko, interview with author, November 1998.
12. Gunjishi Gakkai [The Military History Society of Japan], ed., *Kimitsu Sensō Nisshi (Ge)* [Supreme Headquarters Secret War Diary] (Tokyo: Kinseisha, 1998), p. 683.

Unrestricted Submarine Victory
The U.S. Submarine Campaign against Japan
JOEL HOLWITT

Most of the commerce raiding campaigns described in previous chapters had debatable impacts on the greater wars. This is not to say that these campaigns were trivial. But the British probably would still have lost at Yorktown without American privateers, and there is no guarantee the Union would have won any sooner if CSS *Alabama* had never sailed. While there is some argument whether the German U-boat campaigns in either world war could have succeeded against the Allies, there is no question that they in fact failed.

In sharp contrast to these operations, the U.S. submarine campaign against Japan in World War II not only succeeded but played a decisive role in neutralizing the Japanese Empire. The U.S. submarine campaign was no fluke or bizarre anomaly. The victory resulted from decades of war planning, a final year of specific planning for an unrestricted submarine campaign, and innovation and courage in combat by American submariners.

It also owed its success, in no small part, to the manner in which Japan's national leaders mishandled their maritime vulnerability. Despite the unmistakable similarities between Japan's and Great Britain's geographic situations, as well as the success of German U-boats during the two years before Pearl Harbor, the Japanese naval leadership gravely mismanaged the merchant marine and made little effort to protect the empire's economic lifelines before hostilities began. But as the previous chapter has shown, the Japanese merchant marine performed better than expected from 1942 through 1943 in bringing crucial southern resources to the home islands. This early and illusory success lulled the Japanese naval leadership into continuing its inadequate antisubmarine measures. By 1944, when Japan belatedly attempted to organize and adequately protect crucial shipping, it was too late.

Interwar U.S. Submarine Development, Strategy, and Doctrine

From the earliest iteration of the U.S. war plan against Japan, War Plan ORANGE, the United States always intended to achieve victory by preying on Japan's greatest strategic vulnerability—its reliance, as an island nation, on the importation of war materiel by sea.[1] Having only a very few natural resources in abundance, Japan required its merchant marine to import oil, steel, aluminum, and even foodstuffs. Indeed, despite Japan's own supply of coal, it still required significant imports of coal to meet domestic demand, as well as imports of the higher-grade coal necessary for industrial uses. Japan also depended on a strong merchant fleet to supply its many island possessions with the daily necessities. If Japan's supply lines could be cut off, not only would Tokyo be unable to supply its war machine, but even the Japanese home islands would be choked from inexorable economic pressure.[2]

However, the blockade envisioned by War Plan ORANGE did not initially include submarine warfare. Plan ORANGE presumed that sufficient naval forces would be available to destroy the Japanese fleet and gain control of the sea to enforce a regular naval blockade. Surface ships, utilizing cruiser rules of warfare, would maintain that blockade.[3] ORANGE envisioned the submarine force's primary mission as that of a naval combatant. In 1936 the commander in chief of the U.S. Fleet, Adm. Joseph M. Reeves, stated, "The primary employment of submarines will be in offensive operations against enemy larger combatant vessels. . . . No submarines will be assigned in the early stages of the war to operate against enemy trade routes."[4]

American submariners were limited not only by strategy but by law. The United States was a signatory to both article 22 of the London Naval Treaty and the London Submarine Protocol of 1936. The two documents, which in this respect were identical, required a submarine to remove a merchant ship's crew to a place of safety before sinking the ship. A lifeboat on the open sea, furthermore, was not considered to be a place of safety. It did not matter whether the merchant ship belonged to a belligerent nation or a neutral nation. It did not matter whether a merchant ship was arguably in the service of a belligerent nation's war machine. Regardless of origin or ownership, merchant ships simply could not be attacked without warning.[5]

Consequently, the interwar submarine force planned to support Plan ORANGE by scouting ahead of the battle fleet and skirmishing with the Japanese fleet somewhere in the vast expanses of the Pacific Ocean. This was a pretty tall order, given that the predominant type of submarines built at the end of World War I, the S class, did not have the surface speed or endurance needed to make a long transpacific transit and stay ahead of the battle fleet. Therefore, the navy used a 1916 congressional authorization to build nine "fleet submarines" to investigate the characteristics necessary for future submarines.⁶ By 1930 the U.S. submarine force had identified the ideal characteristics of a fleet submarine: long range, high surface speed, and sufficient weaponry.⁷

After a subsequent decade of experimentation, submariners developed 1,500-ton (surface displacement) *Gato*-class fleet submarines. These submarines, as designed, fully met the U.S. Navy's needs in the Pacific. They could regularly make twenty knots on the surface, allowing them to proceed far ahead of American surface forces and maintain contact with enemy battle fleets. With their displacement, fuel capacity, and technological innovations, they had the range to shadow Japanese fleet movements all the way from the Sea of Japan or to stay on station for almost two months at a time. The *Gato* class turned out to be extremely versatile and capable, but prewar doctrine and training differed greatly from the reality of the war that the submarine force eventually fought.

The U.S. Navy Submarine Force tactical doctrine released in April 1939 broke down the strategic missions of scouting and skirmishing into more detailed guidance.⁸ The doctrine explicitly separated attack on fleet units from secondary missions like "patrol."⁹ The doctrine explained that submarine patrolling differed from "attack" in that the former's purpose was to destroy sea lines of communication, not specific fleet units.¹⁰ The doctrine limited commerce destruction to armed merchant ships and convoys:

> Patrol against enemy lines of communication may include the destruction of commerce. It may be expected that the convoy system will be used, especially at focal and terminal points. On the high seas circuitous routing will be employed. Due to the limitations of submarines in exercising the right of visit and search, and the difficulty of distinguishing between enemy and neutral shipping because of the disguise of enemy shipping as neutral, submarine operations against enemy commerce is limited to attacks on convoys, or attacks on positively identified armed enemy shipping, unless unrestricted commerce destruction is directed as a last resort.¹¹

In any case, traditional cruiser warfare was prohibited because of the danger from armed merchant ships: "Under the limitations imposed by the laws of war and as interpreted in the Treaty of London, submarines cannot be used effectively against merchant ships without running undue risk of destruction."¹²

The published doctrine established what the U.S. Navy expected from its submarines, with emphasis on attacking enemy fleet units, particularly capital ships, and prohibitions against cruiser warfare. The submarine force doctrine acknowledged the remote possibility of unrestricted submarine warfare, "in its operational and tactical preparations, [but] the service held a consistent view: the U.S. Navy would not allow its submarine captains to attack merchant shipping without warning."¹³ Submarine commanders like Dick Voge, who commanded USS *Sealion* at the outbreak of hostilities, even believed that "submarines who [sic] violated [article 22] were subject to being 'hunted down and captured or sunk as pirates.'"¹⁴ This view changed dramatically with the beginning of the war.

The Shift to Unrestricted Submarine Warfare

During the early 1940s, the primary mission of the submarine force shifted from naval combat to commerce warfare. This change dated from November 1940, when the U.S. Navy's senior leadership had to confront squarely a number of strategic choices. These strategic choices were encapsulated in a document that would lay the foundation for the American national military strategy in the Second World War—Plan DOG (that is, D).

Unlike many U.S. war plans, Plan DOG was the brainchild of one person, the Chief of Naval Operations (CNO), Adm. Harold R. Stark, who was promoted to this position on 1 August 1939. Stark quickly found himself bringing the U.S. Navy up to wartime readiness when the Germans invaded Poland only a month later. Stark worked hard to enlarge the navy, while also devising an appropriate strategic vision for employing the forces he had. Shortly after the reelection of President Franklin D. Roosevelt to a third term, Stark committed his thoughts to a long memorandum, which was revised by the chief war planners from both the army and the navy. The memorandum was then studied and approved by a joint board that included Admiral Stark and the army chief of staff, Gen. George C. Marshall. From there the memorandum went to the secretaries of state, war, and the navy, who then forwarded it on to the president. Although he never formally approved the plan, Roosevelt agreed with its general principles.[15]

The memo offered several possible scenarios for war involving the United States, as well as several plans to go with those scenarios. The planners listed the advantages and disadvantages of each one before settling on the fourth (thus D) as the most advantageous to the United States no matter how war came, whether from Nazi Germany or imperial Japan. Of course, the planners hoped not to fight a two-ocean war, but no matter the circumstances, they called for immediate aid to Great Britain upon the commencement of hostilities. Almost a full year before Pearl Harbor, therefore, the U.S. Navy had already adopted a plan that called for first winning the war in the European theater, while fighting a delaying action against the Japanese.[16]

Plan DOG's influence on the decision to conduct unrestricted submarine warfare came in its conclusion that to fight a delaying action against the Japanese the U.S. Navy would have to wage an economic war of attrition. In fact, the American war planners did not believe that the United States would be able to defeat Japan totally in a two-ocean war, so "it should therefore settle upon a war having a more limited objective than the complete defeat of Japan. The objective in such a limited war against Japan would be the reduction of Japanese offensive power chiefly through economic blockade."[17]

In a sense, this economic blockade was no different from what War Plan ORANGE already projected. But ORANGE presumed that sufficient naval forces would be available to destroy the Japanese fleet and gain control of the sea to enforce such a blockade. By late 1940, however, the United States simply did not have the overwhelming naval forces necessary to conduct the blockade operation envisioned by the plan.

Without American control of the western Pacific, there was only one force that could still attack Japan's economy—U.S. submarines. Maintaining their stealth and chances of survivability, however, meant abandoning cruiser warfare. Consequently, the campaign of economic strangulation against Japan to be waged before the United States could gain control of the sea implicitly required unrestricted submarine warfare.

Putting the Rainbow Plans into Effect

Admiral Stark and his chief war planner, Rear Adm. Richmond Kelly Turner, subsequently used Plan DOG as the template to write the navy's "Rainbow" war plans, which were drafted to defeat the Axis powers. In particular, the new RAINBOW 3 plan authorized "Strategical Areas . . . from which it is necessary to exclude merchant ships and merchant aircraft to prevent damage to such ships or aircraft, or to prevent such ships or aircraft from obtaining information, which, if transmitted to the enemy, would be detrimental to our own forces."[18]

As soon as RAINBOW 3 was complete, Admiral Stark sent advance copies to his fleet commanders, including the commander in chief of the Asiatic Fleet, Adm. Thomas C. Hart. Admiral Hart immediately noticed the section regarding "strategical areas" in RAINBOW 3, and he queried Stark about how much freedom he had regarding merchant shipping: "The possibilities in raids on Japanese sea communications,—meaning shipping other than naval forces,—would be great if our submarines were free to wage 'unrestricted' war." However, Hart quickly added, "Unless we are otherwise ordered, our submarines will not be directed to depart from the War Instructions now in force."[19]

Admiral Stark responded, "The term 'sea communications' includes all naval as well as merchant shipping. Raids on military and naval supply ships should prove very profitable. The question of inability to sink merchant shipping by submarines, without warning, is unlikely to arise, since it is probable that all shipping within your reach will be under Japanese naval operation or control. . . . The employment of submarines as proposed is considered suitable and highly desirable."[20]

Stark's response highlighted his assumptions about an unrestricted submarine war in the Pacific would not produce the sort of backlash that the German unrestricted submarine war had caused in World War I. First of all, Stark assessed that there would be little or no neutral shipping for U.S. submarines to sink. In his opinion, the remaining merchant ships, all Japanese flagged, would be, although perhaps nominally in the employ of Japan's civilian merchant marine, actually under the control of the Japanese military. As such, they would be legitimate military targets, not civilian targets.

The strategic reasoning of Plan DOG evolved over 1941, particularly after the U.S. Naval War College, in Newport, Rhode Island, strongly recommended against a formal blockade in the event of a war with Japan and instead recommended the creation of "war zones." Before the Naval War College's recommendations, whenever Admiral Stark had discussed "strategical areas" he had ostensibly been describing "strategic war zones" for the defense of the fleet. From this point on, "strategical areas" would come to mean war zones for unrestricted air and submarine warfare. This became clear in subsequent correspondence by Stark and his chief war planner, Rear Admiral Turner, who wrote that strategical areas, if established, would actually be for unrestricted warfare and that the United States would establish them "immediately upon the outbreak of war."[21]

By the end of November 1941 that moment appeared to have arrived. Negotiations between the United States and Japan had reached a critical impasse. Beginning on 27 November Admiral Stark issued a number of dispatches to prepare the Pacific Fleet and the Asiatic Fleet for war. Among his first dispatches was a message, lengthy but direct and to the point, to Admiral Hart: "If formal war eventuates between US and Japan . . . instructions for the Navy of the United States governing maritime and aerial warfare May 1941 . . . will be placed in effect but will be supplemented by additional instructions including authority to CINCAF [Commander in Chief, Asiatic Fleet] to conduct unrestricted submarine and aerial warfare."[22]

Two weeks later, between 0753 and 0755 Hawaii time on 7 December 1941, the first Japanese bombs fell on Oahu, bringing the stunned U.S. Pacific Fleet and U.S. Army units straight into battle. After informing President Roosevelt of the attack, Admiral Stark issued his orders to conduct unrestricted submarine warfare at 1752 Washington time, which was 1222 at Pearl Harbor: "Execute against Japan unrestricted air and submarine warfare."[23]

Overcoming Prewar Mistakes

The smoke had not even begun to clear over Pearl Harbor when Admiral Stark issued his orders to destroy all Japanese shipping. These orders, however, turned out to be easier to transmit than to execute. For almost two years, numerous self-imposed problems hamstrung the U.S. submarine force.

The most vexing and complicated problem facing the U.S. submarine force turned out to be the submariners' own torpedoes. Just before the war began, the navy's Bureau of Ordnance revealed its top secret warhead for the Mark XIV steam-driven torpedo, with the Mark VI magnetic exploder. By all accounts, the weapon was truly remarkable. The warhead sensed the magnetic field around an enemy ship and was designed to detonate at the point of maximum magnetism, directly underneath the target. The resulting detonation of over six hundred pounds of explosive would snap the target's keel like a toothpick.[24]

But unbeknownst to the submariners, the Bureau of Ordnance, as both a cost-saving measure and a misguided effort to maintain secrecy, never live-tested the Mark VI warhead. Instead, the bureau presented the untested warhead to the U.S. submarine force and claimed that torpedoes carrying it would need only one shot against a target. As it turned out, submariners could fire six shots directly at a target, and the torpedoes still would not work. Instead, torpedoes ran too deep, exploded prematurely, or reached the target but did not explode. Consequently, American submariners would pursue daring attacks only to see their torpedo wakes bubble under a target or prematurely detonate, giving away their own position.[25]

If the failure of the American torpedoes was not extraordinary enough, even more incredible was the Bureau of Ordnance's reaction to criticism, steadfastly insisting that the problem was not the Mark VI exploder but the aim of American submariners. Eventually, submarine force leaders carried out their own tests using fishing nets, underwater cliffs, and even cherry pickers, the latter used to drop torpedo warheads onto the ground to see why they failed. At the forefront was Vice Adm. Charles A. Lockwood, who would become the commander of the Pacific Fleet Submarine Force from February 1943 until the end of the war. Throughout the war, Lockwood and his staff persistently pursued the torpedo problem. They discovered that the depth-excursion defect was due to a combination of the weight of the Mark VI warhead and a poorly designed depth-sensing mechanism. Lockwood ultimately pulled the problematic magnetic exploder out of service after determining it was too complex, eventually determining that the contact exploder was improperly constructed as well. Lockwood and his staff finally fixed the torpedoes, but it was a painfully prolonged process. American submariners could not put to sea knowing that their torpedoes would actually work until October 1943, over twenty-one months after the start of hostilities.[26] Even so, a few torpedo malfunctions continued to plague the American submarine force for the rest of the war, including circular runs that could have been responsible for the sinking of as many as eight U.S. submarines with all hands.[27]

The torpedo issue was the most serious problem facing the submarine force, but it was hardly the only one. Timid commanding officers and unrealistic tactics developed in the interwar period constrained the submarine force just as much as did the torpedoes. When the test of war came, neither the tactics nor the commanders shaped up. Some U.S. submarine commanding officers simply could not handle the stress of combat. Others were relieved out of hand for lack of aggressiveness. Simultaneously, American submariners were forced to reinvent their tactics and learn how to fight while in combat, an unenviable task for any combatant.[28]

Younger and more aggressive American submarine commanders eventually proved equal to the task. Without a doubt, the one submarine commander who most instilled aggressiveness and tenacity into the force was Lt. Cdr. Dudley W. "Mush" Morton, commanding officer of USS *Wahoo*. Starting in January 1943, Morton's ferocity transformed U.S. submarine warfare. He audaciously took *Wahoo* into a Japanese-controlled harbor in Wewak, New Guinea, using only an enlarged almanac map as his chart. Although he was in water that was in depth less than a third of *Wahoo*'s length, Morton attacked a Japanese destroyer, sinking it at point-blank range. Later during the same patrol he daringly attacked and completely destroyed an entire convoy (earning *Wahoo*, when it reentered port, a broomstick on its periscope for a "clean sweep"). Out of torpedoes and finding another convoy, Morton once more attempted to strike, using only his small deck gun. His plan derailed when the escorting destroyer discovered *Wahoo* and shelled it; *Wahoo* barely escaped.[29] This sort of tenacity and determination inspired the entire force. After the war, Cdr. (and author) Edward L. Beach praised Morton, who "more than any other man . . . showed the way to the brethren of the Silent Service."[30]

Men like Morton energized the submarine force, but new and reliable equipment was necessary as well. In addition to the improved Mark XIV, new types of torpedoes appeared, including the wakeless Mark XVIII electric torpedo and the acoustic Mark XXVII.[31] Also, in the last years of the war the Americans gained even greater technological edge over the Japanese with the new SJ radar and its plan position indicator, the improved target bearing transmitter, and a bathythermograph to find thermal layers, which allowed U.S. submarines to evade Japanese sonar.[32]

The Momentum Shifts

As a result of the myriad of equipment and leadership problems plaguing the U.S. submarine force, American submariners did not get much chance to shine during the first year of the war. By the end of 1942 they had only sunk 180 ships in exchange for seven American submarines. It was a start, but since the total number of Japanese ships sunk by all American submarines equaled the number of Allied ships sunk by German U-boats in only two months of 1942, it was disappointing.[33]

But even this small beginning was enough, because Japanese military leaders undervalued the protection of their merchant marine. At the beginning of the war, Japan had only about six million tons of merchant shipping, and of that, only 525,000 tons of tankers. Even though Japan went to war over raw materials in Southeast Asia, the Japanese military command saw no inherent contradiction in requisitioning almost two-thirds of Japan's merchant marine solely for military transportation and supplies. Thus, even as the war began, Japanese military leaders had already drastically cut the vital importation of raw materials to supply the Japanese war machine and economy. Moreover,

Japan's leaders spared little thought to building up the merchant marine. Further, the ships that were afloat were used so inefficiently that they might as well have been on the bottom—empty merchant ships passed other empty merchant ships leaving ports to which they were themselves heading. If that was not enough, the Japanese navy essentially chose to ignore commerce protection, disregarding the lessons of the First World War and interwar Japanese submarine exercises. Consequently, despite the numerous troubles plaguing the U.S. submarine force, the amount of Japanese tonnage sunk in 1942 exceeded the amount Japan constructed.[34]

In 1943, the momentum began to shift even more to the U.S. submarine force, thanks to the aggressiveness of commanders like Mush Morton and the correction of the numerous torpedo problems. By the end of 1943, 335 Japanese ships had been sunk in exchange for fifteen submarines. But the Japanese had focused on one important slice of their tonnage in which U.S. submarines had not made enough of a dent—oil tonnage. The Japanese started off the war with few tankers, but Japan's shipbuilding industry quickly ramped up to supply more. Despite the rising success of U.S. submarines, the Japanese replaced their tanker losses in both 1942 and 1943.[35] The torpedo problem explains in part why Japan's tanker fleet seemed to have remained so far relatively unscathed—tankers were hard targets to sink, and even being holed by an unexploded torpedo was no great emergency. Indeed, Japanese merchant mariners claimed that "a tanker would not sink if torpedoed."[36] If the Japanese believed that their momentary success with tankers was decisive, however, they were completely mistaken. Japan still lost twice as much shipping as it constructed in 1943.[37]

As 1944 began, Japan's leaders finally began to awaken to the mortal danger they had faced since the beginning of the war. Ironically, Japan's awakening was probably slowed by the miserable performance of American torpedoes, which lulled Japanese naval leaders into a false sense of security arising from the apparent impotence of U.S. submarines. Toward the end of 1943, however, Japanese naval leaders suddenly "realized that some innovation had come to the American torpedoes . . . [and the] sinking rate of our torpedoed ships suddenly began to increase." The date that the Japanese sensed that the Americans had solved their torpedo problems was reportedly 20 August 1943, about a month before the Americans officially considered that to be the case.[38]

The Japanese finally began systematic convoying in March 1944 and attempted to establish and equip an effective antisubmarine force, but it was too little too late. Even if the resources had been present to create such a force, the rest of the Japanese military would have greedily seized them, as happened to the few air components of the Japanese antisubmarine effort. Consequently, 1944 turned out to be the halcyon year of the U.S. submarine force. Finally equipped with reliable torpedoes and equipment and manned by experienced crews, it chewed into the Japanese. American submariners sank 603 ships in 1944, at the cost of only nineteen U.S. submarines. Importantly, they annihilated the Japanese tanker fleet, quadrupling the number of tankers sunk. By the beginning of 1945 virtually no oil from the oil fields of Southeast Asia, for which Japan had gone to war, was reaching the home islands.[39]

As 1945 went on, American submarines found fewer and fewer targets to sink. In a quest for what little remained of Japanese shipping, Admiral Lockwood approved Operation BARNEY, the invasion of the mined Sea of Japan by submarines specially equipped with antimine sonar. But even that once-protected haven offered little shipping to sink. By the end of the war Japan had only 700,000 tons of "serviceable" merchant tonnage remaining.[40]

A Decisive Factor in Victory

The U.S. submarine force carried out its mission to strangle Japan with devastating efficiency. By the end of the war American submarines had sunk 1,113 Japanese merchant ships and 201 warships. That came out to 4,779,902 tons of enemy commerce and 540,192 tons of naval warships. The commerce figures were particularly impressive, since Japan had started the war with only 6,337,000 tons of commercial shipping. In terms of casualties, the Japanese lost virtually their entire prewar merchant marine; out of 122,000 sailors, 27,000 were killed and 89,000 were wounded or "otherwise incapacitated."[41]

But the true effectiveness of the U.S. submarine *guerre de course* did not lie at sea. Commerce raiding severely affected the Japanese military throughout the Pacific, as well as the population on the home islands. In particular, the U.S. unrestricted campaign dramatically reduced the nutritional intake of most Japanese soldiers and civilians. Instead of combat, it was starvation—as well as related illnesses, such as beriberi—that ended up killing many Japanese soldiers overseas.[42] On the home islands, the Japanese population felt the pangs of hunger from a very early stage of the war. Even before U.S. bombers destroyed Japanese industrial centers, a large percentage of the Japanese workforce suffered from malnutrition and related illnesses. By the end of the war the food situation was so bad that authorities in Osaka recommended that civilians add such items as acorns, rose leaves, silkworm cocoons, grasshoppers, and sawdust to their diet. Even after the surrender, as many as six people a day died from starvation in just one center for the homeless in Tokyo. In October 1945 the Japanese minister of finance told the United Press that as many as ten million people would starve to death without immediate American food aid. Although this number was perhaps exaggerated, it reflected the desperate situation facing Japan. The exact toll on the Japanese military and population due to starvation and privation during and immediately after the war may never be fully known, but the number is probably staggering.[43]

How important was the U.S. submarine contribution to the Allied victory against Japan? By some accounts, American submarines single-handedly broke the Japanese war machine. For instance, submarine historian Clay Blair claims, "Many experts concluded that the invasions of the Palaus, the Philippines, Iwo Jima, and Okinawa, and the dropping of fire bombs and atomic bombs on Japanese cities were unnecessary. They reasoned that despite the fanatical desire of some Japanese to hang on and fight to the last man, the submarine blockade alone would have ultimately defeated that suicidal impulse."[44] Other historians have argued vigorously against this extreme conclusion. D. M. Giangreco points out that while many ordinary Japanese citizens would have starved by 1946, military leaders had hoarded food collected from farming areas.[45] These leaders and die-hard military units would have hardly caved in under the pressure of the submarine attacks, particularly given how vigorously and fanatically many of Japan's military leaders refused to consider surrender even after the atomic bombings. On the basis of this premise, it seems safe to say that the U.S. submarine force alone did not achieve Japan's unconditional surrender.

But few people, on either side of the war, would dispute that U.S. submarines were devastatingly effective. Mark Parillo, the foremost American expert on the Japanese merchant marine in the Second World War, writes, "The submarine had stopped Japan's industrial heart from beating by severing its arteries, and it did so well before the bomber ruptured the organ itself."[46] After the emperor decided to surrender, the Japanese cabinet reported to the Diet that "the greatest cause of defeat was the loss of shipping," a remarkable admission given the Japanese navy's earlier extraordinary nonchalance toward antisubmarine warfare.[47]

Conclusions

It is undeniable that unrestricted submarine warfare played an essential role in defeating Japan. Indeed, unrestricted submarine warfare's impact went far beyond the economic holding action envisioned by Plan DOG, to contribute significantly to the overall ORANGE strategy to advance across the Pacific and encircle Japan. But it is important to note that the United States did not rely solely on the unrestricted submarine campaign to achieve its strategic goals. Rather, the unrestricted submarine war was just one part of a much larger and cohesive strategy that overwhelmed Japan's defenses. As Edward S. Miller concludes, "The old concept of blockade by surface vessels could not have been made effective until late in the war. The decision for undersea predation magnified the success of one of the ORANGE Plan's most basic prescriptions."[48]

The American victory is even more remarkable given the small size of the U.S. submarine force. Including all rear-echelon personnel, the submarine force amounted to only fifty thousand officers and men, about 1.6 percent of the entire U.S. Navy's personnel.

Of those fifty thousand, only sixteen thousand actually went to sea. Of those submariners, 3,500 never returned, amounting to a 22 percent casualty rate, the highest of any combat branch in the U.S. armed forces during the Second World War.

Yet despite the high casualty rate and extremely low number of personnel serving in the U.S. submarine force, American submarines sank 55 percent of all Japanese ships destroyed in World War II.[49] In terms of sheer magnitude and cost-effectiveness, it is hard to argue with the conclusions of Japanese naval historian Masanori Ito, who writes, "U.S. submarines . . . proved to be the most potent weapon . . . in the Pacific War."[50]

Notes

The thoughts and opinions expressed in this essay are those of the author and are not necessarily those of the U.S. government, the Navy Department, the Naval War College, or the submarine force. Segments of this essay are adapted from the author's book *"Execute against Japan": The U.S. Decision to Conduct Unrestricted Submarine Warfare* (College Station: Texas A&M Univ. Press, 2009).

1. Edward S. Miller, *War Plan Orange: The U.S. Strategy to Defeat Japan, 1897–1945* (Annapolis, Md.: Naval Institute Press, 1991), pp. 21–38, 363–69.

2. Mark P. Parillo, *The Japanese Merchant Marine in World War II* (Annapolis, Md.: Naval Institute Press, 1993), pp. 32–35; Clay Blair, Jr., *Silent Victory: The U.S. Submarine War against Japan* (Philadelphia: J. B. Lippincott, 1975), p. 17; John W. Dower, *Embracing Defeat: Japan in the Wake of World War II* (New York: W. W. Norton / New Press, 1999), p. 91. Before Pearl Harbor, Japan imported 31 percent of its rice, 92 percent of its sugar, 58 percent of its soybeans, and 45 percent of its salt from Korea, Formosa, and China.

3. Miller, *War Plan Orange*, pp. 152, 295–97, 319–20.

4. Adm. J. M. Reeves, Commander in Chief, United States Fleet, to the Chief of Naval Operations, Subject: Employment of BLUE Submarines—Orange War, 20 January 1936, Enclosure: "A Study on the Initial Employment of the BLUE Submarine Force in an ORANGE War," December 1935, series 3, pp. 1–2 (microfilm), Strategic Plans Division Records, Records of the Office of the Chief of Naval Operations, Washington, D.C. [hereafter CNO Records], Record Group [hereafter RG] 38, National Archives at College Park, College Park, Md. [hereafter NARA].

5. "Solution of a Situation, issued 3 December 1938: Solution of a Situation II—Inter. Law, 3 Dec., 1938," p. 11, box 86, nos. 2197-IL2 through 2201-IL2, Publications, RG 4, Archival Records, Naval Historical Collection, Naval War College, Newport, R.I.; see also Emily O. Goldman, *Sunken Treaties: Naval Arms Control between the Wars* (University Park: Pennsylvania State Univ. Press, 1994), p. 317. William Phillips, Acting Secretary of State, to President Franklin D. Roosevelt, 20 March 1936, pp. 2–3, Dept. of State January–August 1936, box 4, OF 20, Department of State, Official File, Franklin D. Roosevelt Presidential Library, Hyde Park, N.Y. A printed version of this letter, with other correspondence related to this topic, is in *Foreign Relations of the United States: Diplomatic Papers, 1936*, vol. 1, *General, The British Commonwealth* (Washington, D.C.: U.S. Government Printing Office [hereafter GPO], 1953), pp. 160–64.

6. John Alden, *The Fleet Submarine in the U.S. Navy: A Design and Construction History* (Annapolis, Md.: Naval Institute Press, 1979), p. 10.

7. Gary E. Weir, *Building American Submarines, 1914–1940*, Contributions to Naval History, no. 3 (Washington, D.C.: Naval Historical Center, 1991), pp. 40–43.

8. Rear Adm. C. S. Freeman, Commander Submarine Force, to the Chief of Naval Operations, Subject: Submarines—Employment of in a Pacific War, 27 July 1938, p. 1, 420-15 1938, box 112, Subject File 420-15, General

Board, Subject File 1900–1947, General Records of the Department of the Navy, RG 80, National Archives Building, Washington, D.C.

9. "Current Doctrine, Submarines, 1939, U.S.F. 25, Revised, Prepared by Commander Submarine Force, April 1939," p. 7, NRS 1977–86, Current Tactical Orders and Doctrine—Submarines, 1939–1944, Microfilm Reel, Operational Archives, Naval History and Heritage Command, Washington, D.C. [hereafter NHHC].

10. Ibid., p. 11.

11. Ibid.

12. Ibid.

13. Randy Papadopoulos, "Between Fleet Scouts and Commerce Raiders: Submarine Warfare Theories and Doctrines in the German and U.S. Navies, 1935–1945," *Undersea Warfare* 7, no. 4 (Summer 2005), p. 31.

14. "Submarine Operational History World War II, Prepared by Commander Submarine Force, U.S. Pacific Fleet, Volume 1 of 4," p. 2, Submarines Pacific Fleet, Operational History, vol. 1 of 4, box 357, Submarines, Pacific Flt. History—Bulletins, Submarine vol. 2, 1945, Type Commands, World War II Command File, Operational Archives Branch, NHHC.

15. Joint Planning Committee to the Joint Board, Subject: National Defense Policy for the United States, 21 December 1940, cover letter, p. 16, Plan DOG, box 85, Series XIII: Pearl Harbor Investigations, Papers of Admiral H. R. Stark, Operational Archives Branch, NHHC; Miller, *War Plan Orange*, pp. 269–70.

16. Joint Planning Committee to the Joint Board, p. 15.

17. Ibid., p. 8.

18. "W.P.L.-44, Naval Basic War Plan—Rainbow No. 3, United States Navy, December 1940," p. 42, box 33, WPL-44, W.P.L. series, War Plans Division, Strategic Plans Division Records, CNO Records, RG 38, NARA.

19. Adm. Thomas C. Hart to Adm. Harold R. Stark, memo re: Substance of "Rainbow 3," 18 January 1941, p. 3, War Plans, box 86, Series XIII: Pearl Harbor Investigations, Papers of Admiral H. R. Stark, Operational Archives Branch, NHHC.

20. Adm. Harold R. Stark, Chief of Naval Operations, to the Commander in Chief, U.S. Asiatic Fleet, Subject: Instructions Concerning the Preparation of the U.S. Asiatic Fleet for War under "RAINBOW No. 3," 7 February 1941, pp. 3–4, War Plans, box 86, Series XIII: Pearl Harbor Investigations, Papers of Admiral H. R. Stark, Operational Archives Branch, NHHC.

21. Rear Adm. Richmond K. Turner, memo for War Plans Files, 29 September 1941, p. 3, WPL-46 Letters (1939–1945), box 147J, WPL-46 through WPL-46-PC, Part III: OP-12B War Plans and Related Correspondence, Plans, Strategic Studies, and Related Correspondence (Series IX), Strategic Plans Division Records, CNO Records, RG 38, NARA; Chief of Naval Operations to the Commander in Chief, U.S. Pacific Fleet, and the Commander in Chief, U.S. Asiatic Fleet, Subject: Action by Submarines against Merchant Raiders, 21 October 1941, WPL-46 Letters (1939–1945), box 147J, WPL-46 through WPL-46-PC, Part III: OP-12B War Plans and Related Correspondence, Plans, Strategic Studies, and Related Correspondence (Series IX), Strategic Plans Division Records, CNO Records, RG 38, NARA.

22. OpNav to CinCAF, 271422, 28 November 1941, box 4, Decodes of Confidential and Secret Dispatches, Sept. 1941–Apr. 1942, Records Relating to the Asiatic Fleet and the Asiatic Defense Campaign 1933–1942, Naval Historical Center, CNO Records, RG 38, National Archives Building, Washington, D.C. The date-time group of the message is "271422," which indicates the message was probably sent at 0352 Pearl Harbor time, 27 November/0922 Washington time, 27 November/1422 Zulu time, 27 November/2222 Manila time, 27 November. This presumes that messages from Washington carried a Zulu date-time group.

23. CNO to CinCPac, Com Panam, CinCAF, Pacific Northern, Pacific Southern, Hawaiian Naval Coastal Frontiers, 072252, 7 December 1941 (available on microfilm), Operation Orders, 7 December 1941–2 April 1942, Operations Orders, box 37, reel 2, Military Files series 1, Map Room Army and Navy Messages, December 1941–May 1942, Map Room Files of President Roosevelt, 1939–1945 [Franklin D. Roosevelt Library, Hyde Park, N.Y.].

24. Blair, *Silent Victory*, pp. 20, 84; Vice Adm. Charles A. Lockwood, *Sink 'Em All: Submarine Warfare in the Pacific* (New York: E. P. Dutton, 1951), p. 76.

25. Edward L. Beach, *Salt and Steel* (Annapolis, Md.: Naval Institute Press, 1999), pp. 126–27; Blair, *Silent Victory*, pp. 20, 61–62, 84; Lockwood, *Sink 'Em All*, pp. 21, 75.

26. The story of American torpedo problems has been told in many studies. The most comprehensive discussion is now found in Anthony Newpower, *Iron Men and Tin Fish: The Race to Build a Better Torpedo during World War II* (Westport, Conn.: Praeger Security International, 2006), pp. 22–32, 59–73, 87–111, 131–96; and in Blair, *Silent Victory*, pp. 273–78, 429–31, 435–39. For Admiral Lockwood's discussion of his steady and prolonged fight to fix the faulty Mark XIV torpedo, see his *Sink 'Em All*, pp. 20–22, 75–76, 85–86, 88–89, 103–104, 111–14.

27. Beach, *Salt and Steel*, p. 127.

28. Blair, *Silent Victory*, pp. 18–19.

29. Although the legendary story of *Wahoo*'s third patrol has been told in virtually all books about the U.S. submarine force, the most complete version remains Dick O'Kane's first-person account: Rear Adm. Richard H. O'Kane, *Wahoo! The Patrols of America's Most Famous World War II Submarine* (Novato, Calif.: Presidio, 1987), pp. 109–72.

30. Edward L. Beach, *Submarine!* (Annapolis, Md.: Naval Institute Press, 1952), p. 36.

31. Keith Wheeler, *War under the Pacific* (Alexandria, Va.: Time-Life Books, 1980), p. 167.

32. Ibid., p. 70.

33. Blair, *Silent Victory*, pp. 359–60.

34. Parillo, *Japanese Merchant Marine in World War II*, pp. 6–31, 63–83, 94–124; John Ellis, *Brute Force: Allied Strategy and Tactics in the Second World War* (New York: Viking, 1990), pp. 468–76; David C. Evans and Mark R. Peattie, *Kaigun: Strategy, Tactics, and Technology in the Imperial Japanese Navy, 1887–1941* (Annapolis, Md.: Naval Institute Press, 1997), pp. 430–31.

35. Blair, *Silent Victory*, pp. 551–52; Ellis, *Brute Force*, pp. 470–71.

36. Oi Atsushi, "Why Japan's Antisubmarine Warfare Failed," in *The Japanese Navy in World War II: In the Words of Former Japanese Naval Officers*, trans. and ed. David C. Evans, 2nd ed. (Annapolis, Md.: Naval Institute Press, 1986), p. 397.

37. Ellis, *Brute Force*, pp. 470–71.

38. Oi, "Why Japan's Antisubmarine Warfare Failed," pp. 397–98.

39. Blair, *Silent Victory*, pp. 816–18; Ellis, *Brute Force*, pp. 470–71, 474–75; Oi, "Why Japan's Antisubmarine Warfare Failed," pp. 401–14; Parillo, *Japanese Merchant Marine in World War II*, pp. 6–31, 63–73, 94–124.

40. Blair, *Silent Victory*, pp. 844, 857–65; Ellis, *Brute Force*, pp. 471–74.

41. Of this number, Blair credits the U.S. submarine force with 16,200 deaths and 53,400 casualties. See Blair, *Silent Victory*, pp. 877–79, 900; Oi, "Why Japan's Antisubmarine Warfare Failed," p. 392; Wheeler, *War under the Pacific*, pp. 21–23; and Parillo, *Japanese Merchant Marine in World War II*, pp. 207, 243–44.

42. Dower, *Embracing Defeat*, p. 91; United States Strategic Bombing Survey, *Summary Report (Pacific War)* (Washington, D.C.: GPO, 1946), pp. 20–21; Parillo, *Japanese Merchant Marine in World War II*, pp. 213–15.

43. Dower, *Embracing Defeat*, pp. 90–93; Parillo, *Japanese Merchant Marine in World War II*, pp. 215–21.

44. Blair, *Silent Victory*, pp. 17–18.

45. D. M. Giangreco, *Hell to Pay: Operation Downfall and the Invasion of Japan, 1945–1947* (Annapolis, Md.: Naval Institute Press, 2009), pp. 117–18.

46. Parillo, *Japanese Merchant Marine in World War II*, p. 225.

47. Oi, "Why Japan's Antisubmarine Warfare Failed," p. 414.

48. Miller, *War Plan Orange*, p. 352.

49. Blair, *Silent Victory*, pp. 877–79; Wheeler, *War under the Pacific*, pp. 21–23.

50. Masanori Ito, *The End of the Imperial Japanese Navy* (New York: Jove Books, 1956), p. 17.

Guerre de Course in the Charter Era
The Tanker War, 1980–1988
GEORGE K. WALKER

The Tanker War in many ways typified post–World War II conflicts. This war was regional in geographic scope and so did not have a global military impact. But for the antagonists, Iran and Iraq, it was a major war. By war's end they had spent the equivalent of their combined oil revenues since World War II. For Iraq, the Tanker War led to the 1990–91 first Gulf War, to the 2003 second Gulf War, and to a nine-year postwar U.S. occupation.

These "small wars" were not small in the belligerents' perspectives, and they raised issues of merchant ship interdiction similar to those in World Wars I and II. There were new variables, however, including more modern technologies and changes in oceangoing shipping. After 1945 all ships had radar and radio communications, and Internet usage later became increasingly common. Vessels also became larger, with smaller crews, thanks to greater automation. Increasingly warships used missiles instead of traditional powder-based guns. Men-of-war became larger in size, smaller in number, extraordinarily expensive, and more "fragile," in the sense that an unerring missile or torpedo could send one to the bottom, instead of a possibility of slight damage from gunfire or a bomb. Sea mines had always been able to sink a ship, and newer mines—such as the Mark 60 Encapsulated Torpedo, or CAPTOR—were more sophisticated.

Following Iraq's invasion of Iran on 22 September 1980, each state tried to interfere with the other's international trade, especially in petroleum, in order to hamstring its military procurement programs and undermine its economy. This chapter will examine the 1980–88 Tanker War, in particular with regard to how new technologies and ship practices profoundly affected the way countries could conduct commerce-raiding operations.

The Origins of the Tanker War, 1980

Iraq invaded Iran in September 1980, claiming self-defense. Soon afterward an Iranian notice to mariners (NOTMAR) declared the waterways near its coast a war zone; established new shipping lanes for vessels passing the Strait of Hormuz, at the Persian Gulf's southern end; abjured responsibility if a ship did not follow those lanes; refused access to Iraqi ports, including those in the Shatt al-Arab waterway; and warned of retaliation if Gulf states gave Iraq port facilities.[1] Iran later called the access refusal a "blockade" of the Iraqi coast. There were Iranian attacks on shipping in the Shatt in the war's early days, and some of those hulks still remain in the waterway, which divides Iran and Iraq. Seventy neutral-flagged ships were trapped in the Shatt and, despite UN good offices in seeking a cease-fire to allow them to leave under a United Nations or Red Cross flag, Iraq refused, citing "full" sovereignty over the Shatt. Iran accepted the UN proposal, but most of these ships remained trapped until war's end.

The European Community (EC) quickly endorsed an Arab League cease-fire appeal calling for freedom of navigation in the Gulf. Meanwhile, the UN Security Council called for an end to hostilities, with Iraq this time accepting the resolution but Iran calling first for an end to Iraqi aggression.[2] Japan and the United States stressed the importance of freedom of navigation of the Gulf, and the United States furthermore pledged neutrality, while emphasizing maintaining unhindered passage through the strait as a matter of national policy. The Soviet Union (USSR) also declared neutrality.

Iran confirmed its commitment to freedom of the seas by keeping the Strait of Hormuz open for navigation and said it had never extended its war zone to the strait. But in October 1980, Iraq declared all of the Gulf north of 29° 30' north latitude a prohibited war zone. Iran began shuttling merchant convoys under naval protection down its coast through its Gulf maritime exclusion zone (GMEZ) to the lower Gulf. In response, Iraq began using pipelines to export oil and imported war-sustaining goods through nearby third-state Gulf ports.

In November 1980, Iran's NOTMARs directed ships entering or leaving Iranian ports to get Gulf travel coordinates from its navy and to inform the relevant Iranian port of their hourly positions. Inbound ships had to give their estimated times of arrival at Bandar Abbas and be cleared. If not cleared, they had to anchor at Bandar Abbas. Early in 1981 another Iranian NOTMAR further directed very large crude carriers or ultra-large crude carriers not inbound for Iranian ports but intending to cross the Iranian restricted zone to contact Iran's naval headquarters and provide travel information forty-eight hours before departure.

The belligerents did not declare contraband lists, but because both governments attacked neutral-flagged oil carriers both loaded and in ballast, it is clear that they regarded oil as contraband. Whether considered to be absolute or conditional contraband, foreign-made armaments, which were paid for by either selling or bartering oil, were indispensable to the war efforts. Neither state established prize courts until the very end of the war, when Iran finally published its rules. These still excluded detailed contraband lists, which did not satisfy the international community.

The International Reaction to the Tanker War

From almost the very beginning of the conflict, the international response to the Tanker War was both quick and highly critical. Britain rapidly established the Armilla Patrol, designed to escort and protect its shipping in the Gulf. By mid-October 1980 at least sixty Australian, French, British, and American warships were in the Indian Ocean to protect this oil route, and twenty-nine Soviet vessels were also present. In 1981, the Islamic Conference Organization offered a peace plan, but it was rejected. UN mediation also failed. The Gulf Cooperation Council (GCC), comprising all Gulf states except the belligerents, was established.

In May 1981 tensions increased after Iran seized a Kuwaiti survey ship and the Danish-flagged *Elsa Cat,* bound for the United Arab Emirates (UAE) and Kuwait and carrying military equipment destined for Iraq. Baghdad protested the seizures, and both ships were released. Beginning in 1981 and continuing through 1984, however, Iraq frequently attacked commercial shipping in the northern Gulf, usually tankers and cargo ships calling at Bandar Khomeni or Bushire after being convoyed through Iranian territorial waters. In 1982 it was reported that Iraq had mined the Bandar Khomeni–Bandar Mashahr channel to the open sea.

During 1982 Iraq tried to invoke the Arab League mutual defense treaty, analogous to the NATO agreement, to receive military aid from league members. But Syria warned that if Egypt, a league member, lined up with Iraq, Syria would join with Iran. The result was a political deadlock. By late 1982 all the Gulf states except Kuwait and Saudi Arabia, which favored Iraq, had declared neutrality. In August 1982 Iraq proclaimed its own GMEZ, announcing it would attack any ship within the zone and that tankers of any nationality docking at Iran's main export terminal, at Kharg Island, would be legitimate targets. In announcing its GMEZ and the Kharg "blockade," Iraq stressed that its war zones were designed to cope with the difficulty of distinguishing among vessel nationalities in the Gulf. Later that month, Iran declared it would protect foreign shipping, began escorting foreign vessels, and deployed ships with surface-to-air missiles to Kharg. Iran also began providing naval protection to convoys of Iran-flagged and neutral merchantmen transporting oil from Iranian northern Gulf ports to others farther down its coast for world export.

Iraq modified its GMEZ in November 1982 by advising companies and tanker owners that their ships would be subject to attack upon entering the zone. In general, however, Iraq attacked virtually all ships in the GMEZ through March 1984—the only aspect of the war where the initiative lay with Iraq. The London-based War Risks Rating Committee raised marine cargo insurance rates in 1982 and again in 1984 because of Iraqi attacks on Gulf shipping. Also, the United States redefined its freedom of navigation policy in 1982, making nonbelligerent access to the Gulf a top priority. These third-party decisions undoubtedly influenced local governments to consider taking action to maintain freedom of navigation.

An October 1983 UN Security Council Resolution again called for a cease-fire, condemning violations of international humanitarian law, in particular the 1949 Geneva Conventions. The resolution affirmed the right of free navigation and commerce in international waters, called on states to respect this right, and urged the belligerents to cease hostilities in the Gulf region, including in sea-lanes, navigable waterways, harbors, terminals, offshore installations, and ports with sea access. The GCC endorsed the resolution and went on record as supporting Gulf freedom of navigation. As a result, the French, British, Soviet, and American naval presence in the Indian Ocean adjacent to the Gulf continued unchanged.

A Turning Point in the War

The year 1984 was a Tanker War turning point. A January U.S. notice to airmen (NOTAM) and NOTMAR proclaimed a "cordon sanitaire" around American warships and aircraft, warning of possible defensive action if a ship or aircraft ventured inside the zone.[3] Iran protested this decision and opposed transits by U.S. Navy ships of its territorial sea, but the American response was that these measures had been adopted only in self-defense. The United Kingdom decided not to declare a security envelope around its Armilla Patrol.

In February 1984, Iraq extended its GMEZ to fifty miles around Kharg, warning that ships approaching Bandar Khomeni or Bushire would be sunk. The United Kingdom protested an Iraqi attack on a convoyed cargo ship in the Khomeni approaches; Indian and Turkish vessels were also attacked. Tankers were hit in Iraqi attacks on Kharg. Iraq also destroyed Saudi tankers steaming outside its GMEZ. Iraqi forces appeared to have devoted only minimal effort to obtaining visual identification before launching missiles.

In response, Iran attacked Kuwaiti and Saudi tankers for the first time in April and May 1984. Iran mainly used rockets, not missiles, and seemed to do a better job than Iraq of identifying targets. An Arab League summit in May condemned the attacks on Kuwaiti and Saudi ships. The GCC complained to the UN Security Council. Many states, including open-registry countries, raised freedom-of-navigation concerns in council debates. The resulting Council Resolution 552 called on all states to respect freedom of navigation, reaffirmed the right of freedom of navigation in international waters and sea-lanes, and condemned attacks on ships en route to and from Kuwaiti and Saudi ports. Significantly, the council decided that if there were future noncompliance with Resolution 552, it would meet again to consider measures to ensure freedom of navigation. This warned belligerents of the possibility of binding council decisions, perhaps involving force. International nongovernmental organizations (NGOs)—the International Transport Workers Federation, the International Chamber of Shipping, and the International Shipping Federation—expressed concern over the deteriorating situation. These were cited in the UN secretary-general's report, and they are an example of how NGO protests can contribute to binding rules.

In 1985, a temporary truce in the land war was broken. Iraq renewed attacks on Kharg and Iranian tankers, and Iran restarted a desultory campaign against neutral tankers. In June 1985, Iran intercepted and detained a Kuwait-flagged ship, *Al-Muharaq,* which was Kuwait-bound but carrying Iraq-destined merchandise. Iraq had used Kuwait as an entry port since 1980. Iran's ex post facto prize-court law justified its seizing *Al-Muharaq* and other ships headed to Kuwait. In September, Iran stepped up its visit-and-search procedure. If enough shipping were warned off, it thought, that might tip the scales, since oil sales financed Iraq's war effort. To be sure, oil left Iraq by international pipeline, but all nations save Turkey had refused Mediterranean Sea access for Iraqi oil. Meanwhile, Iran's crude was being ferried in Iranian tankers from Kharg to Sirri Island in the lower Gulf, where it was stored in "mother" ships for transfer to customers' tankers. Iranian tankers also shuttled between Kharg and Lavan Island in the lower Gulf. By the end of 1985 the Tanker War had become the Iran-Iraq conflict's most important feature.

During late 1985 the United States issued a NOTMAR "special warning" advising of ship visit and search and occasional seizure within the strait and the Gulf of Oman in the lower Gulf. American mariners were advised to avoid Iranian or Iraqi ports and coastal waters and to remain outside declared zones. The NOTMAR added that the United States did not recognize the validity of any foreign rule, regulation, or proclamation. While asserting freedom of the seas and straits transit rights, the United States offered to work with the GCC and to help militarily if publicly requested.

In October 1985, France began defending French-flagged ships, and a French warship moved between a merchantman and an Iranian warship, warning that it would use force if the Iranian tried to intercept the merchantman. France's rules of engagement (ROE) declared that its warships could fire on forces refusing to break off attacks on neutral merchant ships. The end result was a sudden drop in attacks near French men-of-war.

The U.S. Navy Enters the Fray

In January 1986 Iran boarded and searched a U.S.-flagged vessel. The United States recognized belligerent rights to board and search but cautioned against overstepping these rights. Later that month the United Kingdom also justified Iranian interceptions and seizures of British-flagged ships as self-defense. The Netherlands recognized board-and-search rights, but only for ships proceeding to or from belligerents' ports. Finally, in February 1986 UN Security Council Resolution 582 called for a cease-fire, deploring all attacks on neutral shipping.

During February 1986, Iraq extended its zone up to an area near Kuwait's territorial waters. In April 1986 a U.S. destroyer warned off an Iranian warship from what may have been a planned boarding of a U.S.-flagged merchantman. In May, after more Iranian strikes on shipping, the United States reaffirmed a commitment to Saudi self-defense, freedom of navigation, free flow of oil, and open access through the strait. In response, Iran warned that its naval forces would attack U.S. warships escorting or convoying cargo ships carrying material for Iraq or those that tried to interfere with Iran's interception procedures.

A U.S. NOTMAR advised of additional cordon-sanitaire precautions in force for U.S. ships in the Gulf, the strait, the Gulf of Oman, and the northern Arabian Sea. These measures would also apply to U.S. forces transiting the strait or in innocent passage in foreign territorial waters and when operating in such waters with coastal-state approval. The NOTMAR added that its publication served to advise that U.S. forces would exercise self-defense and that freedom of navigation of any ship or any state should not be impeded.

In August Iraq bombed the Sirri terminal, badly damaging a British-flagged, Hong Kong–owned tanker. By then Iraq had hit five of the eleven shuttle ships running between Kharg and Sirri. Later that year it struck the Larak and Lavan terminals. In September, Iranian warships fired on, stopped, and searched a Soviet-flagged Kuwait-bound ship with arms on board destined for Iraq. This was but one of a thousand Iranian ship inspections during 1985–86. In November Iraq bombed the UAE's Abu al-Bukosh offshore installations.

The 1986 attacks reduced Iranian oil production considerably, and a fall in world oil prices aggravated Iran's economic problems. Iraq was also in financial trouble, but its creditor states rescheduled debts while supporting increased military aid. Meanwhile, American arms sales to Iran came through Israel; one shipment even came from Eilat, an Israeli port in the Gulf of Aqaba, to Bandar Abbas aboard a Danish-flagged ship.

By 1987 the war had become more internationalized. In April Iran delivered a note on strait transit passage. The U.S. response rejected Iran's claim that straits-passage rights under the 1982 UN Convention on the Law of the Sea (UNCLOS) were contractual and not customary international law, arguing that UNCLOS rules recited long-standing custom. The United States also rejected Iran's claim of a right to interfere with any vessel's lawful transit passage in a strait used for international navigation. In May the Kuwaiti Oil Tanker Company reregistered eleven tankers under the U.S. flag, and three others went under British registry. The USSR chartered three more to Kuwait.

Iran tried to convince Kuwait to stop reflagging, but when this failed Iran concluded that Kuwait had for all practical purposes turned itself into an Iraqi province, placing its resources at the disposal of France, the USSR, and the United States. Iran declared it could not allow Iraq to receive oil income to build its war machine through Kuwaiti tankers' flying other flags. Then an Iranian warship fired on a Soviet merchantman. In mid-May, a Soviet-flagged tanker hit a mine that the USSR claimed the Iranian navy had laid. Another Kuwait-bound tanker hit a mine in June. Sea mines were detected in approaches to the channel leading to Kuwait's Mina Ahmadi terminal. Mines, apparently laid by Iranian Revolutionary Guards using small boats, began appearing throughout the Gulf. The Saudi and U.S. navies cleared a channel to Kuwait, and the USSR sent three more minesweepers to the Gulf.

On 17 March 1987 two Iraqi fighter-bombers launched Exocet missiles that severely damaged a frigate, USS *Stark,* and killed thirty-seven sailors.[4] The United States ordered its forces to a higher state of alert, revising its ROE for possible interaction between American and Iraqi forces or against anyone displaying hostile intent or committing hostile acts. British rules continued to reflect the view that the UN Charter permitted self-defense, as an attack on merchantmen would trigger the self-defense clause. U.S. NOTAMs and NOTMARs from July and September reflected a stronger self-defense policy, including anticipatory self-defense if a warship were illuminated by a weapon fire-control radar. However, these measures were to be implemented so that they would not unduly interfere with freedoms of navigation and overflight.

In July 1987 the U.S. Navy began convoying reflagged tankers that carried no contraband from Iraq. On 24 July the reflagged *Bridgeton,* and on 10 August *Texaco Caribbean,* chartered to an American company, hit mines. Immediately the U.S. Navy began mine protection. The *Bridgeton* incident opened a new chapter of direct U.S.-Iranian naval confrontation. Mines began appearing all over the Gulf, outside the Gulf in the strait and the Gulf of Oman, and in Kuwaiti and Omani territorial waters. French and British naval operations expanded to meet the threat in the latter areas.

The Armilla Patrol began "accompanying" but not escorting or convoying British merchantmen. As a result, foreign ships were attracted to British registry to get protection in the lower Gulf. British seafarer unions opposed arming merchantmen, a plan that was reminiscent of the Q-ships of World War I, but this soon became British and Italian policy. Some merchant ships began carrying chaff canisters to confuse incoming missiles, while others were repainted dull, nonreflective gray for the same reason.

In August 1987 the U.S. Navy, claiming self-defense, attacked an Iranian minelayer. Iran countered that this was overt aggression, that self-defense could be claimed as a response only to an armed attack. The American action effectively halted Iranian minelaying for six months, but by mid-1987 there had been over a hundred mine attacks on ships of thirty nationalities. Meanwhile, Iraq had attacked over two hundred ships, mostly Iran-flagged or -chartered.

A June 1987 Vienna Economic Summit had reaffirmed freedom of navigation and free, unimpeded flow of oil and other traffic through the Gulf. Also in July 1987, UN Security Council Resolution 598 again "*Deplore*[d] . . . attacks on neutral shipping[,]" "*Demand*[ed] an immediate cease-fire," and "*Call*[ed] *upon* all other States to exercise the utmost restraint and to refrain from any act which may lead to further escalation and widening of the conflict." Iraq accepted Resolution 598, but Iran refused. In September the EC also supported the resolution, reiterating firm support for freedom of navigation, "which is of the utmost importance to the whole international community."[5]

On 3 August 1987 Iran announced planned naval maneuvers in its territorial waters and the Gulf of Oman, warning ships against approaching those waters. Iraq protested, noting that Iran's territorial waters included part of the strait and correctly claiming that under the 1958 Territorial Sea Convention and 1982 UNCLOS a country could not suspend passage through territorial straits. Contemporaneous with a Gulf buildup among the U.S., Saudi, and European navies, the Western European Union declared that Europe's vital interests required that Gulf freedom of navigation be assured at all times.[6] In November an Arab League Extraordinary Summit supported Resolution 598 and called on Iran to do so.

Interestingly, from 1980 Iran and Iraq had maintained diplomatic relations; these finally ended only in October 1987.[7] On 8 October Iranian speedboats fired on U.S. helicopters. In accordance with American self-defense principles and ROE, the helicopters returned fire, sinking one boat and damaging others. U.S. forces, claiming self-defense, responded to an Iranian Revolutionary Guards attack on a U.S.-flagged tanker by destroying Iran's Rostum offshore oil platform in the lower Gulf. Iran claimed the attacks were aggression and, again, that self-defense could only be asserted in response to armed attack.[8]

The United States did not respond to a similar attack on *Sungari*, a Liberian-flagged, U.S.-owned tanker. At that time Washington did not consider open-registry ships, even if owned by American interests, to have enough U.S. connection to merit protection. It also followed a long-standing law of armed conflict rule that the flag flown and only the flag flown—not ownership—counts, as distinguished from law of the sea rules. These Iranian attacks seemed aimed at tankers in Kuwait's al-Hamadi port, where Kuwaiti and Saudi oil donated to Iraq was transported to pay for munitions shipped to Iraq through neutral ports. Three days after the American attack on Rostum, Iran hit Kuwait's deepwater Sea Island Terminal. In November Iranian speedboats shot up three tankers carrying Saudi oil.

In December 1987 a U.S. warship helped rescue a Cypriot tanker's crew after an Iranian gunboat attack set it ablaze. Although the conflict was outside territory covered by the NATO agreement, the NATO Council supported Resolution 598 in December, recalling the importance of security of navigation in the Gulf. The GCC urged the Security Council to implement the resolution and approved a comprehensive security strategy approaching the level of a collective self-defense pact. But as the year ended it appeared that some permanent council members who held the veto (China, France, and the USSR) under UN Charter articles 23 and 27 would vote against a U.S.-sponsored sanctions resolution.

An early 1988 U.S. NOTMAR summarized the perilous situation in the Gulf, warning of the belligerents' apparent intentions and a possibility of mine attacks and of visit, search, and possible seizure or diversion of nonbelligerent merchantmen. Iran published its Prize Law, effective in January 1988, of which article 3 declared these to be war prizes:

(a) All goods, merchandise, means of transport and equipment belonging to a State or to States at war with ... Iran.

(b) Merchandise and means of transport ... belonging to neutral States or their nationals, or to nationals of the belligerent State if they could effectively contribute to increasing the combat power of the enemy or their final destination, either directly or via intermediaries, is a State at war with ... Iran.

(c) Vessels flying the flag of a neutral State as well as vehicles belonging to a neutral State transporting the goods set out in this article.

(d) Merchandise, means of transport and equipment which ... Iran forbids from being transported to enemy territory.

The language in article 3(b), "effectively contribute to increasing the combat power of the enemy," echoes the language of current views on neutral merchant ships carrying military materials.[9] The law declared that article 3(a) property would become Iranian property and that articles 3(b) and 3(c) property of neutrals would be confiscated and adjudicated. Article 3(d) property would "become the property of ... Iran or be confiscated according to circumstances. Any person contesting this must appear before the [prize] Tribunal."[10]

Winding Down the Tanker War

Iraqi tanker attacks resumed in February 1988. The U.S. government was willing to consider a UN Gulf naval force if a collective action plan were spelled out clearly. But the United States would not support a UN force replacing U.S. and American-aligned forces. The United Kingdom was also unenthusiastic, even though Italy and the USSR supported the idea. The USSR, in particular, wanted to replace the large Western naval presence with a UN flotilla.

In April, however, an Iranian mine severely damaged the frigate USS *Samuel B. Roberts*. Four days later the United States responded by engaging Iranian warships, sinking or damaging them, and destroying the Sassam and Sirri oil platforms, which had been the speedboat bases. Iran branded the attacks as aggression, but some later saw these actions, plus Iran's simultaneous loss of the Fao Peninsula to Iraq, as a turning point in the war.[11] Even more warships now crowded the Gulf. Unprecedented international concern within the United Nations and within NATO ushered in a new phase of the war for neutral countries.

In April, after Iranian gunboats attacked a Saudi tanker off Dubai, the United States announced it would begin assisting, upon their request, "friendly, innocent neutral vessels flying a nonbelligerent flag outside declared war exclusion zones that [were] not carrying contraband or resisting legitimate visit and search" by a belligerent, if the U.S. warship's or aircraft's mission allowed rendering such aid.[12] This offer, more expansive than a British policy of protecting foreign-flagged ships with a clear majority British ownership interest, was partly a response to Saudi, UAE, and American oil shippers navigating under foreign flags. The British policy was really a distinction without a difference, since British warships gave humanitarian assistance to neutral ships after attacks and were prepared to interpose between an attacker and a target ship. French warships were "available to assist [merchantmen] according to circumstances."[13] Italy's escort was limited to Italian-flagged ships, although its ROE promised a military response if a belligerent committed a hostile act; however, these did not contemplate "repressive acts" on bases of operation, such as oil platforms.[14] Mine clearance became more cooperative. In May 1988, Iraq damaged the Liberian-registered *Seawise Giant*, the world's largest supertanker, among five ships at Iran's Larak terminal in the strait.

During July 1988, tragedy struck when the United States accidentally shot down a civilian airplane. The United States claimed self-defense in USS *Vincennes*'s shooting down of an Iranian airbus in July.[15] A week later U.S. helicopters attacked Iranian gunboats that had set ablaze a Panamanian-registered, Japanese-owned tanker with American nationals in the crew, thus implementing the new U.S. policy. By the war's end the U.S. Navy had conducted over a hundred convoys in the Gulf, and other states had also been engaged in many escort operations.

International organizations like the Arab League, the EC, the GCC, and the Toronto Economic Summit continued to support Resolution 598. In July 1988 Iran finally accepted the resolution, perhaps prompted by the airbus incident. On 8 August the UN secretary-general announced a cease-fire effective 20 August 1988. Iran announced on that day that it would continue inspecting vessels during the cease-fire—largely a theoretical gesture. Nevertheless, Iraq protested the announcement. U.S. convoy operations ended in October, and in January 1989 "deflagging" procedures began, reverting tankers to Kuwaiti from U.S.-flag status.

The Importance of the Tanker War

The Tanker War was the most important single theater of naval warfare during the Iran-Iraq conflict. Over two hundred mariners died in attacks by Iran and Iraq on over four hundred ships, almost all of which flew neutral flags. The attacks resulted in over forty million deadweight tons of damaged shipping, thirty-one sunk merchantmen, and another fifty damaged ships eventually declared total losses. By the end of 1987, write-off losses stood at nearly half the tonnage of all merchant shipping sunk in World War II. One reason was that ships had become larger in size and fewer in number. The relatively low figure for lives lost reflected the fact that modern vessels' crews are smaller, owing to automation. Ships under the flags of more than thirty countries, including UN Security Council permanent members, were attacked.

Only about 1 percent of Gulf voyages involved attacks. Nevertheless, in terms of percentages of losses due to maritime casualties worldwide the statistics were staggering. During 1982 alone, 47 percent of all Liberian-flag tonnage losses due to maritime casualty occurred in the Gulf; in 1986 it was 99 percent; in 1987 it was still over 90 percent. Most Gulf tankers were open registry, but American nationals owned a third of them, while American nationals chartered another substantial portion. The U.S. financial loss was therefore substantial. Insured losses reached thirty million dollars in one month, and there were tremendous increases in war-risk insurance premiums, which drove up shipping costs. The total price of the war and the direct or indirect damages it caused was estimated to be nearly $1.2 trillion.

Although nothing about the war can be considered truly positive, there were two indirect "benefits." Because of the large number of ship sinkings and the extensive damage to ships, there was a sharp reduction in what was seen as an oversupply of available tanker tonnage. As a result, Western reliance on the strait as an oil lifeline declined from twenty million barrels of petroleum a day in 1978 to 6.4 million in 1985. Meanwhile, pipelines were built to transport Gulf oil through Saudi Arabia. Increased production from other fields, including those in the North Sea, may have prevented a worldwide oil shortage during 1980–88.

Convoying or escorting merchantmen, a tactic that the United States and Great Britain had used in both world wars and that Britain employed during the Icelandic fishing disputes of the 1960s and 1970s, turned out to be a big factor during the Tanker War. A good question is whether traditional rules for escorting neutral convoys during war, practiced as long ago as the Spanish silver fleets, still applied during the charter era. Practice thus far seems to say yes to this question.

One of the continuing issues for self-defense of naval platforms and their convoys or of escorted ships is whether anticipatory self-defense, or reactive self-defense, as Iran seemed to have advocated, appears to be the rule rather than the exception during the charter era. The U.S. position, which was supported by many other states, allowed anticipatory self-defense to be invoked so that a ship need not "take the first hit" before there is a response to a threat.

Collective self-defense through prior agreement, except perhaps among the GCC members, was never an issue. On the other hand, states with Gulf naval presence cooperated to a greater or lesser degree. The legal basis for this cooperation was "informal" collective self-defense, which was analogous to coalition warfare to defeat a common enemy. A further question is whether this was a "war" as traditional international law would have it. Iraq declared war, but Iran never did, which raised many issues under self-defense and aggression doctrines flowing from charter articles 2(4) and 51.

Conclusions

Tanker War statistics clearly show the trend toward transporting oil on larger vessels with smaller crews. The number of ships lost was small, but the tonnage sunk or damaged was huge. Proliferation of open-registry shipping and factors like containerization meant that there were more private players from more nationalities. Chartering has always presented a possibility of more private interests, but today there are subcharters and sub-subcharters. Containerization aboard ever-larger cargo ships means potentially more claimants among consignors and consignees. Entry of governments through national shipping lines and problems in private law, like sovereign immunity, remain important factors. Reflagging under the law of the sea can result in different practices for belligerent interception and attack. Additionally, crews are multinational in origin. Because of the problem of determining ownership of cargoes and ships, diverting for inspection instead of boarding and prize taking has become a new modality, accepted in international law. Diversion has been widely practiced in wars since 1945, especially during the Tanker War.

New weapons and weapon-delivery systems have resulted in different methods for boarding-and-search operations. States often use helicopters instead of warships' boats because of merchantmen's size and the relative speed of helicopters, allowing a warship to stand off at a safer distance, given the risk of a missile response. Use of guided missiles, whereby one unerring projectile can do great damage to or even sink a warship, has become nearly universal, and some of these can be carried aboard relatively small warships.

The possible roster of national and international players has also increased dramatically. There are over two hundred entities claiming statehood status today. Besides states' traditional protests and individual state actions, decision makers must contend with a spectrum of intergovernmental organizations—ranging from those with direct maritime interests in a situation, like NATO, the GCC, or the Arab League, to those on the geographic periphery, like the European Union / EC—as well as nongovernmental organizations, including shipping associations, international maritime insurance interests, labor organizations, and human rights or humanitarian law organizations like the International Committee of the Red Cross. These were all important factors during the Tanker War. Some NGOs may have agendas that conflict with those of the shipping companies, such as Greenpeace campaigns against the whaling and tuna fishing industries.[16]

At the top of the law and policy pyramid is the United Nations and its lawmaking potential. This has been a growing factor in most confrontations since 1945. For example, belligerents' acceptance of UN Security Council Resolution 598 ultimately ended the Tanker War, subject to a UN-brokered cease-fire. This may have legal ramifications different from armistices, which ended many post–World War II conflicts, including the Korean War and some Arab-Israeli conflicts. Sometimes UN law even supplied legal decisions that differed from customary law, which proved to be a major feature of the 1990–91 Gulf War.

Notes

This chapter is based on three earlier studies: George K. Walker, *The Tanker War, 1980–88: Law and Policy,* International Law Studies, no. 74 (Newport, R.I.: Naval War College, 2000); "State Practice following World War II, 1945–1990," in *Targeting* *Enemy Merchant Shipping,* ed. Richard J. Grunawalt, International Law Studies, no. 65 (Newport, R.I.: Naval War College, 1993); and "The Crisis over Kuwait, August 1990–February 1991," *Duke*

Journal of Comparative & International Law (1991), p. 25.

The thoughts and opinions expressed in this publication are those of the author and are not necessarily those of the U.S. government, the U.S. Navy Department, or the Naval War College.

1. An October 1980 U.S. NOTMAR warned of these risks, cautioning mariners to be alert to unusual, abnormal, or hostile actions in the Gulf.

2. Only after the 2003 coalition invasion and the end of Saddam Hussein's regime did Iraq admit that it had been the aggressor in the war. Sabrina Tavernise, "Iraqi Government, in Statement with Iran, Admits Fault for 1980's War," *New York Times,* 20 May 2005, p. A8.

3. Stanley F. Gilchrist, "The Cordon Sanitaire— Is It Useful? Is It Practical?," *Naval War College Review* 35, no. 3 (May–June 1982), p. 60.

4. In 1989 Iraq paid U.S. claims for the *Stark* attack without admitting liability, a standard feature of all claims settlements—for example, for car accidents.

5. *Statement by Member States of the European Community,* 3 September 1987, *European Political Cooperation Documentation Bulletin* 3, no. 2, p. 93, in Andrea de Guttry and Natalino Ronzitti, *The Iran-Iraq War (1980–1988) and the Law of Naval Warfare* (Cambridge, U.K.: Grotius, 1993), p. 554.

6. Organized by the Treaty of Economic, Social & Cultural Collaboration & Collective Self-Defence, 17 March 1948, 19 UNTS 51. The Western European Union later negotiated protocols to subordinate military matters to NATO.

7. When conflicts began with a war declaration before the charter era, diplomatic relations typically ended with war declarations—which was not necessarily the case with "imperfect" wars like the 1798 French-U.S. Quasi-War.

8. Oil Platforms (Iran v. U.S.), 2003 ICJ 161, 218–19 (6 November) held that the U.S. actions were not measures necessary to protect American security interests under a 1955 Iranian-U.S. friendship, commerce, and navigation treaty with respect to these actions and similar ones in 1988 and that there was no violation of a right of freedom of commerce and navigation between the parties' territories. The court denied Iran's claim that the actions were a breach of the treaty. The case is not a primary source of law and offers no precedent for other cases; see ICJ Statute arts. 38(1), 59, 26 June 1945, 33 UNTS 993.

9. Compare Iran Law Regarding Settlement of Disputes over War Prizes, 17 November 1987, art. 3, reprinted in de Guttry and Ronzitti, *Iran-Iraq War,* not in force until 31 January 1988, with "International Lawyers & Naval Experts Convened by the International Institute of Humanitarian Law," *San Remo Manual on International Law Applicable to Armed Conflicts at Sea,* ed. Louise Doswald-Beck (San Remo, It.: International Institute of Humanitarian Law, 1995), paras. 67(f), 67(g), cmts. 67.1–67.27; A. R. Thomas and James C. Duncan, eds., *Annotated Supplement to* The Commander's Handbook on the Law of Naval Operations, International Law Studies, no. 73 (Newport, R.I.: Naval War College, 1999), paras. 7.5, 7.5.2, 8.2, 8.2.2, whose predecessor, U.S. Navy Dept., *The Commander's Handbook on the Law of Naval Operations,* NWP 9 (1987), annotated in U.S. Navy Dept., *Annotated Supplement to* The Commander's Handbook on the Law of Naval Operations: *NWP 9 (Rev. A)/FMFM 1-10* (1989), applied during the Tanker War.

10. Merchandise that would rapidly deteriorate or was "not worthwhile preserving" would be sold, the funds put in an account pending tribunal disposition. Iran Law, note 10, arts. 4–5.

11. The U.S. action was also a subject of the International Court of Justice *Oil Platforms* case.

12. As quoted in George K. Walker, *The Tanker War, 1980–88: Law and Policy,* International Law Studies, no. 74 (Newport, R.I.: Naval War College, 2000), pp. 70, 100–101n447.

13. Ibid., pp. 70, 101n451.

14. Ibid., pp. 70–71, 101n453.

15. As in Iraq's response in the *Stark* case, the United States settled states' claims on behalf of their nationals. *Oil Platforms* did not resolve U.S. counterclaims for its nationals in the *Roberts* attack.

16. Paul Greenberg, "Tuna's End," *New York Times Magazine,* 27 June 2010, pp. 28, 32.

Twenty-First-Century High-Seas Piracy off Somalia
MARTIN N. MURPHY

Previous chapters have shown how countries with stable governments have adopted commerce raiding strategies mainly to injure opponents, even while enjoying the profits that could be made from these endeavors. An entirely different type of commerce raiding by failed states uses similar methods mainly to obtain profits, both for the pirates and for the government or local officials backing the pirates' efforts. Between 1993 and 2005, for example, over seven hundred piracy incidents were reported in the waters off Somalia, and there was also a dramatic increase in kidnap and ransom crimes.

This problem appeared not just because Somalia was a failed state but because of geography: for much of recorded history, significant trade routes have passed through Somalia's coastal waters, which also contain rich and underexploited fishing grounds. Piracy arose after 1991 with the spread of conflict and disorder and the rise to power of elites who countenanced a type of behavior that would be regarded as criminal—not by them but by much of the developed world. These elites gradually realized that what worked so well for them on land, kidnapping and extortion in particular, worked even better at sea, where targets worth hundreds of thousands and even millions of dollars were at risk.

The pirates' primary goal is to obtain ransom for captured ships and crews. The initial ransoms were small. In 2005 the Haradheere-based pirates called "Somali marines" demanded only $300,000 for the return of their first major hijack. By 2006, however, they were signaling that they were looking for much more when they demanded a million dollars for the return of *Dongwon-ho*, a South Korean tuna-fishing vessel, which they accused of illegally fishing in Somali waters, even though they eventually settled for half this amount.[1] As of September 2011 Somali pirates held "at least 49 vessels and more than 500 hostages," and the average ransom had increased to five million dollars per ship.[2]

The failure of the Somali state is due to many causes, including poverty and domestic chaos. However, the political dimensions of the problem are perhaps the most relevant for the spread of state-sponsored piracy. The activities and fortunes of pirates are linked to a small number of Somali clans and their political leaders who have struggled for control of the country since the fall of the Mohammed Siad Barre regime in 1991. The problem of piracy will be resolved only once some form of political stability is achieved. The complex situation in Somalia is not pure and simple piracy, therefore, but a new form of commerce raiding, in which the pirates, the Somali people, and Somali officials share in the proceeds of these criminal endeavors.

The Origins of Somali Piracy: 1989–2006

Piracy, whose spectacular success has been splashed across the world's headlines, has occurred both off Somalia's Indian Ocean coastline and in the Gulf of Aden between Somalia's northern coast and Yemen, a large expanse of water that narrows gradually toward the Bab el-Mandeb choke point, through which ships must pass to reach the Suez Canal. Ship hijackings began in the gulf long before President Barre's overthrow. During the 1950s yachts were seized occasionally there and held for ransom, while British colonial records note incidents of piracy against dhows and fishing vessels.[3] After 1989, however, the pace and scope of maritime predation gradually increased.

The initial incidents came after the Somali National Movement (SNM), dominated by the Isaaq clan in what is now Somaliland, lost the support of its Ethiopian backers. It needed to capture weapons, even while preventing supplies from reaching government forces, a goal that led it to warn all shipping agencies "not to cooperate with the dying regime of Mogadishu, because they are not able to ensure the safety of ships and their crews against any dangers that they may be exposed to."[4] As a result, according to the National Geospatial-Intelligence Agency (NGA), the "SNM Coast Guard" seized on 5 December 1989 a Panamanian-flagged ship on its way to Berbera, a port then controlled by Barre's regime.[5]

But not all hijackings even then were obviously related to political struggles. From the start of the civil war in 1989, captains of the dhows and small freighters that form the majority of ships carrying cargoes around the Arabian Peninsula alerted each other about attacks in Somali waters.[6] However, the problems got worse after the Barre regime collapsed at the end of January 1991. In the same month pirates attacked MV *Naviluck*, outbound from Mombasa to Jeddah off Cape Guardafui, in what is now Puntland. Three boatloads of attackers set the ship on fire. They reportedly took some of the crew members ashore, where they killed three of them; the remainder were forced overboard and were later rescued.

A small number of attacks took place up and down the Somali coast in 1993, but the most extraordinary occurred off Yemen in May 1994, when two German vessels, *Norasia Samantha* and *Glucksburg*, reported being fired on by missiles that landed close to both ships but hit neither one. No one claimed responsibility for either attack. In September 1994 a party of twenty-six pirates posing as members of the Somali Coast Guard, North East Region, hijacked MV *Bonsella* and used it for six days as a base of operations to attack other ships.[7] Experience from other pirate-prone areas around the world indicates strongly that many if not most pirate incidents go unreported.[8] The level of organization displayed by this attack suggests, therefore, that the pirates were too well practiced for it to have been anything other than part of a pattern.[9]

Some of these early attacks, such as that on *Bonsella* and another on the motor vessel *Full City* in 1995 during which currency and drink were stolen, share strong similarities with pirate attacks the world over.[10] However, many other features now distinguish Somali piracy, such as the use of "mother ships," the prevalence of kidnapping, the targeting of foreign fishing boats and aid ships, the great distance from shore where the incidents took place, and the involvement of corrupt political figures.[11] In 1997, for example, the International Maritime Bureau (IMB) reported that an armed faction seized MV *Baharihindi* and sailed it to Gara'ad on the east coast.[12] In the same year a Taiwanese trawler, MV *Shen Kno II*, was captured by the Somali Salvation Democratic Front (SSDF), the political organization of the Majarteen clan, which demanded a "fine" of $800,000 to release the ship—forty thousand dollars for the captain and ten thousand for each member of the crew. It threatened that if payment was not made, each man would be imprisoned for ten years for "stealing maritime products."[13]

Fishing craft, however, were never the only targets. In 1998, seven armed men stole a freighter, the 299-ton *Noustar*, from dockside at Boosaaso. They sailed the vessel to Ras Hafoon, near Socotra, where they released half the crew. The remaining crew reported that heavy guns had been placed on board and that the ship was then used to prey on small ships and dhows in the Gulf of Aden.[14] In fact, by 1998 fully two-thirds of all maritime abductions worldwide took place in the Gulf of Aden. During 2000 there were twenty-three piracy incidents recorded in the Red Sea–Aden–Somalia region, a third of the African total.[15] Most importantly, the IMB warned seafarers that attacks by men in small speedboats firing rifles and rocket launchers were taking place up to forty nautical miles (seventy-four kilometers) offshore and advised ships to sail at least fifty nautical miles (almost ninety-three kilometers) off the coast. Most attacks were taking place off Puntland, but some were occurring at unspecified locations along the Somali coastline farther south.[16]

After 9/11, Combined Task Force (CTF) 150 was assembled under the authority of United Nations Security Council Resolutions 1368, 1373, and 1378 to patrol the northern

Arabian Gulf from the Pakistani coast across to Somalia. In addition, a U.S. military presence known as Combined Joint Task Force–Horn of Africa (CJTF-HOA) was established in Djibouti.[17] This appeared to make the pirates more cautious while they assessed the new situation. The number of reported piracy incidents declined to a low of ten in 2004, although the respite was brief. Incidents resumed in 2005 to the extent that the U.S. State Department even advised ships traveling though the Gulf of Aden to do so in convoy.[18] In 2004 the IMB advised ships to sail at least fifty nautical miles (almost ninety-three kilometers) from the coast and preferably farther. Nonetheless, and despite the presence of coalition warships attached to CTF 150, fifteen attacks occurred in January to July 2005, including eight in July alone.

Off the east coast, a new pirate gang based in Haradheere demonstrated that it was prepared to mount attacks far out to sea, which, while not unique, is something that few other modern pirate groups have proved willing to do. Although pirates were on occasion able to locate distant targets, during this period their abilities and the frequency of such incidents appear to have been exaggerated. There are grounds for believing that some of the attacks, which were reported to have taken place two hundred nautical miles or more from the coast, could have been erroneously attributed to Somali pirates as a result of misidentification.

Owing to the continued domestic chaos in Somalia, cases of piracy continued to climb, reaching thirty-two vessels in just the nine months between March and November 2005. During the entire year Somali waters "topped the piracy high risk areas," with thirty-five attacks taking place—sixteen actual and nineteen attempted attacks. As a result of these seizures, over 130 crew members were being held hostage as of November 2005; negotiations for their release were pending.[19] On 5 November 2005 two pirate speedboats even attempted to attack a luxury cruise ship, *Seabourn Spirit,* which fought them off with a sonic boom gun. The cruise ship had 151 passengers and 161 crew members on board, of whom forty-eight were American citizens.

In response to this failed November 2005 attack, the International Maritime Organization (IMO) issued Resolution A 979(24), "which called on all seafarers and other involved parties to work within international law to ensure that further acts of piracy in the region were prevented and current ones terminated." The UN Security Council did not adopt this resolution, however, but instead issued its own antipiracy statement warning local navies to "take appropriate action to protect merchant shipping, in particular the transportation of humanitarian aid, against any such act, in line with relevant international law. . . . The Council further urges cooperation among all States, particularly regional States, and active prosecution of piracy offences."[20]

While the United Nations urged regional states to solve the problem, many newly independent countries—especially in Africa—did not have sufficient naval units to patrol their own territorial waters. This was especially true for countries divided by civil war. In the case of Somalia, what the international community has referred to as "piracy" could also be seen as Somalia's attempts to protect its natural resources in its exclusive economic zone (EEZ). For example, when the U.S. Navy's guided-missile destroyer USS *Gonzalez* located and subsequently captured several Somali pirates in March 2006, the pirates claimed "to be defending local fishermen by 'taxing' illicit foreign trawlers." Nevertheless, and though the pirates considered themselves to be commerce raiders of a sort, because these men were armed and were operating from unmarked ships on the high seas they exactly fit the standard international definition of "piracy."[21]

Piracy and Illegal Fishing

Some observers consider Somalia's huge increase in maritime depredation as a response to widespread illegal fishing in its sovereign waters. Somalia's coastline, which is almost equivalent in length to the Eastern Seaboard of the United States, is the longest in Africa. Thanks to the periodic upwelling of the nutrient-laden Somali Current, its waters are rich in fish and shellfish. In defiance of the 1982 United Nations Convention on the Law of the Sea (UNCLOS), the Barre regime claimed that its territorial waters extended to two hundred nautical miles rather than the twelve that are allowed. This claim has never been recognized and has no standing in international law. Nonetheless, because UNCLOS has broadly been accepted as customary international law, Somalia has been deemed to have a two-hundred-nautical-mile EEZ, meaning that foreign fishing there without permission is illegal. Unfortunately, the lack of an effective successor government has made enforcement impossible. After 1991, in particular, there was no force to protect these fishing grounds. Foreign boats moved in aggressively to catch tuna, shark and ray for their fins, and lobster, in the process destroying reef habitats.

During the civil war in Somalia, fishing boats from Egypt and Yemen and whole fishing fleets from such distant-water fishing nations as Belize, South Korea, and Taiwan moved in, exploiting Somalia's offshore resources with "near impunity." French and Spanish fishermen have reportedly been observed using the Belizean flag to circumvent European Union (EU) rules prohibiting fishing by member states in Somalia's maritime areas. Although it is difficult to estimate exactly, the value of foreign annual fish catches in Somali waters may vary from $90,000,000 to as much as $300,000,000.[22]

The size and sophistication of the foreign vessels hurt the local fishermen. Estimates suggest that in 2000 there were around thirty thousand full-time and sixty thousand part-time Somali fishermen who sold most of their catch for export. In the years that followed reports became increasingly frequent of local boats being crushed by the larger foreign boats, their nets destroyed, and catch stolen.[23] Local fishermen, who were not licensed, reported being fired on by the foreign fishermen and by Somali militiamen on board foreign vessels; on occasion they retaliated, attacking foreign vessels in attempts to drive them off.[24] There have been reports of foreign fishermen destroying gear on Somali boats. Attempts have also been made to ram and sink or disable local fishing vessels, so as to chase local fishermen away from lucrative fishing grounds. In response to this unfair foreign competition, Somali fishermen began to use force to protect their interests. This Somali response has been described as the effective "decentralization of fisheries enforcement to the grass-roots level."[25]

It has been suggested that Somali piracy, aided and abetted by Somali political interests on land, evolved out of this defensive response to foreign exploitation, but to what extent Somali fishermen turned their skills to piracy remains unclear. According to one theory—which might be termed the "evolutionary explanation"—once fishermen realized how much money foreign interests were prepared to pay for the return of fishing boats and crews, they abandoned fishing to exploit this more lucrative line of work. "It's true that the pirates started to defend the fishing business," a Somali diplomat explains, but then, as he puts it, "they got greedy."[26] Over time these groups turned their attention to nonfishing vessels, which in some cases commanded even higher ransoms.

According to a second theory, which might be labeled the "adaptive explanation," once the warlords began selling fishing concessions and in some cases providing foreign boats with armed guards, the self-protection groups turned their attention to unarmed commercial vessels as the only option left open to them to make a living.[27]

According to a third theory, which might be dubbed the "complicity explanation," some fishermen, recognizing economic reality, sold their services as boat handlers to warlord groups, which then supplied the equipment and the men who actually boarded and captured the foreign ships.[28]

The fourth theory might be described as the "unemployed coast guard explanation." The Puntland administration first hired a British private security company, Hart Security, and then, after it left, entered into a joint venture with a Canadian company named SomCan to train and equip a fishery-protection force. When this ended in 2005, men trained by both companies apparently used their interception and ship-boarding skills as pirates.

Whatever the reality, Somali pirate groups have not hesitated to claim that they are defending the country's fishing grounds and preventing toxic-waste dumping. Time and again they have justified their activities as restitution for the theft and destruction foreigners have wrought on Somalia's natural resources.[29] Consequently, the Somali fishermen may be called pirates, regardless of whether their actions are more in line with state-sponsored commerce raiding than with traditional piracy. Understanding this ongoing problem requires an examination of the political backing for the pirates, backing that originated not so much in government institutions as in the clan-based structure of Somali society.

Somalia: A Clan-Based Society

Reaching any understanding of Somali piracy—or indeed, any understanding of Somali society and politics generally—demands the recognition that Somalia is a lineage-based society where almost everyone is identified by membership in a clan. Clan affiliation (or clanism) has been the organizing principle in Somali life and the basis of most social and political institutions, including "personal identity, rights of access to local resources, customary law *(xeer)*, blood payment *(diya)* groups and social support systems," since the precolonial era.[30] Critically, clans are the principal source of individual and family security.[31] Restitution is addressed by blood payment, but criminal acts are deterred by the threat that an entire subclan will retaliate if one of its members is attacked.[32] Consequently, the clan tends to be the institution that people turn to in times of violence and danger.

However, it is important to note that only in situations of conflict do these clans mobilize as actual groups. They form what anthropologists term a "segmentary political system," in which the constituent parts are relative, constituting what is perhaps best thought of as a collection of interlocking, emergent groupings akin metaphorically to Russian nesting dolls.[33] Such systems are decentralized, highly individualistic, and democratic in nature. Clans and subclans are led by "elders," generally senior, adult males, but the idea of "elder" is not synonymous with "chief."[34] "Clan heads (commonly styled 'Sultan') have little established authority. At every level of political division, the elders *(oday, duk)* make policy, meeting in ad hoc councils *(shir)* in which every adult male traditionally has the right to speak."[35]

Clans also provide the critical political context. "A Somali genealogy," Ioan Lewis writes,

> is not a mere family tree.... [I]t represents the social divisions of people into corporate political groups. By reference to his ancestors, a man's relations with others are defined, and his position in Somali society as a whole determined. Thus an understanding of political relations between groups requires a knowledge of their genealogical relationships. At the same time, the range of agnatic relationship recognized on one occasion need not be the same as that on another, so that the corporate kinship group in which an individual has political status varies with the context. Thus although political and legal affiliation might be elastic it fluctuates largely within the range of agnatic connection "defined in the genealogies."[36]

As Moshe Terdman puts it, in Somalia "one does not have a permanent enemy or a permanent friend—only a permanent context."[37] There are six major clan families. Of these, the Hawiye, Darood, Isaaq, and Dir are traditionally pastoralist groupings. They have a higher status than the two other, agro-pastoralist, groupings, the Rahanweyn and the Digil, known collectively as the "Sab."[38] Each grouping is made up of numerous subclans, lineages, and extended families.[39] The Darood are the largest and geographically most widely distributed group; the Isaaq are derived historically from the Dir and linked to the Hawiye in a genealogical grouping known as the "Irir."[40]

Notions of government and statehood can quickly wither in the face of clan-based conflict, much of which has been driven since 1991 by the desire for the wealth and power revolving around commercial opportunities, including ports and airports, access to land and natural resources, jobs, and contracts with aid agencies. All the rather modern-sounding political groupings that emerged to confront Barre, such as the United Somali Congress (USC) and the SNM, were in the main vehicles for individual clans—in these examples the Hawiye and the Isaaq, respectively.[41] Although the SNM cooperated closely with the Hawiye USC, it retained its own Isaaq character and focus on the liberation of what is now Somaliland.[42] The Somali Salvation Democratic Front, led by Abdullahi Yusuf, was always an essentially Majarteen organization whose purpose, at least in the view of the SNM, was the restoration of Darood (and Majarteen) hegemony throughout Somalia.[43]

Somalia's deep-seated clan loyalties have impeded the creation of any sort of unifying political entity, let alone a "state" in the Western sense. Creating such structures may not be easy anywhere but appears to be particularly problematic in Somalia, where social positions can be felt acutely and defended fiercely. This state of affairs has spurred the belief that the creation of clan homelands may well offer the most practical solution to Somalia's political problems. Northern Somalia is divided among three clan families: the Dir, in the west, who reside partly in Somaliland and partly in Djibouti; the Isaaq, in the center; and the Darood, in the east. The borders of Somaliland largely coincide with those of the Isaaq. Those of Puntland embrace the territory of the Harti clan of the Darood, which consists of the Warsangeli, Dulhabante (who live partly in Somaliland), and the Majarteen but is dominated by the Majarteen.[44]

In southern Somalia, however, decades of migration and settlement, some of it forced, have produced an intermingling of clan areas that makes the formation of clan homelands much more difficult, although this did not prevent the Marehan from exercising power in the Gedo region at the expense of the Rahanweyn, and the Rahanweyn declaring nonmembers outsiders in the south-central Bay region. Nor have affiliations prevented clans from forming what usually prove to be fluid but mutually beneficial alliances.[45] For example, the Marehan and Habir Gedir Ayr formed the "Jubba Valley Alliance" to control the lucrative traffic through the port of Kismayo.[46]

Clan affiliations are not immutable, as Lewis makes clear. Peter Little too argues that clans are not the rigid, easily identifiable, and internally homogeneous groupings beloved in Western analysis.[47] While several political and social organizing factors can weaken clan affiliation, clanism remains the principal organizing force in Somali society.[48] Islam might be one such nonclan organizing factor, but any move to unify the country on the basis of religious allegiance is likely to depend on its effect on clan interests.

The clan structure of Somalia makes government corruption particularly potent, and "corruption is the main vehicle, and likely the most socially damaging activity, by which criminal gangs achieve their aims."[49] Poverty comes a close second; it is perhaps the major reason why pirate gangs find it so easy to attract recruits. For these reasons, in addition to civil war and the lack of proper maritime supervision, Somalia has seen a rapid increase in piracy incidents. Experience elsewhere in the world suggests that the number of actual attacks exceeds the number reported, by an unknown margin, and it is therefore quite possible that Somali piracy attacks are more numerous than publicly available figures suggest.

The Financial Backing for Somali Piracy

Piracy quickly became a multimillion-dollar business in the midst of what many consider to be a failed state. Funding for these ventures now comes from a variety of sources, ranging from individuals, who provide the equipment and consumables; and investment groups, some of them almost certainly based overseas, that take shares in individual ventures much as merchants used to buy shares in ship cargoes in the early days of sail; to owners of boats who allocate places to men who contribute food or guns and whose eventual reward reflects these contributions. Profits have not been simply spent on indulgences; some limited reinvestment is made in GPS systems, satellite phones, and even night-vision goggles.

Although the pirates' main goal is to capture ships and their crews for ransom, there is a symbiotic relationship between the internal political situation in Somalia and the maritime security situation offshore. In the north of the country, the self-proclaimed Republic of Somaliland has effective control over its territory and has established a coast guard to combat piracy off Somaliland's shores. Because of effective policing on and off shore, there have been no reports of piracy attacks off Somaliland, even as the number of incidents has escalated elsewhere along the coast. That escalation has been most marked off Puntland in the northeast and the neighboring regions of Mudug and Galguduud to the south. Coast guards alone, however, are not the solution and may, if their activities are not sustained, as indicated already, contribute to the problem. In 1999 the Puntland authorities contracted (as noted above) a British company, Hart Security, to establish such a force. The fees for fishing licenses that it would enforce would pay for its services. In 2001, however, President Abdullahi Yusuf refused to step down at the conclusion of his constitutionally mandated term, sparking a conflict with his rival, who was eventually forced to flee. The conflict caused the coast guard to split into factions that supported the rival candidates, forcing Hart to withdraw.[50]

In 2002 Hart's role was taken over by a new venture, which took its name SomCan from its Somali-Canadian joint ownership. The Somali partners were members of the Taar, a *diya*-paying group aligned with members of the same Oman Mohammed subclan as Yusuf. The same exploitative model as Hart has used funded the new company but on a greater scale—Hart had deployed one fisheries enforcement vessel with between seventy and eighty men, while SomCan deployed six together, with a four-hundred-man force, the members of which Jay Bahadur, who interviewed two Taar leaders, described as "marines." Not all of the licenses the company issued originated in the Ministry of Fisheries; some were of its own make. Given, however, its close connection with Yusuf and the corrupt nature of his regime, this issue may be regarded as a quibble. In a crucial difference from Hart, SomCan required license holders to obtain Somali agents, with whom it could deal directly. In return it not only mounted patrols using its own ships but placed militiamen on board foreign vessels to protect them, in the main, from local Somali fishermen, who suffered from the foreign boats' rapacious fishing methods. Between 2002 and 2005 SomCan operated what was effectively a protection racket for privileged foreign (and some domestic) concerns. Its own boats and those it was protecting regularly moved close to the coast, driving off local artisanal fishermen and tearing up their gear.[51]

It is likely that the main domestic operator who benefited from their protection was Hassan Munya. Munya had been the manager of the Somali High Seas Fishing Company (SHIFCO), a joint venture between an Italian company and the Barre regime prior to Barre's overthrow in 1991. Equipped with five trawlers and a freezer-equipped mother ship, it supplied fish to Italy and the European Economic Community (as the EU was then known).[52] When the Barre regime collapsed, Munya took over the ships and sailed them to Aden. It becomes something of a moot point to describe the continued use of SHIFCO's vessels as illegal, but their actions were enough to provoke local fishermen into holding them for ransom on at least four occasions.[53] In addition to fishing without due regard for local interests, Munya's operation was also accused of piracy against local and international shipping; he was said to have equipped his ships with heavy weapons and to have had between fifty and sixty men under his command. His men fired on the Hart-operated protection vessel at least once. When confronted with evidence of his crew's misbehavior, Munya agreed to purchase licenses from Yusuf's regime. The deal was reached despite his apparent threat to kill anyone who worked for Hart Security.[54]

The official position of the Yusuf administration was that it had no evidence with which to prosecute him.[55]

Many other pirate groups appear to be closely linked with the ruling clans. The "Somali marines" began operating in 2005 from Haradheere, a village that lies outside Puntland in the part of the Mudug region inhabited by the Suleiman subclan of the Hawiye. This is an area that even in Somalia is regarded as little more than a no-man's-land between the more populous regions to the north and south. It was founded by Mohamed Abdi Hassan "Afweyne," a Suleiman, on a cross-clan understanding. The Suleiman, while they are Hawiye, had no particular affiliation with the main clan and reportedly little in common with the politically active Habir Gedir and Abgaal subclans to the south that dominated Hawiye affairs.[56] They certainly had no lasting disagreements with the Majarteen. According to the pirate leader "Boyah," Afweyne handpicked his pirate group carefully, taking pirates and pirate trainers from Puntland, including "Boyah" himself, plus Mohamed Garad and Farah Abdullahi.[57] The remoteness of the base left it isolated from the ravages of Somalia's civil war and underpopulated with public officials needing to be bribed. The result was "an entrepreneurial alliance" between the Suleiman and the Majarteen, which remain the dominant forces in Somali piracy even though it has included other clans subsequently, the Saad in particular.[58]

It seems unlikely that such an operation could have arisen, survived, and then thrived, however, without powerful political protection or without drawing on the reservoir of piratical and commercial expertise that resided within Puntland.[59] The Haradheere group announced its presence with the hijacking of a liquefied–petroleum gas carrier, *Feisty Gas,* fifty nautical miles off the coast in April 2005, about six months after Yusuf became president of the Transitional Federal Government and only one month after Yusuf's successor Muhamed Muse Hersi terminated SomCan's contract. The evidence may be circumstantial, but during his time as a civil servant Afweyne was a close associate of Yusuf's, particularly during his struggle to secure the Puntland presidency in 2001.[60] Senior Majarteen political figures were used to taking shares of the proceeds in recognition of their positions: "You can't have that much money coming in or going out without the top clan people being involved," as Abdirahman Ibrahim, a Puntland academic, puts it.[61] It seems inconceivable they would have allowed such a lucrative source of revenue to slip beyond their control. Meanwhile, the Somali Salvation Democratic Front, the political front organization for the Majarteen clan, led by then colonel Abdullahi Yusuf, is suspected of being behind the attacks on *Bonsella* and *Baharihindi.*[62] Yusuf is believed to have financed his rise in Puntland politics from the proceeds of the huge ransom extorted for the release of MV *Shen Kno II* in particular.[63]

Although the "Somali marines" were openly recognized as a pirate group, they emphasize their role in protecting Somali resources and the livelihoods of coastal communities from foreign exploitation. They portray any fees collected or cargoes expropriated as connected with legitimate defensive efforts.[64] The evolution over time of Somali piracy from individual attacks to organized group attacks, often having more in common with commerce raiding than piracy, has resulted in the adoption of new, more successful, strategies.

Evolving Piracy Strategies

The Haradheere high-seas piracy ring used the traditional tools available to Somali fishermen to increase its success rates. From the outset it employed small motorized boats made of fiberglass with Styrofoam cores. Its members used multiple skiffs, with a larger skiff acting as a "mother ship" to support one or more smaller boats. The smaller skiffs, each with a crew of four or five pirates, would come alongside a vessel, one to starboard and the other to port, with the larger skiff astern in pursuit. They then placed one or more of their number on board the target vessel to intimidate the crew, allowing the rest of the boarding party to bring the captured vessel into port.[65] This technique, first implemented in early 2005, resulted in some huge captures.

The diversion of Combined Task Force 150 elements, especially after the *Seabourn Spirit* incident during 2005, perhaps prompted a change in pirate habits. For example, the Haradheere group soon began using captured low-value vessels as mother ships. During 2006 the Council of Islamic Courts (CIC) tried to halt the pirate activities. The CIC briefly reopened the port of Mogadishu and began to gather port entry fees and profits. However, this new regime lasted only a very short time. Even before the Ethiopian National Defense Force had defeated the CIC in 2006, the pirates were back at sea hunting for new targets.

Beginning in 2007, the number of pirate attacks quickly increased. A small cargo ship, MV *Rozen,* was hijacked in February and another cargo ship, MV *Nimattulah,* on 1 April. The influence of the Afweyne family had temporarily declined during the CIC interlude, allowing several new groups to emerge.[66] These ranged in size from father-and-son combinations to large organizations of two hundred pirates or more. Although the older pirate organizers like Afweyne and Mohamed Garad remained active, Stig Hansen reports that many of the newer groups were often "a loose constellation around a pirate leader, usually a veteran pirate that re-invests funds in new pirate missions and who often functions as a fund-raiser."[67]

One of the most astonishing confirmations of the commercial nature of Somali piracy was the report in December 2009 of the existence in Haradheere of a cooperative, or "stock exchange," that was attracting a wide range of potential investors. According to one pirate, the cooperative had proved to be an excellent way of involving the local community: "The shares are open to all," he told a reporter, "and everyone can take part, whether personally at sea or on land by providing cash, weapons or useful materials[;] . . . we've made piracy a community activity." A local official appeared to confirm this: "Piracy-related business has become the most profitable economic activity in our area and as locals we depend upon their output. The district gets a percentage of every ransom from ships that have been released, and that goes on public infrastructure, including our hospital and public schools."[68] In actuality, such public infrastructure investment, as compared to private construction, has been hard to find.

The international dimension remains murky. Members of the Somali expatriate communities in Kenya and Dubai figure regularly in reports.[69] Initially money was probably paid to interests based outside the country, but much of what entered Somalia never came out again.[70] What money did come out appeared to be invested in Dubai and in Kenya, where the most obvious consequence was a property boom in Eastleigh, a suburb of Nairobi dominated by Somali refugees and expatriates.[71] In May 2009 the head of Interpol, Ronald Noble, pressed for a global alliance of criminal investigators to track pirates through the ransom-money trail, while at an Interpol meeting held in Singapore later in the year Australia's inspector of transport security, Mick Palmer, suggested there was "clear evidence" of increased organization based on the use of more sophisticated weapons and the gangs' ability to locate ships farther and farther from the coast.[72] This would certainly fit the classic pattern of criminal gang evolution, but the evidence for these specific assertions remained thin.[73]

Attempts by Western investigators to track the money once it has entered Somalia have to date borne little fruit.[74] Somalis traditionally base everything on trust.[75] Their society remains closed and xenophobic, while as one maritime analyst pointed out, the degree to which payments needed to be split among various domestic groups to enable the gangs to continue to operate probably left little margin for international syndicates to make money.[76] Whether or not this remains the case now is questionable. Gangs still need to pay off their political protectors and local officials and to spread the proceeds among the members of their clans, but most sources now agree that pirate financiers could take between 20 and 30 percent of the reward. Thus, enormous potential returns on investment suggest that overseas investors have become much more prominent in the gangs' activities.

Because of the very success of these pirate organizations, financial links among the pirates, the Somali clans, and government officials are almost certain. What portion would go to the pirates' government supporters is unclear. However, at least some of the pirates are undoubtedly working hand in hand with clan leaders and local administrations—or at the very least, corrupt officials in those administrations—which makes that portion of Somalia's piracy problem similar to government-sponsored piracy, or commerce raiding.

Conclusions

Although attacks off the Somali coast appear to be pure piracy on the surface, they should in fact be considered as a new form of government-sponsored, at least government-complicit, commerce raiding. Clan-based political groups, such as the SSDF, needed from their inceptions to fund their operations. The UN intervention in Somalia in 1993–95 had taught them that foreign kidnap victims could readily be turned into cash. The armada of foreign fishing vessels that entered Somalia's unprotected waters presented tempting targets that warlord groups up and down the coast preyed on, arguing all the while—some more genuinely than others—that they were "legitimate coastguards protecting the waters from unlawful fishing or contamination."[77]

But years of successful boat seizures and kidnappings, which have led to ever larger ransoms, have made this highly lucrative form of commerce raiding even more attractive to a wide range of backers, including most importantly clan leaders, local government officials, and shadowy financial figures overseas. As a result, cases of piracy in the waters off Somalia have increased rapidly during recent years, to the point where the international community seems to be at a loss for a strategy to deal with the problem.

Any effective strategy would first need to recognize that Somali piracy is a rational response that satisfies an economic need by exploiting a security weakness. That security weakness will never be closed by using naval forces operating only at sea. To exert control the security forces would need to raid ashore. But Somali piracy is not a criminal fraternity hiding in the midst of an otherwise largely law-abiding society. It constitutes a significant part of that society. It has a human and geographic hinterland. It furthermore displays features of a commercial system that shows signs of turning into a permanent way of life.

Navies will have a clear and purposeful role in the solution of this riddle, but only in combination with strategies conducted on land. Better intelligence from land-based sources will improve chances of interdiction. Those caught can be returned with relative ease to face Somali, not international, justice. Ports and stretches of coastline can be closely patrolled and even effective "exclusion zones" established, because ships can be supported by land-based policing. Over time, expensive multimission warships could be replaced by more specialized and less expensive coastal patrol craft and converted civilian ships that can gradually be turned over to local coast guards as their capacities and capabilities increase. Illegal fishing in Somalia's EEZ can be curtailed, perhaps by the navies of the nations from whence some of the illegal boats come. Unity of effort, which has proved elusive so far, might thus be achieved.

But the solution is not solely naval. Suppressing piracy means securing the land, which in turn demands engagement with the Somali people, thus requiring a realistic political strategy. If the Somali pirates were to be treated not as simple pirates but as commerce raiders, a whole new range of options might open to those groups attempting to halt the attacks. It would be most important, of course, to find and eliminate the real operational and financial masterminds behind the pirate organizations. These backers include clan leaders and government officials locally, and financiers, who often reside abroad—just as the merchants who underwrote seventeenth-century piracy often lived in London or New York. The assets of those who persist in piracy, however, by spurning attempts to draw Somalia into the international order, will need to be squeezed, and their money needs to be confiscated wherever it is held. Only when the piratical activities become unprofitable for the clans, for members of the Somali government engaged in piracy, and for international backers will the Somali people as a whole desire to put a stop to these activities.

Notes

Parts of this chapter have been adapted from other published works by the author, including *Somalia: The New Barbary? Piracy and Islam in the Horn of Africa* (New York: Columbia Univ. Press, 2011).

The thoughts and opinions expressed in this publication are those of the author and are not necessarily those of the U.S. government, the U.S. Navy Department, or the Naval War College.

1. See Martin N. Murphy, *Contemporary Piracy and Maritime Terrorism: The Threat of International Security,* IISS Adelphi Paper 388 (London: Routledge, 2007), p. 30.
2. "Vietnam Firm Pays 'Millions' to Free Pirated Ship," Agence France-Presse, 26 September 2011.
3. Rob de Wijk, David M. Anderson, and Steven Haines, "The New Piracy: Three Contexts," *Survival* 52, no. 1 (February–March 2010), p. 45.
4. National Geospatial-Intelligence Agency, "Anti-shipping Activity Message 1989-19," 5 December 1989, Springfield, Va. [The series hereafter referred to using the format NGA ASAM, reference number, date.]
5. Ibid.
6. Brian Scudder, "Somalia Pirates' Last Stand," *African Business,* 1 March 2000.
7. On the *Norasia Samantha* incident see NGA ASAM 1994-25, 31 May 1994; on the *Glucksburg* incident see NGA ASAM 1994-26, 31 May 1994; on the *Bonsella* incident see IMO MSC/Circ. 698, 30 June 1995, p. 3.
8. For a fuller discussion of piracy underreporting, see Martin N. Murphy, *Small Boats, Weak States, Dirty Money: Piracy and Maritime Terrorism in the Modern World* (New York: Columbia Univ. Press, 2010), pp. 59–72.
9. Some support for this suggestion is provided by the notorious and successful pirate leader Farah Hirsi Kulan, known as "Boyah," who told Stig Jarle Hansen in an interview in 2009 that organized piracy had started in 1994, on the basis of a fishery self-protection group that had first come together in 1992. Stig Jarle Hansen, *Piracy in the Greater Gulf of Aden: Myths, Misconceptions and Remedies* (Oslo: Norwegian Institute for Urban Regional Research, 2009), p. 20, en.nibr.no/uploads/publications/26b0226ad4177819779c2805e91c670d.pdf.

10. On MV *Full City*, see NGA ASAM 1995-48, 3 April 1995.

11. On the issue of distance, see NGA ASAM 1995-91, 3 May 1995, recounting the attack on MV *Liliana Dimitrova* a hundred nautical miles off Socotra.

12. International Maritime Bureau, *Piracy and Armed Robbery against Ships: A Special Report* (London: International Chamber of Commerce, July 1997), p. 16.

13. Anthony Foster, "An Emerging Threat Shapes Up as Terrorists Take to the High Seas," *Jane's Intelligence Review*, July 1998, p. 43.

14. NGA ASAM 1998-60, 27 September 1998.

15. Peter Chalk, *Non-military Security and Global Order: The Impact of Extremism, Violence and Chaos on National and International Security* (London: Macmillan, 2002), p. 75.

16. Scudder, "Somalia Pirates' Last Stand."

17. "Horn 'Anti-terror Alliance,'" *BBC News*, 13 January 2003.

18. On the convoy recommendation, see "Regional Security: East Africa," *Fairplay*, 26 September 2005.

19. International Maritime Bureau, *Piracy and Armed Robbery against Ships: Annual Report, 1 January–31 December 2006* (London: International Chamber of Commerce, 2007), pp. 5, 24.

20. See Murphy, *Contemporary Piracy and Maritime Terrorism*, pp. 17, 29, citing "Somalia: Security Council Urges Action over Piracy off the Coast of Somalia in Line with IMO Assembly Resolution," *Cargo Security International*, 17 March 2006.

21. Murphy, *Contemporary Piracy and Maritime Terrorism*, p. 31.

22. Peter Lehr and Hendrick Lehmann, "Somalia: Pirates' New Paradise," in *Violence at Sea: Piracy in the Age of Global Terrorism*, ed. Peter Lehr (London: Routledge, 2006), p. 13, quoting UN figures.

23. Abdulkadir Khalif, "How Illegal Fishing Feeds Somali Piracy," *East African* (Nairobi), 15 November 2005; Horand Knaup, "The Poor Fishermen of Somalia," *Spiegel Online*, 4 December 2008; Shreya Roy Chowdhury, "Piracy Fallout of Illegal Fishing," *Times of India*, 27 January 2009; Gabobe Hassan Musse and Mahamud Hassan Tako, "Illegal Fishing and Dumping Hazardous Wastes Threaten the Development of Somali Fisheries and the Marine Environments" (paper delivered to Tropical Aquaculture and Fisheries Conference 99, Terengganu, Malaysia, 7–9 September 1999).

24. United Nations, Office for the Coordination of Humanitarian Affairs, *Somalia: Fishermen Appeal for Help over Foreign Fishing Ships* (New York: 2006); this report is also available at Reuters AlertNet, 9 March 2006. United Nations, *Report of the Monitoring Group on Somalia Pursuant to Security Council Resolution 1630 (2005)*, S/2006/229 (New York: 2005), p. 25, describes how following the fall of Barre "regional authorities" used "armed local militias to protect [fishing interests against] what they considered to be . . . a threat to their respective self-defined interests." In some cases this involved seizing vessels.

25. Scott Coffen-Smout, "Pirates, Warlords and Rogue Fishing Vessels in Somalia's Unruly Seas," accessed 6 July 2006, www.chebucto.ns.ca/~ar120/somalia.html.

26. Jeffrey Gettleman, "Somali Pirates Tell All: They're in It for the Money," *New York Times*, 1 October 2008.

27. Khalif, "How Illegal Fishing Feeds Somali Piracy"; Coffen-Smout, "Pirates, Warlords and Rogue Fishing Vessels in Somalia's Unruly Seas."

28. The International Expert Group discussed the possible involvement of fishermen in piracy, including the possibility that instead of providing their services voluntarily they could have been bullied into providing them. International Expert Group on Piracy off the Somali Coast, *Piracy off the Somali Coast*, Workshop Commissioned by the Special Representative of the Secretary General of the UN to Somalia Ambassador Ahmedou ould-Abdallah (Nairobi: 21 November 2008), p. 17, available at www.asil.org/files/SomaliaPiracyIntlExpertsreportconsolidated1.pdf.

29. As the UN Monitoring Group commented in its 2010 report, "Exploitation of Somali marine resources is a reality, but it is by no means a preoccupation of Somali pirates or their backers. In 2009, only 6.5 per cent of Somali pirate attacks were aimed at fishing vessels, of which only one . . . is confirmed to have been in Somali territorial waters at the time." United Nations, *Report of the Monitoring Group on Somalia Pursuant to Security Council Resolution 1853 (2008)*, S/2010/91 (New York:

10 March 2010), p. 37; International Expert Group, *Piracy off the Somali Coast*, p. 15.

30. Harmony Project, "Al-Qaida's (Mis)Adventures in the Horn of Africa" (Combating Terrorism Center at West Point, n.d.), p. 29, accessed 4 November 2008, available at ctc.usma.edu/aq/pdf/Al-Qa%27ida%27s%20MisAdventures%20in%20the%20Horn%20of%20Africa.pdf; Kenneth John Menkhaus, "Somalia and Somaliland: Terrorism, Political Islam and State Collapse," in *Battling Terrorism in the Horn of Africa*, ed. Robert I. Rotberg (Washington, D.C.: Brookings Institution, 2005), p. 26.

31. Ioan Lewis, *A Pastoral Democracy* (Oxford, U.K.: James Currey for the International African Institute, 1999), pp. 162–67.

32. "Al-Qaida's (Mis)Adventures in the Horn of Africa," p. 30.

33. Ken Menkhaus, "Local Security Systems in Somali East Africa," in *Fragile States and Insecure People? Violence, Security and Statehood in the Twenty-First Century*, ed. Louise Andersen, Bjørn Møller, and Finn Stepputat (New York: Palgrave Macmillan, 2007), p. 80.

34. Ioan Lewis, correspondence with the author, November 2009.

35. Ioan Lewis, *Understanding Somalia and Somaliland* (London: Hurst, 2008), p. 28.

36. Lewis, *Pastoral Democracy*, p. 2.

37. Moshe Terdman, *Somalia at War: Between Radical Islam and Tribal Politics*, S. Daniel Abraham Center for International and Regional Studies, Research Paper 2 (Tel Aviv: Tel Aviv Univ., March 2008), pp. 32–34, available at www.e-prism.org/images/somalia_-_Mar08.pdf, pp. 36–37.

38. Members of the Sab traditionally followed what in Somali eyes were degrading crafts, such as leatherworking, haircutting, and metalworking; Ioan Lewis, *Blood and Bone: The Call of Kinship in Somali Society* (Lawrenceville, N.J.: Red Sea, 1994), p. 127. The Rahanweyn translate the term as "large crowd," which gives some idea of its mixed composition; ibid., p. 162.

39. Lewis, *Pastoral Democracy*, pp. 1–10. Lewis also provides a summary in *Understanding Somalia and Somaliland*, pp. 3–4, 49–50.

40. Lewis, *Understanding Somalia and Somaliland*, p. 5.

41. Shaul Shay, *The Red Sea Terror Triangle* (New Brunswick, N.J.: Transaction, 2005; repr. 2007), p. 68. For a list of the factions together with their clan linkages, see Lewis, *Blood and Bone*, pp. 234–35.

42. Lewis, *Blood and Bone*, p. 214.

43. Ibid., pp. 207–208.

44. Kenneth John Menkhaus, *Somalia: State Collapse and the Threat of Terrorism* (Oxford, U.K.: Oxford Univ. Press, 2004), p. 20; Lewis, *Blood and Bone*, p. 100. This domination is reflected in the fact that the Warsangeli and the Dulhabante maintained their own political organizations separate from the SSDF. Moreover, the SNM's victory was assured when the Dulhabante, who occupy what is now the Somaliland-Puntland border area, took its side against Barre late in the civil war. Lewis, *Blood and Bone*, p. 214.

45. Lewis, *Pastoral Democracy*, pp. 193–95; Lewis, *Understanding Somalia and Somaliland*, p. 77.

46. Menkhaus, "Somalia and Somaliland," p. 26.

47. Peter D. Little, *Somalia: Economy without State* (Oxford, U.K.: James Currey for the International African Institute, 2003).

48. William Reno, *Somalia and Survival: In the Shadow of the Global Economy*, QEH Working Paper Series 100 (Oxford, U.K.: Oxford Department of International Development, February 2003), p. 2, available at www3.qeh.ox.ac.uk/RePEc/qeh/qehwps/qehwps100.pdf. (This series is now known as the ODID Working Papers.)

49. Murphy, *Contemporary Piracy and Maritime Terrorism*, p. 86, citing Kimberley L. Thachuk and Sam J. Tangredi, "Transnational Threats and Maritime Responses," in *Globalization and Maritime Power*, ed. Sam J. Tangredi (Washington, D.C.: National Defense Univ. Press, 2002), p. 60.

50. See Martin N. Murphy, *Somalia: The New Barbary? Piracy and Islam in the Horn of Africa* (New York: Columbia Univ. Press, 2011), pp. 22–23.

51. Jay Bahadur, *The Pirates of Somalia: Inside Their Hidden World* (New York: Pantheon, 2011), pp. 64–65. See also Christopher Paul Kinsey, Stig Jarle Hansen, and George Franklin, "The Impact of Private Security Companies on Somalia's Governance Networks," *Cambridge Review of International Affairs* 22, no. 1 (March 2009), pp. 152–54; and Jonathan

Gatehouse, "This Cabbie Hunts Pirates," *Macleans,* 12 January 2009.

52. De Wijk, Anderson, and Haines, "New Piracy," pp. 46–47.

53. Abdirahman Jama Kulmiye, "Militia vs. Trawlers: Who Is the Villain?," *East African Magazine,* 9 July 2001.

54. Brian Scudder, "Pirate King Turns Law Enforcer," *African Business,* 1 July 2000. See also an intriguing reference to an interchange between Lord Bethell, Hart's founder, and a man he calls a pirate who shot and injured one of his men: Charlotte Parsons, "A Modern Mercenary's Tale," *BBC News,* 14 February 2002.

55. Scudder, "Somalia Pirates' Last Stand."

56. Interview with insurance negotiator, December 2006.

57. Hansen, *Piracy in the Greater Gulf of Aden,* pp. 16–17.

58. Ibid., p. 17.

59. See Murphy, *Somalia: The New Barbary?,* p. 59.

60. J. Peter Pham, interview, September 2009.

61. Jonathan Clayton, "Millions Are Starving in Somalia, but in Eyl Piracy Is Big Business," *Times,* 19 November 2008. See also Roger Middleton, "Piracy in Somalia: Threatening Global Trade, Feeding Local Wars," Chatham House Briefing Paper (October 2008), p. 5, www.chathamhouse.org.uk/files/12203_1008piracysomalia.pdf.

62. See Murphy, *Somalia: The New Barbary?,* pp. 12–13.

63. J. Peter Pham, "Putting Somali Piracy in Context," *Journal of Contemporary African Studies* 28, no. 3 (July 2010), p. 333.

64. Andrew McGregor, "The Leading Factions behind the Somali Insurgency," *Jamestown Foundation,* www.jamestown.org/terrorism/news/article.php?articleid=2373348.

65. For example, see the action of USS *Porter* in 2007 near the Horn of Africa: Barbara Staff, "U.S. Destroyer Pursuing Hijacked Ship in Somali Waters, Military Says," CNN.com, 29 October 2007; and "Pirates Attack UAE Ship off Somalia," *Yahoo,* 25 January 2006, asia.news.yahoo.com/060125/3/2eolb.html.

66. Hansen, *Piracy in the Greater Gulf of Aden,* p. 27.

67. Ibid., pp. 34–35.

68. Mohamed Ahmed, "Somali Sea Gangs Lure Investors at Pirate Lair," Reuters, 1 December 2009.

69. Mohamed Olad Hassan and Elizabeth Kennedy, "Somali Piracy Backed by International Network," Associated Press, 10 December 2008.

70. International Expert Group, *Piracy off the Somali Coast,* p. 20.

71. Shashank Bengali, "Kenya Awash in Somali Pirates Cash," *Seattle Times,* 23 May 2009; "Pirate Cash Suspected Source of Kenya Property Boom," Associated Press, 1 January 2010.

72. Doreen Carvajal, "Interpol Takes Aim at Somali Pirates," *New York Times,* 29 May 2009; "Interpol: Crime Syndicates Helping Somali Pirates," *VOANews,* 14 October 2009; "Foreigners Involved in Somalia Piracy," Agence France-Presse, 14 October 2009.

73. For a discussion of organized-crime development and how this might apply to piracy gangs, see Murphy, *Small Boats, Weak States, Dirty Money,* pp. 124–26.

74. John Cassara, "Following the Pirates Money? Here We Go Again," Johncassara.com, 22 April 2009; Tabassum Zakaria, "US Chase of Somali Pirate Assets Faces Rough Seas," Reuters, 21 April 2009.

75. Robyn Hunter, "How Do You Pay a Pirate's Ransom?," *BBC News,* 3 December 2008.

76. Mary Harper, "Chasing the Somali Piracy Money Trail," *BBC News,* 24 May 2009.

77. "Freed South Korean Trawler Arrives to Cheers in Kenya," *Khaleej Times,* 5 August 2006.

Conclusion: *Guerre de Course* in the Modern Age

Guerre de course—or more generically, commerce raiding—is a means to contribute to the achievement of national or nonstate-actor goals. What kinds of commerce raiding have been employed? What types of operational goals has commerce raiding furthered most effectively? What types of strategic effects have various kinds of commerce raiding produced? What circumstances are most and least conducive to their operational and strategic success? How have other parties—both belligerents and neutrals—responded? What kinds of enemy adaptation and counterstrategies have been most effective?

These questions will be examined here in terms of time, space, force, operational versus strategic goals, countermeasures, and overall operational and strategic effectiveness. *Time* includes both the rate of implementation and duration. *Space* focuses not just on the area under attack but also on the sea and land lines of communications of both sides. *Force* refers to all available instruments of national power. *Operational goals* concern the intended first-order effects of the commerce raiding, while *strategic goals* mean the objectives of the conflict. *Countermeasures* concern enemy adaptation. Finally, *effectiveness* is measured at both the operational and strategic levels.

Commerce raiding types can be categorized in a number of ways: rapid, intermittent, tightening, or loosening (in terms of implementation); short, medium, and long (in terms of duration); close or distant (in terms of the distance of the theater from the territory of the victim); near or far (in terms of the distance of the theater from the territory of the perpetrator); joint (when different military services of one country cooperate) or combined (when militaries of allied countries coordinate); and partial or total (in terms of porosity). Over time technological breakthroughs have greatly influenced the cost, execution, and feasibility of all types of commerce raiding, as shown initially by the development of instruments to locate and target specific ships in World War II, and most recently by the ability of small Somali skiffs to hijack huge oil tankers with the aid of handheld GPS tracking devices.

Time: Implementation and Duration

Both the rate of implementation and duration of a commerce raiding campaign can influence its effectiveness. Implementation can be rapid, intermittent, tightening, or loosening, while the duration can be short, medium, or long. For instance, the French quickly adopted commerce raiding operations during the French Wars, but over the long term they conducted them only intermittently; meanwhile, because of faulty torpedoes and poor leadership, American commerce raiding against Japan during World War II tightened only gradually over several years.

Table 1, "Time," shows how these factors played out in the sixteen case studies in this volume. The five rapidly implemented commerce raiding campaigns comprise the Confederate side of the U.S. Civil War, the *Jeune École* (in theory, at least), the Japanese in the First Sino-Japanese War, the American campaign upon entry in World War I, and the Fascists in the Spanish Civil War. In all of these conflicts rapid commerce

TABLE 1
Time

CONFLICT	IMPLEMENTATION	DURATION	OPERATIONAL EFFECTIVENESS	STRATEGIC OUTCOME
Seven Years' War	British tightening	long	yes, but creates trade with enemy	win
American Revolution	American intermittent	long	yes, undermines stronger Royal Navy response	win
French Wars	French intermittent	long	no, failed seriously to hurt Britain	lose
1812	American intermittent	medium	yes, ties up Royal Navy units, halts trade, helps force peace talks	draw
U.S. Civil War	Confederate rapid	medium	yes, sinks commercial ships, diverts Union response	lose
Jeune École	rapid (theory)	short (theory)	yes, imposes disproportionate costs (theory)	win (theory)
First Sino-Japanese War	Japan rapid	short	yes, cuts China's troop movements, immobilizes navy	win
Russo-Japanese War	Russia loosening	short	no, creates hostile neutral response	lose
WWI Germany	Germany tightening	medium	hurts cross-Atlantic and Mediterranean trade, but also creates hostile neutral response	lose
WWI USA	America joins war rapid	medium	yes, in combination with blockade results in German malnutrition to pressure home front	win
Spanish Civil War	Fascist rapid	medium	yes, cuts key arms imports, undermines Republican morale	win
WWII Germany	Germany tightening	long	hurts trade, but also creates hostile neutral response	lose
WWII Japan	Japan loosening	medium	Japan uses ASW too late, poor coordination of shipping	lose
WWII USA	America tightening	medium	yes, cuts trade, ruins Japanese economy, immobilizes expeditionary forces	win
Tanker War	Iran and Iraq intermittent	long	hurts oil trade, but also creates hostile neutral response	draw
Somali pirates	Somalia intermittent	long	yes, gains huge revenue stream, but creates hostile neutral response	unclear

raiding operations were relatively short (one to two years) to medium (three to four years) in duration and were usually waged by the victorious side—but not always, as the American Civil War showed. In the First Sino-Japanese War, the Chinese never contested Japanese command of the sea after suffering the loss of a troop transport and then defeat in one naval engagement. Once the United States entered World War I, its convoys and antisubmarine warfare campaign began rapidly and soon neutralized German commerce raiding. In line with the expectations of Jeune École theorists, the rapid introduction of commerce raiding seemed to force the Spanish Republic to capitulate. Nevertheless, commerce raiding was not the sole determining factor in the outcome of these wars.

Likewise, in all four cases of gradually tightening commerce raiding—Britain in the Seven Years' War, Germany in both world wars, and the United States against Japan in World War II—the raiding did not determine the outcome of the wars. Rather, commerce raiding worked in conjunction with other strategies: the attrition of ground forces, blockade, naval combat, alliance, and (in the World War II case of Japan) the American use of atomic bombs. The extensive size of these theaters perhaps explains why implementation was gradual versus rapid. Strategies of gradually tightening raiding worked for the dominant sea power but not for land powers, or even for a secondary sea power such as Germany in both world wars. Britain and the United States successfully combined commerce raiding with other strategies to win, variously, the Seven Years' War and both world wars, while unrestricted submarine warfare by Germany in World War I cost that nation the war by transforming a great naval power, the United States, from a neutral into a belligerent.

Five cases of intermittent commerce raiding—the American Revolution, the French Wars, the War of 1812, the Tanker War, and Somalia—produced mixed results: loss, draw, or victory due partly to other factors (such as French intervention at Yorktown, in the case of the American Revolution). These campaigns were perhaps most strategically effective in what they prevented from happening—in other words, in their deterrent or diversionary effects. For example, U.S. commerce raiding in the War of 1812 tied down many Royal Navy ships that might otherwise have attacked conventional American targets. However, when intermittent commerce raiding negatively affected the international community, such as during the French Wars, the Tanker War, and Somalia, it spurred the intervention of neutral powers.

There were only two cases in this book of loosening campaigns—Russia in the Russo-Japanese War and Japan in World War II—both powers whose attention turned to massive land battles that they were losing. In Russia's case, threats of neutral intervention convinced Russia to terminate commerce raiding from neutral Chinese ports. By contrast, huge Japanese naval and merchant ship losses after 1943 precluded effective countermeasures.

Six commerce raiding campaigns were long (five years or more)—the Seven Years' War, the American Revolution, the French Wars, Germany during World War II, the Tanker War, and Somali piracy. Two of these six gradually tightened over time, while the others were intermittent. Short and medium-length commerce raiding campaigns generally concentrated on an operational center of gravity, such as the Japanese sinking of the Chinese troopship *Kowshing* or the Confederate focus on Union shipping.

Only six of the cases examined (not counting Jeune École theorists) ended in clear victories for the sides with the more robust commerce raiding campaigns. In the Seven Years' War, the American Revolution, the First Sino-Japanese War, the American campaigns during both World Wars I and II, and the Spanish Civil War, the ability to conduct a commerce raiding campaign or to protect vital imports by sea appears to have been crucial for victory. Even so, it was but one of multiple critical factors that together determined the outcome. In eight cases, however, a side engaging in commerce raiding either lost or the conflict ended in a draw. All were continental powers, with the exception of Japan in World War II, and Japan in that war chose a continental strategy rather than a purely maritime strategy that could have led to a more productive use of its navy and merchant marine.

In other words, generally commerce raiding seems more strategically effective for a naval power than a continental power. Land powers that pursue the strategy can often achieve the operational effect of imposing far greater financial losses on their enemy than their raiding operations cost. Although the ships sunk are not available for future passages and their replacement requires a vigorous shipbuilding capacity, dominant naval powers typically have such capacity, either domestically or through allies. So the costs from lost ships are significant and cumulative but potentially disastrous only for a country lacking the capacity to replace lost ships and dependent on crucial war materiel delivered by sea. Usually, most of the trade still gets through, with the result that the financial losses constitute a small fraction of total trade. Moreover, commerce raiding can become a morale-enhancing catalyst for an angry victim. Although it might seem that overseas commerce would be a critical vulnerability for a maritime power, foreign trade has actually been most vulnerable when targeted by the dominant maritime power against a continental adversary. In fact, overseas commerce turns out to be an even more important critical vulnerability for continental powers, let alone secondary maritime powers, which lack the means to protect their trade. Dominant maritime powers tend to combine blockade with commerce raiding to cut off the victim's overseas trade virtually in its entirety—as exemplified by Germany's fate in both world wars.

Space: The Nature of the Theater

In most cases—ten of the sixteen—land powers adopted commerce raiding operations, and in eight they targeted sea powers. Only two of the campaigns were part of victorious wars: the American Revolution (targeting a sea power) and Spanish Civil War (targeting a land power). In each of these two cases the outcome depended in part on conventional military aid supplied by great-power allies—France for the decisive battle of Yorktown and the Axis powers for the Spanish Fascists. In half of the ten cases, the land powers engaged in commerce raiding had significant navies, and yet only one case resulted in victory—the Fascists in Spain, whose maritime assets were allied navies. The Spanish case was also the only one of the five in which the victim was a land, not a sea, power. In other words, commerce raiding conducted by land powers against sea powers has not generally resulted in a victorious war.

Five of the campaigns were conducted by sea powers, which won all but one of the wars. The only loss was Japan's in World War II, in which it took on the dominant naval powers, the United States and Great Britain. All the victorious campaigns save one took place in theaters distant from the victims and far from the perpetrators, the dominant sea power roaming the seas in search of targets. These were global wars—the Seven Years' War, World War I, and World War II—in which the global order was at stake and fighting took place around the planet. The exception, the First Sino-Japanese War, was a regional war between only two belligerents, not between global coalitions, and geography dictated a close-near theater.

The nature of the theater of operations can help determine success or failure. Commerce raiding distances can vary greatly. The terms "close" (for roughly a hundred to 150 nautical miles) and "distant" (more than 150 nautical miles) refer to the distance of the theater from the victim country, while "near commerce raiding" and "far commerce raiding" refer to the distance of the theater from the commerce raiding country. As shown in table 2, "Space," most cases included operations both distant from the shores of the victim and far from the shores of the perpetrator, such as in the Seven Years' War, the American Revolution, the French Wars, the War of 1812, the American Civil War, World War I for the United States, World War II for Japan and the United States, and more and more so in Somalia. Prior to technological improvements in the ability to locate hostile ships, nations engaged in far raids tended to pursue targets close to enemy shores. Technological improvements, however, made raids adjacent to enemy shores increasingly dangerous, so close commerce raiding campaigns have become rare. For a limited time submarines changed this dynamic in World War II, when Germany sank U.S. merchant ships along the Eastern Seaboard, but only until the United States implemented countermeasures. In the case of the Spanish Civil War and the Tanker War, the belligerents lacked significant navies, permitting commerce raiding close to

TABLE 2
Space

CONFLICT	COMMERCE RAIDER	VICTIM	DISTANCE FROM VICTIM	DISTANCE FROM RAIDER
Seven Years' War	Britain dominant sea power	France land power with large navy	close + distant	near + far
American Revolution	America land power	Britain dominant sea power	close + distant	far
French Wars	France land power with large navy	Britain dominant sea power	close + distant	near + far
1812	United States land power	Britain dominant sea power	close + distant	far
U.S. Civil War	Confederacy land power	Union land power with small navy growing to navy second only to that of Britain	close + distant	near + far
Jeune École	sea power (theory)	sea power (theory)	close + distant	near + far
First Sino-Japanese War	Japan sea power with large army	China land power with large navy	close	near
Russo-Japanese War	Russia land power with large navy	Japan sea power with large army	close + distant	near + far
WWI Germany	Germany land power with large navy	Britain, United States dominant sea powers	close + distant	far
WWI USA	United States dominant sea power	Germany land power with large navy	distant	far
Spanish Civil War	Fascists land power coalition with navy	Republic land power	close	far
WWII Germany	Germany land power with large navy	United States, Britain dominant sea powers	close + distant	far
WWII Japan	Japan sea power with large army	United States dominant sea power	distant	far
WWII USA	United States dominant sea power	Japan sea power with large army	distant	far
Tanker War	Iran and Iraq land powers	third-party neutral sea powers	close	near
Somali pirates	failed state land power	third-party neutral sea powers	distant	near + far

shore. Only three cases entailed commerce raiding near to the raider's shores—Japan in the First Sino-Japanese War, Russia in the Russo-Japanese War, and the Iran-Iraq Tanker War—cases where the belligerents bordered on each other or were separated by a narrow sea. Normally, merchant ships would not be sent near enemies intent on raiding, unless geography offered no alternative.

In virtually all of these case studies the country victimized by commerce raiding was too far away to retaliate effectively against the home territory of the perpetrator. The exclusively "distant" commerce raiding was also "far" commerce raiding. In the exceptional case, the Spanish Civil War, the commerce raiding occurred far from the perpetrator but close to the victim, mainly because the Republic lacked an adequate navy. Meanwhile, in the two cases of raiding exclusively close to the shores of the victim, the raiding was also near the shores of the perpetrator: the First Sino-Japanese War and the Tanker War. Both were regional, not global, wars, and the close-near factors reflect the constricted geography of the theater. In the Russo-Japanese War, which occurred in much the same theater as the First Sino-Japanese War, Russia took advantage of its geography to engage both in near commerce raiding, with its locally based naval assets, and also in far commerce raiding in the Red Sea, with its European-based naval assets. As for the situation in Somalia, as shipping companies ordered their vessels to sail farther away from Somalia, the "pirates" would also venture out onto the "high seas" to find prey.

Thus history suggests that countries that are primarily land powers have a very small chance of successfully using commerce raiding operations, except perhaps for a deterrent effect or to improve leverage for a peace settlement in the rare case when the costs from commercial losses are sufficiently disproportionate to promote negotiations, such as U.S. commerce raiding operations during the American Revolution and the War of 1812. Commerce raiding has been most significant in global wars as one of many elements of national power necessary to defeat a great power, such as Germany in both world wars and Japan in World War II.

Force: Joint and Combined Operations

After time and space, force constitutes a critical dimension of commerce raiding operations. Commerce raiding violates the "commons," historically defined in an international context as the oceans, which under international law are open to the common use of all. Commerce raiding normally attempts to transform passage through the commons into a gauntlet that imposes heavy costs on the enemy by diverting, restricting, or eliminating traffic. With the advent of aircraft, submarines, and now satellites, the commons have expanded from the surface of the oceans to their depths and to the air and space above them. Prior to the development of aircraft and submarines, commerce raiding required mainly surface ships; in cases of raiding far from the raider's home territory it has also required "mother ships" or ports in friendly countries to service and replenish ships. With the development of technology, commerce raiding has relied increasingly on submarine, air, satellite, and intelligence assets to locate targets.

Over time, commerce raiding has moved away from purely naval operations to joint operations—entailing close cooperation among air, sea, and intelligence. In past eras, commerce raiding was often conducted by privateers and judged by prize courts. However, the 1856 Paris Declaration Respecting Maritime Law made privateering illegal, transforming privateers into pirates. In the era of modern international law, professional navies, not individuals, conduct commerce raiding, with the Somali pirates an obvious exception. In practice, naval attacks on trade have been most effective for sea powers in global wars, which by their nature are coalition wars, so that coordinated (if not combined) operations have also figured prominently in these cases.

According to table 3, "Force," surface patrols were crucial in virtually all of the sixteen case studies. The main exceptions are represented by the predominant roles played by submarines in Germany's commerce raiding operations during both world wars, plus the U.S. Navy's unrestricted submarine-warfare campaign against Japan during World War II. Airpower, after its advent, also played a prominent role, particularly to locate German submarines in World War II. Land-based missiles became an important instrument of force in only one case study—the Tanker War—but their potential for use in future commerce raiding is great, especially in restricted waters like the Persian Gulf.

TABLE 3
Force

CONFLICT	PATROLS	SUBS	LAND OPS	AIR OPS	ALLIES
Seven Years' War	X		X		X
American Revolution	X		X		X
French Wars	X		X		X
1812	X		X		
U.S. Civil War	X		X		
Jeune École	X (theory)				
First Sino-Japanese War	X		X		
Russo-Japanese War	X		X		
WWI Germany	X	X	X	X	X
WWI USA	X		X	X	X
Spanish Civil War	X	X	X	X	X
WWII Germany	X	X	X	X	X
WWII Japan	X	X	X	X	X
WWII USA	X	X	X	X	X
Tanker War	X		X	X	
Somali pirates	X				

The most successful commerce raiding operations worked in combination with simultaneous land campaigns and often in combination with blockades. All three were evident in the American Civil War, the First Sino-Japanese War, the Entente strategy in World War I, the Spanish Civil War, and the Allied strategy in World War II. Commerce raiding was usually at best an important but secondary means to pressure an adversary to capitulate. Usually land campaigns exerted far more pressure than did commerce raiding. The only exception was the U.S. unrestricted submarine warfare against Japan, which targeted not only commerce but also the Imperial Japanese Navy and army troop transports. Although the United States never invaded the Japanese home islands prior to its surrender, a devastating air campaign including atomic bombs leveled Japan's cities, and a Soviet land invasion loomed had the war protracted further.

Commerce raiding was cheap to execute, at the operational level, for all sides, and eventually it imposed greatly disproportionate costs on the enemy, both from trade losses and from countermeasures to end the raiding. U.S. submarines in the Pacific theater of World War II inflicted by far more damage against Japanese forces per dollar of American investment than any other military service or branch.[1] These disproportionate costs were evident in the Seven Years' War, the American Revolution, the War of 1812, the U.S. Civil War, the Russo-Japanese War, World War I for all sides, the Spanish Civil War, World War II for all sides, the Tanker War, and Somalia. The only exception might be the French Wars, where France might have suffered more than Britain did, not because of the privateering, but because of the British blockade of French ports and the British policy of warehousing thousands of detained commerce raiders in prison hulks.

At the strategic level, however, the costs could also become enormous, as Germany discovered with its unrestricted submarine-warfare campaign that brought the United States into World War I. Commerce raiding that threatens neutral shipping risks escalation and retaliation. Russia recognized these costs and cut short its campaign in the Russo-Japanese War rather than suffer British and American intervention. Conversely, however large the economic costs, they may be insufficient to alter the outcome of the war—for example, the American Revolution or the U.S. Civil War.

Finally, commerce raiding can be executed unilaterally or in combination with allies. Combined operations have figured prominently in commerce raiding. Global wars usually depend on allies—the Seven Years' War, the French Wars, and both world wars. Weak powers also often depend on allies—the colonists in the American Revolution and the Fascists in Spain. However, commerce raiding of neutral shipping is not to be undertaken lightly, since it is likely to produce an opposing alliance. The conduct of diplomacy was made complicated thereby for the American colonists targeting British trade, for revolutionary France, for the Confederates targeting Union trade, for Russia in the Russo-Japanese War, for Germany in World War I, for the Fascists in Spain, and for both sides in the Tanker War. The United States largely escaped political ramifications from destroying Japanese trade in World War II because Japan's simultaneous attacks on all of the neutral powers had left it with no friends in Asia.

Operational and Strategic Goals

Commerce raiding is a means to an end. At the operational level, it provides a means to impede enemy trade, transportation, and communications through such actions as the destruction of merchant ships, the elimination of land forces while on shipboard, and the destruction of enemy naval forces; strategic effects can range from sanction enforcement, cost escalation, and bottleneck creation to full economic strangulation, if applied in combination with blockade. At the strategic level it can contribute—sometimes alone, but more often in combination with other strategies—to the achievement of war aims, whether limited or unlimited. Since it tends to work slowly, *guerre de course* has been most important in protracted coalition wars, which are often fought for unlimited objectives. World War I, for example, left Germany blockaded and hungry, its trade from the sea cut; World War II left Japan in even worse shape.

Commerce raiding can be total or partial. Total commerce raiding operations are designed to halt prohibited traffic completely, while partial campaigns, by intent or by default, allow either a percentage or certain categories of trade to continue. Commerce raiding strategies that are effective at sea but fail to cut alternative land routes are still partial, even though they may make critical contributions to victory. In practice total campaigns are rare, because they require specific circumstances to become feasible—for instance, China's unwitting cooperation with Japanese designs by failing to contest command of the sea in the First Sino-Japanese War, or the unusual oceanic theater of the World War II Pacific, marked by widely scattered islands and long distances separating Japan from key resources. Such factors allow a dominant naval power totally to cut off a secondary naval power by sea.

Finally, the commerce raider's goals can be unlimited, meaning the overthrow of the enemy government, or something less, such as a negotiated peace. Hence the terms "unlimited" and "limited" commerce raiding, defined in terms of strategic objective, not the quantity of resources devoted to the operation. As shown in table 4, "Operational and Strategic Goals," in seven of the sixteen cases the original strategic objectives were unlimited, and in at least one case—the German attack on Great Britain during World War II—the original, limited goal escalated to an unlimited goal for at least a time. These were mainly global wars. Two additional cases were Russia in the Russo-Japanese War and the Fascists in the Spanish Civil War.

TABLE 4
Operational and Strategic Goals

CONFLICT	OPERATIONAL GOAL	STRATEGIC GOAL	FOCUS	PARTIAL/ TOTAL	GOAL L/U	ENEMY GOAL
Seven Years' War	British stop French trade	defeat France	trade	partial	U	U
American Revolution	fight Royal Navy; cut British trade	gain independence	navy + trade	partial	L	U
French Wars	cut British trade	defeat Britain	trade	partial	U	U
1812	disrupt British trade; stop impressment	maintain the country	trade	partial	L	L
U.S. Civil War	undermine Union trade	negotiated peace	trade	partial	L	U
Jeune École	disrupt enemy economy	impel change of government policy	trade	total		
First Sino-Japanese War	sink troopships; bottle up Chinese navy	negotiated peace	troop transport	total	L	L
Russo-Japanese War	sink enemy ships; stop trade	defeat Japan	navy + trade	partial	U	L
WWI Germany	deny resources; bottle up Royal Navy	defeat Britain	navy + trade	partial	U	U
WWI USA	stop German U-boat attacks	defeat Germany	navy	partial	U	U
Spanish Civil War	cut trade and supplies	defeat Republicans	trade	partial	U	U
WWII Germany	cut British imports	defeat Britain	trade	partial	L-U-L	U
WWII Japan	keep supplies flowing to Japan	protect empire	navy	partial	L	U
WWII USA	cut trade/ petroleum imports to Japan	defeat Japan	trade	total	U	L
Tanker War	halt oil exports	weaken enemy	trade	partial	L	L
Somali pirates	increase revenue stream	tax illegal fishing	trade	partial	L	L

U—unlimited   L—limited

Limited strategic objectives include U.S. goals in both the American Revolution and the War of 1812, neither of which involved seeking regime change in London. Likewise, the Confederacy in the Civil War did not seek to destroy the North and reunite the country under its own government but merely to achieve a negotiated settlement establishing its own independence. Other limited wars include the First Sino-Japanese War, which ended in a negotiated settlement, and the current situation off Somalia, which is more about maximizing revenue rather than overthrowing any particular country or challenging the global trade system. Japan tried to fight a limited war against the United States in World War II but became the object of an unlimited counterattack that overthrew the imperial government in Tokyo.

Not surprisingly, most commerce raiding operations focus on an enemy's trade, although the enemy's military force—including both naval vessels and troop transports (for instance, Japan's sinking of the Chinese troopship *Kowshing*)—can also be targeted. Wars for unlimited objectives always targeted both civilian and military vessels, and many (but not all) wars for limited objectives also targeted both.

Adaptations and Countermeasures

Countermeasures against commerce raiding include reconnaissance, patrols, and interdiction of the raiders from the surface, under water, from the air, and now even from space. In the Age of Sail, surface ships conducted mainly search-and-destroy operations and such defensive measures as sailing in convoys. As submarines and aircraft became more available and dependable, they too were used for patrol and search-and-destroy duty. Thus, joint operations have played an increasingly important role, with joint sea-air operations substituting for joint land-sea operations in the modern period.

Enemies can make commerce raiding extraordinarily expensive in terms of money, personnel, prestige, and strategic effect, so much so that the costs can ultimately outweigh the benefits. For instance, Napoleon's strategy of commerce raiding hurt France more than Britain, because Britain was far better positioned to cut off French overseas trade than the reverse. Germany's unrestricted submarine campaign eventually cost it World War I by spurring the entry of the United States. The Tanker War, which sank nearly half the merchant tonnage lost in all of World War II, produced crippling economic and political effects for both sides. Moreover, the Tanker War ushered in an era of intrusive great-power intervention in the Middle East, which continues to this day. Such outcomes can easily become nightmare scenarios for countries engaged in commerce raiding operations. However, if the victim either lacks effective countermeasures or does not incorporate them in time (such as Republican Spain or Japan in the Pacific theater of World War II) and the victim requires goods delivered by sea (critical war materiel, in the cases of Spain and Japan), commerce raiding can have critical "dream scenario" effects. But these situations are rare. Most commerce raiding operations fall somewhere between these extremes.

As table 5, "Adaptation and Countermeasures," shows, of the sixteen case studies examined there are arguably five nightmare scenarios for the country first to adopt commerce raiding operations—Russia in the Russo-Japanese War, Germany in both World War I and World War II, Japan in World War II, and both parties in the Tanker War. In the first case, Russian naval vessels attempted to capture all commercial ships supplying Japan, but in the process prompted an opposing neutral reaction by Britain and to a lesser degree the United States, which was an emerging sea power at the time. German attacks on neutral commerce in World War I triggered a fatal third-party intervention. Initially, Germany's U-boat campaigns in World Wars I and II were operationally successful, but once the Allies ramped up production, especially of new ships, organized convoys, and fine-tuned intelligence assets to locate the raiders, the long-term tonnage trends worked against Germany. Japan met a similar fate in World War II, when it underestimated the damage American submarines could do to its commercial fleet and all attempts to reform its shipping system proved too little too late. Likewise, both Iran and Iraq experienced huge financial losses and intrusive third-party interventions in the Tanker War.

TABLE 5
Adaptation and Countermeasures by the Victim of Commerce Raiding

CONFLICT	COUNTER-COMMERCE RAIDING	SEARCH & DESTROY	BASE RAIDS	BLOCKADE	CONVOY	THIRD-PARTY ADDITION	NIGHT-MARE SCENARIO	DREAM SCENARIO
Seven Years' War	X	X	X	X	X	potential		
American Revolution	X	X	X	X	X	X		
French Wars	X	X	X	X	X			
1812	X	X		X	X			
U.S. Civil War	X	X	X	X		potential		
Jeune École	X (theory)							
First Sino-Japanese War						potential		X
Russo-Japanese War	X	X				potential	X	
WWI Germany	X	X		X	X	X	X	
WWI USA	X	X						
Spanish Civil War	X	X		X		X		
WWII Germany	X	X		X	X	X	X	
WWII Japan		X		X	X		X	
WWII USA	X	X	X					X
Tanker War	X	X	X		X	X	X	
Somali pirates		X				X		

In the first of the two dream scenarios—the First Sino-Japanese War and the U.S. unrestricted submarine campaign in World War II—the theater of hostilities was small and the victim (China) took no countermeasures, while in the second the theater was huge and the victim (Japan) took few countermeasures. In both cases the theater ideally suited the capabilities of the commerce raider. In the first war, the Japanese had a navy with regional capabilities facing a Chinese navy under incompetent command. In the second war, Japan's reliance on numerous overseas resources in combination with the dispersion of its troops over scattered islands allowed the dominant naval power, the United States, to freeze the movements of Japan's goods and troops. In the dream scenarios, the victims largely followed the scripts anticipated by the commerce raiders, effectively becoming "cooperative" adversaries that could have imposed far higher costs on their enemies had they taken countermeasures and, in World War II, had Japan realized that its military codes had been compromised. Indeed, China could have won the First Sino-Japanese War had it targeted Japanese troop transports, contested their landings, and drawn any remaining Japanese forces inland for a long winter on low rations before delivering on them an annihilating counterattack during the spring.

Most of the sixteen commerce raiding case studies involved neither dream nor nightmare scenarios. Usually both sides adapted to the other's strategies. For example, in the U.S. Civil War the Union merchant-marine companies quickly sold their ships to foreign countries, mainly Great Britain, to protect them from Confederate attack. Likewise, France sold off much of its merchant marine during the Seven Years' War. During the Russo-Japanese War the Japanese quickly halted Russian attempts to use neutral Chinese ports to conduct commerce raiding attacks. But not all adaptations were effective. For example, in the American Revolution the U.S. government initially formed its own fleet of ships to attack the British, before belatedly deciding to grant letters of marque to private commerce raiders instead.

In one set of paired cases—the German U-boat campaigns in the two world wars—the commerce raiding country used the same basic strategy twice, with greater initial success the second time. However, in World War II not only did the Allies eventually obtain an Enigma machine and break the German codes, but the introduction of new technologies, including radar and aerial antisubmarine patrols, offset the greater capabilities of the German U-boats. Just as the war was ending, however, the Germans were about to introduce a new class of submarines that might have in turn offset these Allied advantages. Thus the speed of adaptation can be crucial.

Over the years, the countermeasures to commerce raiding have become more effective as the technology for locating and targeting has improved. Search-and-destroy missions undermined commerce raiding in the Seven Years' War, the French Wars, the U.S. Civil War, World War I, the Battle of the Atlantic in World War II, and the Tanker War. The targeting of bases proved most effective by Britain in the French Wars, by the Union in the U.S. Civil War, and by the United States in World War II. Convoys greatly reduced the damage in the Seven Years' War, the French Wars, the War of 1812, World

War I, World War II (Europe), and the Tanker War. Note that these countermeasures of search-and-destroy, base raids, and convoys are more accessible to the dominant than to the secondary naval power, and more accessible to maritime than continental powers generally. Base raids are rarely feasible for a secondary naval power, let alone a continental power.

When evaluating the efficacy of commerce raiding, it is important to consider all possible enemy responses, as well as the intervention of third parties. For example, French assistance during the American Revolution and the American intervention in both World Wars I and II cut short what might have otherwise been promising commerce raiding campaigns. In addition, geography determines the availability of alternate land lines. In theaters with alternative land routes available to replace trade by sea, effective commerce raiding requires the cooperation, if not the active support, of the third parties that control the alternative land routes, in order to sever them. The likelihood of operational and strategic effectiveness plummeted for commerce raiders that triggered major additions to the victim's coalition; hence the relatively cautious commerce raiding conducted by the Confederacy in the Civil War, Japan in the First Sino-Japanese War, Russia in the Russo-Japanese War, and the Fascists in the Spanish Civil War. Such precedents for neutral intervention do not bode well for Somalia, which as a pariah state has no allies.

Operational and Strategic Effectiveness

Evaluating the effectiveness of a commerce raiding campaign is a three-part process. Did the commerce raiding operation achieve its operational goals? Did this contribute to the achievement of strategic success? Were the costs entailed worth the benefits delivered? Factors influencing effectiveness include the number of commerce raiders, the size of the theater, the economic or military importance of the targeted goods, and the availability of substitutes to offset bottlenecks. Tight commerce raiding operations do not in theory let any prohibited items through, whereas porous operations stop only a percentage of the traffic.

Rather than a binary choice of targets—naval ship or merchantman—these case studies reveal a wide spectrum of potential targets, ranging from neutral merchantmen to enemy-commandeered merchantmen, enemy merchantmen, enemy-allied naval vessels, and finally enemy naval vessels. Each type of target entailed a different level of operational risk of the attacker surviving the engagement and a different level of strategic risk of precipitating a third-party intervention. For this reason, strategic effectiveness can be difficult to measure with any certainty.

Table 6, "Operational and Strategic Effectiveness," shows that of the six commerce raiding operations conducted by victorious powers (excluding the Jeune École), half were porous (Seven Years' War, the American Revolution, and the Spanish Civil War) and the other half tight (the First Sino-Japanese War, the Entente's destruction of German trade in World War I, and the U.S. destruction of Japanese trade in World War II). With the exceptions of the First Sino-Japanese War and Spanish Civil War, the other four involved large geographic areas, and the tight commerce raiding campaigns focused on interdicting specific war materiel, such as petroleum, or in halting the enemy's commerce raiding efforts. By contrast, all cases resulting in a loss or a draw were porous. Nevertheless, in all eight of these cases commerce raiders managed to reduce the flow of goods, drive up costs, and impose burdens on the enemy. In all cases the outcome of the war turned not solely on commerce raiding but rather on the integration of multiple strategies.

Of the eight commerce raiding campaigns that were either victorious or a draw (again, excluding the Jeune École), only four were conducted by naval powers, while in the four others—the American Revolution, the War of 1812, the Spanish Civil War, and the Tanker War—the perpetrator hardly had a navy, which is also the case for Somalia. This suggests that selected land powers can successfully use commerce raiding operations to

TABLE 6
Operational and Strategic Effectiveness

CONFLICT	THEATER SIZE	BOTTLENECK	W/L	EFFECT
Seven Years' War (Britain)	huge	trade	W	porous
American Revolution (colonies)	huge	trade	W	porous
French Wars (France)	huge	trade	L	porous
1812 (USA)	huge	trade	D	porous
U.S. Civil War (Confederacy)	huge	trade	L	porous
Jeune École	huge (theory)	trade + war materiel	W	tight
SJWI (Japan)	medium	troopships	W	tight
RJW (Russia)	medium	war materiel	L	porous
WWI (Germany)	huge	war materiel	L	porous
WWI (USA)	huge	antisubmarine warfare	W	tight
Spanish Civil War (Fascists)	limited coastline	war materiel	W	porous
WWII (Germany)	huge	war materiel	L	porous
WWII (Japan)	long SLOC	petroleum imports	L	porous
WWII (USA)	huge	petroleum imports	W	tight
Tanker War	Persian Gulf	petroleum exports	D	porous
Somali pirates	long coastline	hijack ships transiting SLOC	N/A	porous

W—commerce raider wins   L—commerce raider loses   D—draw   SLOC—sea line of communication further their objectives, in particular if a draw leading to a negotiated settlement is an acceptable outcome. On the operational level, commerce raiding provides a means to inflict disproportionately high costs on the enemy. For this to translate favorably at the strategic level requires avoidance of hostile third-party intervention and the absence of cost-effective countermeasures.

Of the six cases studied here that ended in defeat for the side engaged in commerce raiding, four involved land powers targeting the dominant maritime power—France targeting Britain in the French Wars, Germany targeting Britain and the United States in both world wars, and Japan also targeting the United States in World War II. Similarly, in the U.S. Civil War the Confederate navy was inferior to that of the Union. Not surprisingly, maritime dominance positions a country to minimize the impact of commerce raiding.

Outside intervention by another great power was the most common reason for strategic failure. For example, in the eighteenth and nineteenth centuries France played a major role in opposing Britain—most successfully in the American Revolution—while in the twentieth century the United States twice came to Great Britain's assistance against Germany. Whenever commerce raiding has affected the global trade system, such as in the Tanker War or potentially in Somalia, great-power diplomatic if not military intervention becomes more likely. Affected third parties actively engaged in diplomacy in the Seven Years' War, the Napoleonic Wars, the American Civil War, the First Sino-Japanese War, the Russo-Japanese War, both world wars, the Spanish Civil War, and the Tanker War. Thus, commerce raiding can entail significant strategic risk.

As naval theorist Alfred Thayer Mahan observed over a century ago in the concluding paragraphs of his classic *The Influence of Sea Power upon History,* commerce raiding has been "a most important secondary operation of naval warfare." Mahan predicted that it was "not likely to be abandoned till war itself has ceased." But he warned against regarding it as the cheap silver bullet, sufficient on its own "to crush an enemy." He called such optimism "a most dangerous delusion," particularly when aimed at a strong sea power with a "widespread healthy commerce and a powerful navy." As he argued and this work has shown, far-flung commerce "can stand many a cruel shock."[2]

Mahan wrote these lines a generation before the First World War, when commerce raiding figured more prominently than he imagined it might, let alone the Second World War, when commerce raiding brought imperial Japan to its knees. Particularly in global wars, commerce raiding in combination with other military strategies and other instruments of national power can produce outcomes lethal to the victims. Although the Jeune École presented commerce raiding as a weapon of the weaker maritime power to defeat the dominant maritime power, in practice it has most often offered a strategically effective way for the dominant naval power to set the conditions for the economic decline of a continental adversary and thereby to put time on its side in a high-stakes attrition war.

Commerce raiding was not strategically decisive but tended to work in combination with other strategies, such as blockade, embargo, invasion, and bombing, and together these strategies were strategically effective. The maritime powers were more financially able to conduct these strategies than their continental adversaries were to endure them. Over time the cumulative effects changed the balance of forces in the favor of the maritime powers by inducing the financial and military exhaustion of the continental adversaries. The British victory in the Seven Years' War, the Entente victory in World War I, and the Allied victory in World War II illustrate this pattern.

Certain geographic conditions are particularly favorable to the strategy. Peninsular or island adversaries dependent on trade to conduct military operations in conflicts against more powerful foes proved particularly vulnerable to commerce raiding. For example, the Spanish Republic did not survive the Spanish Civil War, nor did imperial Japan survive World War II. For them, when commerce raiding cut military supply lines, defeat loomed.

In most of the case studies examined, the weaker naval power, not the stronger, adopted a commerce raiding strategy, since the stronger power, often the guarantor of the international order, was more likely to impose a blockade rather than to put international commerce at risk.[3] This suggests that a weaker naval power could not effectively blockade a stronger power and so fell back on commerce raiding as the only feasible way to attack the enemy's trade, disperse it, and impose costs on the enemy's navy seeking to protect endangered trade. Weaker maritime powers engaging in commerce raiding have included the American colonies in the American Revolution, France in the French Wars, the United States in 1812, the Confederacy in the U.S. Civil War, Russia in the Russo-Japanese War, and Germany in both world wars. All lost or drew except for the American colonies, and in that war a costly insurgency in combination with the French-supported victory at Yorktown, not commerce raiding, accounted for the British change of heart.

Commerce raiding seemed to impose disproportionate costs on the adversary in the cases studied, mainly because it was far cheaper to conduct than to eradicate, but this was insufficient to change the wars' outcomes. In fact, with the possible exception of the Seven Years' War, through the end of World War I commerce raiding actually had only a minor impact on commerce. In the American Revolution, the French Wars, the U.S. Civil War, the Russo-Japanese War, and World War I, most of the traffic arrived safely. Only with the development of technology capable of efficiently locating merchant ships at sea and accurately targeting them did commerce raiding become a potentially lethal strategy for the dominant maritime power. In the Pacific theater, this American strategy sank so many Japanese commercial ships so quickly that it virtually froze Japan's expeditionary forces in place, strangled its economy, and immobilized its fleet for lack of fuel.

The most strategically effective *guerre de course* operations included the Entente elimination of German trade in World War I and the Allied elimination of German, Italian, and Japanese trade in World War II. In these cases the victors combined commerce raiding with blockade. The dominant naval powers allied themselves and wiped out the commerce of their enemies. It was most effective against the one maritime enemy, Japan, that had no internal trade routes. Key conditions for these examples were, first, the technology to find and destroy targets—in World War II, a combination of cryptography and torpedoes; second, commerce raiding in conjunction with blockade to minimize seepage; and third, a global war with no powerful neutrals to ally with the victim and resist the raiding.

In regional wars with powerful neutral nations sitting on the sidelines, commerce raiding is likely to prejudice their interests. Second-order effects against neutrals can trigger a war-changing third-party intervention. For example, British commerce raiding in the Seven Years' War threatened Dutch interests. In the Russo-Japanese War, Russia abandoned its commerce raiding lest Britain and the United States intervene.

Because commerce raiding is inexpensive to conduct but costly to stop, theoretically the strategy would work best in a low-stakes war, to work in combination with other strategies to impose high enough costs on the adversary to induce a negotiated settlement. But the case studies in this volume do not support this conclusion. For the dominant maritime power, naval attacks on commerce threaten the very global commercial order it is intent on preserving; any sustained restriction on commerce quickly ups the stakes from commercial loss to the survival of the global system. Also, such attacks can affect neutral shipping, through higher insurance and freight rates, bringing other interested parties into the conflict. Counterintuitively, then, in limited wars where belligerents seek to minimize escalation, commerce raiding actually turns out to be extremely expensive to conduct, given its potentially alienating effect on others.

As technology has changed, so have commerce raiding operations. Attacking countries have turned to smaller boats, which depend on speed and darkness to succeed. The deployment of large naval ships to oppose these efforts has become increasingly expensive and often ineffective. In "choke point" commerce raiding, such as the Tanker War or Russian commerce raiding in the Red Sea during the Russo-Japanese War, the disruption of trade can impact stock, commodity, and insurance markets, causing numerous second-order effects with potentially global reach.

Owing to the interconnections of the international trading system, commerce raiding by developed countries will probably decline, since these nations have ever larger stakes in the global economy, but commerce raiding operations by failed or pariah states will most likely increase, as they perceive them to be a lucrative business in lands of little economic opportunity. The widespread use of GPS and other modern technologies suggests that future commerce raiding attacks may take place hundreds, or even thousands, of miles from shore. Countermeasures will thus extend farther out to sea as well, which means they will increasingly rely on new forms of aerial and space-based surveillance and even interdiction. As before, navies will remain essential for countering commerce raiding, but the necessary sensing tools will increasingly require the integration of naval, air, and space assets.

Notes

The thoughts and opinions expressed in this publication are those of the authors and are not necessarily those of the U.S. government, the U.S. Navy Department, or the Naval War College.

1. Ronald H. Spector, *Eagle against the Sun* (New York: Free Press, 1985), p. 487; Sadao Asada, *From Mahan to Pearl Harbor: The Imperial Japanese Navy and the United States* (Annapolis, Md.: Naval Institute Press, 2006), p. 181; George W. Baer, *One Hundred Years of Sea Power: The U.S. Navy, 1890–1990* (Stanford, Calif.: Stanford Univ. Press, 1993), pp. 233–34.

2. The quotation comes from the fourth-to-last paragraph of Alfred Thayer Mahan, *The Influence of Sea Power upon History 1660–1783* (1890; repr. New York: Wang and Hill, 1957), p. 481. After the quoted section, Mahan overstates the case for command of the sea. As his rough contemporary and fellow naval theorist Julian Corbett observes, in a naval war "the normal position is not a commanded sea but an uncommanded sea." Mahan's focus on command of the sea, however attractive in theory, is rarely feasible in practice. Corbett developed the more applicable concept of local sea denial in *Some Principles of Maritime Strategy* (New York: Longmans, Green, 1911; repr. Annapolis, Md.: Naval Institute Press, 1988), pp. 87–91.

3. See Bruce A. Elleman and S. C. M. Paine, eds., *Naval Blockades and Seapower: Strategies and Counter-strategies, 1805–2005* (London: Routledge, 2006), for an entire book on the subject of blockade.

Selected Bibliography

Abendroth, Hans-Hening. *Hitler in der spanischen Arena: Die deutsch-spanischen Beziehungen im Spannungsfeld der europäischen Interessenpolitik vom Ausbruch des Bürgerkrieges bis zum Ausbruch des Weltkrieges, 1936–1939.* Paderborn, W. Ger.: Schöningh, 1973.

Acerra, Martine, José Merino, and Jean Meyer, eds. *Les Marines de Guerre européennes, XVII–XVIIIᵉ siècles.* Paris: Presses Universitaires de la Sorbonne, 1985.

Acerra, Martine, and Jean Meyer. *Histoire de la Marine française des origines à nos jours.* Rennes, Fr.: Editions Ouest-France, 1994.

———. *Marines et Révolution.* Paris: Editions Ouest-France, 1988.

Adams, John. *Diary and Autobiography of John Adams.* Edited by L. H. Butterfield. 4 vols. Adams Papers. Cambridge, Mass.: Belknap of Harvard Univ. Press, 1961.

Aizawa Kiyoshi. "Differences Regarding Togo's Surprise Attack on Port Arthur." In *The Russo-Japanese War in Global Perspective: World War Zero,* edited by John W. Steinberg et al. Leiden: Brill, 2005.

Albion, Robert Greenhalgh. *Naval & Maritime History: An Annotated Bibliography.* 4th rev. and exp. ed. Mystic, Conn.: Munson Institute of American Maritime History, 1972.

Albion, Robert Greenhalgh, and Jennie Barnes Pope. *Sea Lanes in Wartime: The American Experience, 1775–1942.* New York: W. W. Norton, 1942.

Alden, John. *The Fleet Submarine in the U.S. Navy: A Design and Construction History.* Annapolis, Md.: Naval Institute Press, 1979.

Allen, Gardner W. *A Naval History of the American Revolution.* Vol. 1. Boston: Houghton Mifflin, 1913.

Anderson, Fred. *Crucible of War: The Seven Years' War and the Fate of Empire in British North America, 1754–1766.* New York: Knopf, 2000.

Audran, Karine. "Les négoces portuaires bretons sous la révolution et l'Empire. Bilan et Stratégies. Saint-Malo, Morlaix, Brest, Lorient et Nantes, 1789–1815." Vol. 1. Unpublished PhD diss., Univ. of Lorient, 2007.

Baer, George W. *One Hundred Years of Sea Power.* Stanford, Calif.: Stanford Univ. Press, 1993.

Bahadur, Jay. *The Pirates of Somalia: Inside Their Hidden World.* New York: Pantheon, 2011.

Bargoni, Franco. *L'Impegno navale italiano durante la Guerra Civile Spagnola (1936–1939).* Rome: Ufficio Storico della Marina Militare, 1992.

Barlow, Jeffrey G. *From Hot War to Cold: The U.S. Navy and National Security Affairs, 1945–1955.* Stanford, Calif.: Stanford Univ. Press, 2009.

Baugh, Daniel A. *British Naval Administration in the Age of Walpole.* Princeton, N.J.: Princeton Univ. Press, 1965.

———. *The Global Seven Years' War, 1754–1763: Britain and France in a Great Power Contest.* Harrow, U.K.: Longman, 2011.

———. "Why Did Britain Lose Command of the Sea during the War for America?" In *The British Navy and the Use of Naval Power in the Eighteenth Century,* edited by J. Black and P. Woodfine. Leicester, U.K.: Leicester Univ. Press, 1988.

Bayly, Admiral Sir Lewis. *Pull Together: The Memoirs of the Late Admiral Sir Lewis Bayly.* London: G. G. Harrap, 1939.

Beach, Edward L. *Salt and Steel.* Annapolis, Md.: Naval Institute Press, 1999.

———. *Submarine!* Annapolis, Md.: Naval Institute Press, 1952.

Beasley, W. G. *A Modern History of Japan.* 2nd ed. New York: Praeger, 1974.

Beatson, Robert. *Naval and Military Memoirs of Great Britain from 1727 to 1783.* 6 vols. London: Longman, Hurst, Rees, and Orme, 1804.

Beesly, Patrick. *Very Special Intelligence: The Story of the Admiralty's Operational Intelligence Centre, 1939–1945.* London: Greenhill Books, 2000.

Bendert, Harald. *Die UB-Boote der Kaiserlichen Marine, 1914–1918*. Berlin: E. S. Mittler, 2000.

———. *Die UC-Boote der Kaiserlichen Marine, 1914–1918*. Berlin: E. S. Mittler, 2001.

Bennett, Frank M. *The Monitor and the Navy under Steam*. Boston: Houghton Mifflin, 1900.

Biddle, Alexander, ed. *Old Family Letters*. Philadelphia: J. B. Lippincott, 1892.

Billias, George Athan. *General John Glover and His Marblehead Mariners*. New York: Henry Holt, 1960.

Binaud, Daniel. *Les corsaires de Bordeaux et de l'estuaire. 120 ans de guerres sur mer*. Biarritz, Fr.: Atlantica, 1999.

Blair, Clay, Jr. *Hitler's U-boat War: The Hunters 1939–1942*. New York: Random House, 1996.

———. *Silent Victory: The U.S. Submarine War against Japan*. Philadelphia: J. B. Lippincott, 1975.

Bōei Kenshūjo Senshishitsu, ed. *Rikugun Gunju Dōin*. Vol. 2. Tokyo: Asagumo Shinbunsha, 1970.

Bogdenko, V. L. "Stranitsy starykh bloknotov." In *Leningradtsy v Ispanii: Sbornik vospominami*. 3rd ed. Leningrad: Lenizdat, 1989.

Bonnel, Ulane. *La France, les États-Unis et la guerre de Course (1797–1815)*. Paris: Nouvelles Éditions Latines, 1961.

Bosscher, Ph. M. *De Koninklijke Marine in de Tweede Wereldoorlog*. Franeker, Neth.: Wever, 1984.

Boulle, Pierre Henri. "The French Colonies and the Reform of Their Administration during and following the Seven Years' War." PhD diss., Univ. of California at Berkeley, 1968.

Bowen, Ashley. *The Journals of Ashley Bowen (1728–1813) of Marblehead*. Edited by Philip C. F. Smith. 2 vols. Boston: Colonial Society of Massachusetts, 1973.

Bowler, R. Arthur. *Logistics and the Failure of the British Army in America, 1775–1783*. Princeton, N.J.: Princeton Univ. Press, 1975.

Boxer, Charles R. *The Dutch Seaborne Empire: 1600–1800*. New York: Knopf, 1965.

Brodie, Bernard. *A Guide to Naval Strategy*. Princeton, N.J.: Princeton Univ. Press, 1944.

Buel, Richard, Jr. *In Irons: Britain's Naval Supremacy and the American Revolutionary Economy*. New Haven, Conn.: Yale Univ. Press, 1998.

Bulloch, James D. *The Secret Service of the Confederate States in Europe, or How the Confederate Cruisers Were Equipped*. 1884; repr. New York: Modern Library, 2001.

Burrell, William. *Reports of Cases Determined by the High Court of Admiralty and Upon Appeal Therefrom, . . . Together with Extracts from the Books and Records of the High Court of Admiralty and the Court of the Judges Delegates, 1584–1839*. Edited by Reginald G. Marsden. London: William Clowes and Sons, 1885.

Butel, Paul. *The Atlantic*. London: Routledge, 1999.

Canny, Nicholas, ed. *The Origins of Empire: British Overseas Enterprise to the Close of the Seventeenth Century*. Vol. 1 of *The Oxford History of the British Empire*, edited by Wm. Roger Louis. Oxford, U.K.: Oxford Univ. Press, 1998.

Capp, Bernard. *Cromwell's Navy: The Fleet and the English Revolution, 1648–1660*. Oxford, U.K.: Clarendon, 1989.

Carter, Alice Clare. *The Dutch Republic in Europe in the Seven Years' War*. London: Macmillan, 1971.

Casse, Michel. "Un armateur en course bordelaise sous la Révolution et l'Empire: Jacques Conte, 1753–1836." In *Bordeaux, porte océane, carrefour européen. Actes du 50ᵉ congrès de la Fédération Historique du Sud-Ouest*. Bordeaux, Fr.: FHSO, 1999.

Cassell's History of the Russo-Japanese War. Vol. 2. New York: Cassell, 1905.

Chalk, Peter. *Non-military Security and Global Order: The Impact of Extremism, Violence and Chaos on National and International Security*. London: Macmillan, 2002.

Chapelle, Howard I. *The History of the American Sailing Navy: The Ships and Their Development*. New York: Norton, 1949; repr. n.d.

Charbonnel, Nicole. *Commerce et course sous la Révolution et le Consulat à La Rochelle: Autour de deux armateurs—les frères Thomas et Pierre-Antoine Chegaray*. Paris: PUF, 1977.

Ciano, Galeazzo. *Diary, 1937–1943*. New York: Enigma, 2002.

Clark, John G. *La Rochelle and the Atlantic Economy during the Eighteenth Century.* Baltimore: Johns Hopkins Univ. Press, 1981.

Clark, William Bell. *Ben Franklin's Privateers: A Naval Epic of the American Revolution.* Baton Rouge: Louisiana State Univ. Press, 1956.

———. *George Washington's Navy: Being an Account of His Excellency's Fleet in New England Waters.* Baton Rouge: Louisiana State Univ., 1960.

Clauder, Anne. *American Commerce as Affected by the Wars of the French Revolution and Napoleon, 1793–1812.* Clifton, N.J.: A. M. Kelley, 1932 [repr. 1972].

Clowes, William Laird. *The Royal Navy: A History from the Earliest Times to the Present.* 7 vols. London: Sampson Low, Marston, 1897–1903.

Colombos, C. John. *The International Law of the Sea.* 6th rev. ed. London: Longman, 1967.

Coltman, Robert, Jr. *The Chinese, Their Present and Future: Medical, Political, and Social.* Philadelphia: F. A. Davis, 1891.

Conway, Stephen. *War, State, and Society in Mid-Eighteenth-Century Britain and Ireland.* Oxford, U.K.: Oxford Univ. Press, 2006.

Cooling, Benjamin Franklin. *Gray Steel and Blue Water Navy: The Formative Years of America's Military-Industrial Complex 1881–1917.* Hamden, Conn.: Archon Books, 1979.

Corbett, Julian S. *England in the Seven Years' War: A Study in Combined Strategy.* 2 vols. London: Longmans, Green, 1907 [repr. 1918].

———. *Maritime Operations in the Russo-Japanese War: 1904–1905.* Vol. 2. Annapolis, Md.: Naval Institute Press, 1994.

———. *Some Principles of Maritime Strategy.* New York: Longmans, Green, 1911; repr. Annapolis, Md.: Naval Institute Press, 1988.

Costello, John, and Oleg Tsarev. *Deadly Illusions.* New York: Crown, 1993.

Couteau-Bégarie, Hervé. *L'histoire maritime en France.* Paris: Economica, 1995.

Coverdale, John F. *Italian Intervention in the Spanish Civil War.* Princeton, N.J.: Princeton Univ. Press, 1975.

Crouzet, François. *L'économie britannique et le blocus continental, 1806–1813.* 2nd ed. Paris: Economica, 1987.

Crowhurst, Patrick. *The Defence of British Trade, 1689–1815.* Folkestone, Kent, U.K.: Wm. Dawson & Sons, 1975.

———. *The French War on Trade: Privateering, 1793–1815.* London: Scolar, 1989.

Cutler, Carl C. *Greyhounds of the Sea.* Annapolis, Md.: Naval Institute Press, 1984.

Dai toa syo [Department of Great East Asia], ed. *Southern Economic Measures.* N.p.: NIDS Archives, 31 July 1943.

Dalzell, George W. *The Flight from the Flag: The Continuing Effect of the Civil War upon the American Carrying Trade.* Chapel Hill: Univ. of North Carolina Press, 1940.

Daniels, Josephus. *Our Navy at War.* New York: G. H. Doran, 1922.

de Guttry, Andrea, and Natalino Ronzitti, eds. *The Iran-Iraq War (1980–1988) and the Law of Naval Warfare.* Cambridge, U.K.: Grotius, 1993.

de Vattel, Emer. *The Law of Nations; or, Principles of the Law of Nature: Applied to the Conduct and Affairs of Nations and Sovereigns.* London: G. G. J. and J. Robinson, 1793.

Delaney, Norman C. *John McIntosh Kell of the Raider Alabama.* Tuscaloosa: Univ. of Alabama Press, 1973.

Dobson, [John]. *Chronological Annals of the War, from Its Beginning to the Present Time.* Oxford, U.K.: Clarendon, 1763.

Documents on British Foreign Policy, 1919–1939. Series 2. Vol. 18. London: HMSO, 1980.

Documents on German Foreign Policy, 1918–1945. Series D. Vol. 3, *Germany and the Spanish Civil War.* Washington, D.C.: U.S. Government Printing Office [hereafter GPO], 1950.

Dodington, George Bubb. *The Diary of the Late George Bubb Dodington.* Salisbury, U.K.: E. Easton, 1784.

Dönitz, Karl. *Memoirs: Ten Years and Twenty Days.* Annapolis, Md.: Naval Institute Press, 1990.

Dorn, Walter L. *Competition for Empire, 1740–1763.* New York: Harper and Row, 1963.

Dower, John W. *Embracing Defeat: Japan in the Wake of World War II.* New York: W. W. Norton / New Press, 1999.

Drea, Edward J. *Japan's Imperial Army: Its Rise and Fall, 1853–1945*. Lawrence: Univ. Press of Kansas, 2009.

Dreifort, John E. *Yvon Delbos at the Quai d'Orsay: French Foreign Policy during the Popular Front, 1936–1938*. Lawrence: Univ. Press of Kansas, 1973.

Ducéré, Édouard. *Les corsaires basques et bayonnais sous la République et l'Empire*. Bayonne, Fr.: A. Lamaignère, 1898.

Dudley, William S., and Michael J. Crawford, eds. *The Naval War of 1812: A Documentary History*. 3 vols. Washington, D.C.: Naval Historical Center, 1992.

Dull, Jonathan R. *The French Navy and the Seven Years' War*. Lincoln: Univ. of Nebraska Press, 2005.

Eastlake, Warrington, and Yamada Yoshi-aki. *Heroic Japan: A History of the War between China & Japan*. 1897; repr. Washington, D.C.: University Publications of America, 1979.

Eckert, Edward K. *Navy Department in the War of 1812*. Gainesville: Univ. of Florida Press, 1973.

Elleman, Bruce A. *Modern Chinese Warfare, 1795–1989*. London: Routledge, 2001.

Elleman, Bruce A., and S. C. M. Paine. *Modern China: Continuity and Change 1644 to the Present*. Boston: Prentice Hall, 2010.

Ellis, John. *Brute Force: Allied Strategy and Tactics in the Second World War*. New York: Viking, 1990.

Epkenhans, Michael. "Technology, Shipbuilding and Future Combat in Germany, 1880–1914." In *Technology and Naval Combat in the Twentieth Century and Beyond*, edited by Phillips Payson O'Brien. London: Frank Cass, 2001.

———. *Tirpitz: Architect of the German High Seas Fleet*. Washington, D.C.: Potomac Books, 2008.

Esthus, Raymond A. *Theodore Roosevelt and Japan*. Seattle: Univ. of Washington Press, 1967.

Evans, David C., and Mark R. Peattie. *Kaigun: Strategy, Tactics, and Technology in the Imperial Japanese Navy, 1887–1941*. Annapolis, Md.: Naval Institute Press, 1997.

Evans, George A. *The Influence of the Sea on the Political History of Japan*. New York: E. P. Dutton, 1921.

"Extracts from an Epitome of the Chino-Japanese War." In *Chino-Japanese War, 1894–95*, edited by N. W. H. Du Boulay. London: typescript, ca. 1903.

Falconer, William. *A New Universal Dictionary of the Marine*. London: T. Cadell and W. Davies, 1769 [repr. 1815].

Fayle, C. Ernest. *Seaborne Trade*. 3 vols. London: John Murray, 1920–24.

Fernow, Berthold, ed. *New York (Colony) Council: Calendar of Council Minutes, 1668–1783*. Harrison, N.Y.: Harbor Hill Books, 1987.

Fletcher, Joseph. "Ch'ing Inner Asia c. 1800." In *Cambridge History of China*. Vol. 10, edited by John King Fairbank. Cambridge, U.K.: Cambridge Univ. Press, 1978.

Foreign Relations of the United States: Diplomatic Papers, 1936. Vol. 1, *General, The British Commonwealth*. Washington, D.C.: GPO, 1953.

Fowler, William M., Jr. *Rebels under Sail: The American Navy during the Revolution*. New York: Scribner's, 1976.

Franco, Lucas Molina, and José María Manrique. *Legion Condor: La historia olvidada*. Valladolid, Sp.: Quirón, 2000.

Frank, Willard C., Jr. "German Clandestine Submarine Warfare in the Spanish Civil War, 1936." In *New Interpretations in Naval History: Selected Papers from the Ninth Naval History Symposium Held at the United States Naval Academy, 18–20 October 1989*, edited by William R. Roberts and Jack Sweetman. Annapolis, Md.: Naval Institute Press, 1991.

———. "The Nyon Arrangement 1937: Mediterranean Security and the Coming of the Second World War." In *Regions, Regional Organizations and Military Power: XXXIII International Congress of Military History, 2007*, edited by Thean Potgieter. Stellenbosch, S.A.: South African Military History Commission / African Sun Media, 2008.

Fuehrer Conferences in Matters Dealing with the German Navy. 7 vols. Washington, D.C.: Secretary of the Navy, 1947.

Fuehrer Conferences on Naval Affairs 1939–1945. Annapolis, Md.: Naval Institute Press, 1990.

Fuehrer Directives and Other Top-Level Directives of the German Armed Forces 1939–1941. Typescript translation [Office of Naval Intelligence], Washington, D.C., 1948. Henry E. Eccles Library, Naval War College, Newport, R.I.

Fuller, William C., Jr. *Strategy and Power in Russia, 1600–1914*. New York: Free Press, 1992.

Gallois, Napoléon. *Les corsaires français sous la république et l'Empire*. 2 vols. Le Mans, Fr.: Julien, Lanier, 1854.

Gardiner, Robert, ed. *The Naval War of 1812*. London: Chatham, 1998; repr. London: Caxton, in association with the National Maritime Museum, 2001.

———, ed. *Navies and the American Revolution, 1775–1783*. London: Chatham, in association with the National Maritime Museum, 1996.

Gemzell, Carl-Axel. *Organisation, Conflict and Innovation: A Study of German Naval Strategic Planning, 1888–1940*. Stockholm: Esselte Studium, 1973.

Gerace, Michael P. *Military Power, Conflict and Trade*. London: Frank Cass, 2004.

Giangreco, D. M. *Hell to Pay: Operation Downfall and the Invasion of Japan, 1945–1947*. Annapolis, Md.: Naval Institute Press, 2009.

Gibson, R. H., and Maurice Prendergast. *The German Submarine War, 1914–1918*. London: Constable, 1931.

Gipson, Lawrence Henry. *The British Empire before the American Revolution*. 15 vols. New York: Knopf, 1939–70.

Glete, Jan. *Navies and Nations: Warships, Navies and State Building in Europe and America, 1500–1860*. Vol. 2. Stockholm: Almqvist and Wiksell International, 1993.

Goldman, Emily O. *Sunken Treaties: Naval Arms Control between the Wars*. University Park: Pennsylvania State Univ. Press, 1994.

Görlitz, Walter, ed. *The Kaiser and His Court: The Diaries, Notebooks and Letters of Admiral George Alexander von Müller, Chief of the Naval Cabinet, 1914–1918*. [English translation.] London: Macdonald, 1961.

Grivel, Richild. *De la guerre maritime avant et depuis les nouvelles inventions*. Paris: Arthus Bertrand, 1869.

Guilliatt, Richard, and Peter Hohnen. *The Wolf*. New York: Free Press, 2010.

Güth, Rolf. *Von Revolution zu Revolution: Entwicklungen und Führungsprobleme der deutschen Marine 1848–1918*. Herford, Ger.: E. S. Mittler, 1978.

Hagan, Kenneth J. "The Birth of American Naval Strategy." In *Strategy in the American War of Independence*, edited by Donald Stoker, Kenneth J. Hagan, and Michael T. McMaster. New York: Routledge, 2010.

———. *This People's Navy: The Making of American Sea Power*. New York: Free Press, 1991.

Halpern, Paul G. *A Naval History of World War I*. Annapolis, Md.: Naval Institute Press / UCL, 1994.

———. *The Naval War in the Mediterranean, 1914–1918*. London: Allen & Unwin; Annapolis, Md.: Naval Institute Press, 1987.

Hansard, T. C., ed. *Parliamentary Debates. From the Year 1803 to the Present Time*. 41 vols. London: HMSO, 1803–20.

Hansen, Stig Jarle. *Piracy in the Greater Gulf of Aden: Myths, Misconceptions and Remedies*. Oslo: Norwegian Institute for Urban Regional Research, 2009.

Harman, Joyce Elizabeth. *Trade and Privateering in Spanish Florida, 1732–1763*. Tuscaloosa: Univ. of Alabama Press, 2004.

Hartwig, Dieter. *Großadmiral Karl Dönitz: Legende und Wirklichkeit*. Munich: Paderborn, 2010.

Hearn, Chester G. *George Washington's Schooners: The First American Navy*. Annapolis, Md.: Naval Institute Press, 1995.

Hervey, Frederick. *The Naval, Commercial, and General History of Great Britain*. 5 vols. London: J. Bew, 1786.

Herwig, Holger H. *"Luxury Fleet": The Imperial German Navy, 1888–1918*. London: Allen & Unwin, 1980.

———. "Total Rhetoric, Limited War: Germany's U-boat Campaign, 1917–1918." In *Great War, Total War: Combat and Mobilization on the Western Front, 1914–1918*, edited by Roger Chickering and Stig Förster. Washington, D.C.: German Historical Institute; New York: Cambridge Univ. Press, 2000.

Hezlet, Vice Admiral Sir Arthur. *The Submarine and Sea Power*. London: Peter Davies, 1967.

Hickey, Donald R. *The War of 1812: A Forgotten Conflict*. Urbana: Univ. of Illinois Press, 1990.

Higginbotham, Don. *The War of American Independence: Military Attitudes, Policies, and Practice, 1763–1789*. Boston: Northeastern Univ. Press, 1971 [repr. 1983].

Hinsley, F. H. *British Intelligence in the Second World War: Its Influence on Strategy and Operations*. 5 vols. Cambridge, U.K.: Cambridge Univ. Press, 1979–90.

Hobson, Rolf. *Imperialism at Sea: Naval Strategic Thought, the Ideology of Sea Power, and the Tirpitz Plan, 1875–1914*. Boston: Brill, 2002.

Howe, Octavius T. *Beverly Privateers in the American Revolution*. Cambridge, Mass.: J. Wilson and Son, [1922?].

Hubatsch, Walther, ed. *Hitlers Weisungen für die Kriegführung, 1939–1945*. Frankfurt, Ger.: Bernard & Graefe, 1962 [repr. 1983].

Hull, Isabel V. *Absolute Destruction: Military Culture and the Practices of War in Imperial Germany*. Ithaca, N.Y.: Cornell Univ. Press, 2005.

Hunter, Janet E., comp. *Concise Dictionary of Modern Japanese History*. Berkeley: Univ. of California Press, 1984.

Ito, Masanori. *The End of the Imperial Japanese Navy*. New York: Jove Books, 1956.

Iwatake Teruhiko. *Nanpō gunseika no keizai shisaku* [Economic Policies under the Southern Army's Administration]. Vol. 1. Tokyo: Ryukei Shosha, 1995.

Jackson, Melvin H. *Privateers in Charleston, 1793–1796: An Account of a French Palatinate in South Carolina*. Washington, D.C.: Smithsonian Institution, 1969.

James, William. *Naval Occurrences of the War of 1812*. 1817; repr. London: Conway Maritime, 2004.

Japan. Army General Staff, ed. *Sugiyama Memo*. Vol. 1. Tokyo: Hara Shobo, 1967.

Japan. Imperial General Staff. *A History of the War between Japan and China*. Vol. 1, translated by Major Jikemura and Arthur Lloyd. Tokyo: Kinkodo, 1904.

Jen Hwa Chow. *China and Japan: The History of Chinese Diplomatic Missions in Japan 1877–1911*. Singapore: Chopmen, 1975.

Jennings, Francis. *Empire of Fortune: Crowns, Colonies & Tribes in the Seven Years War in America*. New York: W. W. Norton, 1988.

Jones, Jerry W. *U.S. Battleship Operations in World War I*. Annapolis, Md.: Naval Institute Press, 1998.

Kaeppelin, Jeanne, ed. *Surcouf dans l'océan Indien: Journal de bord de la "Confiance."* Saint-Malo, Fr.: Cristel, 2007.

Kahn, David. "Codebreaking in World War I and II: The Major Successes and Failures." In *The Missing Dimension: Governments and Intelligence Communities in the Twentieth Century*, edited by Christopher Andrew and David Dilks. London: Macmillan, 1984.

———. *Seizing the Enigma: The Race to Break the German U-boat Codes, 1939–1943*. Boston: Houghton Mifflin, 1991.

Karau, Mark D. *"Wielding the Dagger": The MarineKorps Flandern and the German War Effort, 1914–1918*. Westport, Conn.: Praeger, 2003.

Kell, John M. *Recollections of a Naval Life*. Washington, D.C.: Neale, 1900.

Kelly, Patrick J. *Tirpitz and the Imperial German Navy*. Bloomington: Indiana Univ. Press, 2011.

Kennedy, Paul M. *The Rise and Fall of British Naval Mastery*. New York: Scribner's, 1976.

Kimball, Gertrude Selwyn, ed. *Correspondence of William Pitt*. 2 vols. New York: Macmillan, 1906.

Klachko, Mary, and David F. Trask. *Admiral William Shepherd Benson: First Chief of Naval Operations*. Annapolis, Md.: Naval Institute Press, 1987.

Klooster, Wim. *Illicit Riches: Dutch Trade in the Caribbean, 1648–1795*. Leiden: KITLV, 1998.

Knox, Dudley W. *The Naval Genius of George Washington*. Boston: Riverside, 1932.

Kuo Ting-yee. *Sino-Japanese Relations, 1862–1927*. New York: Columbia Univ. Press, 1965.

Labaree, Benjamin W. *A Supplement (1971–1986) to Robert G. Albion's Naval & Maritime History, An Annotated Bibliography, Fourth Edition*. Mystic, Conn.: Munson Institute of American Maritime Studies, 1988.

Lambi, Ivo Nikolai. *The Navy and German Power Politics, 1862–1914*. Boston: Allen & Unwin, 1984.

Lavery, Brian. *Nelson's Navy: The Ships, Men and Organization, 1793–1815*. Rev. ed. 1989; repr. Annapolis, Md.: Naval Institute Press, 1997.

Le Guellaff, Florence. *Armements en course et droit de prises maritimes (1792–1856)*. Nancy, Fr.: Presses Universitaires de Nancy, 1999.

Leadam, I. S. *The History of England from the Accession of Anne to the Death of George II*. London: Longmans, Green, 1909.

Lecky, William Edward Hartpole. *A History of England in the Eighteenth Century*. 8 vols. New York: D. Appleton, 1888.

Lee, Robert H. G. *The Manchurian Frontier in Ch'ing History*. Cambridge, Mass.: Harvard Univ. Press, 1970.

Legohérel, Henri. *Histoire de la Marine française*. Paris: PUF, 1999.

Lehr, Peter, and Hendrick Lehmann. "Somalia: Pirates' New Paradise." In *Violence at Sea: Piracy in the Age of Global Terrorism*, edited by Peter Lehr. London: Routledge, 2006.

Leighton, John Langdon. *SIMSADUS London: The American Navy in Europe*. New York: Henry Holt, 1920.

Leighton, Richard M., and Robert W. Coakley. *Global Logistics and Strategy 1940–1943*. 2 vols. Washington, D.C.: GPO for Chief of Military History, U.S. Army Dept., 1955. (These two volumes constitute vol. 4, pt. 4 of the multivolume U.S. Army in World War II series, in the "The War Department" subseries.)

Lewis, Ioan. *Blood and Bone: The Call of Kinship in Somali Society*. Lawrenceville, N.J.: Red Sea, 1994.

———. *A Pastoral Democracy*. Oxford, U.K.: James Currey for the International African Institute, 1999.

———. *Understanding Somalia and Somaliland*. London: Hurst, 2008.

Little, Peter D. *Somalia: Economy without State*. Oxford, U.K.: James Currey for the International African Institute, 2003.

Lockwood, Vice Admiral Charles A. *Sink 'Em All: Submarine Warfare in the Pacific*. New York: E. P. Dutton, 1951.

Lone, Stewart. *Japan's First Modern War: Army and Society in the Conflict with China, 1894–95*. London: St. Martin's, 1994.

Luntinen, Pertii, and Bruce W. Menning. "The Russian Navy at War, 1904–05." In *The Russo-Japanese War in Global Perspective: World War Zero*, edited by John W. Steinberg et al. Leiden: Brill, 2005.

Lydon, James G. *Pirates, Privateers, and Profits*. Upper Saddle River, N.J.: Greg, 1970.

Lynn, John A., ed. *Feeding Mars: Logistics in Western Warfare from the Middle Ages to the Present*. Boulder, Colo.: Westview, 1993.

———. *Giant of the Grand Siècle: The French Army, 1610–1715*. Cambridge, U.K.: Cambridge Univ. Press, 1997.

Macpherson, David. *Annals of Commerce, Manufactures, Fisheries, and Navigation*. 4 vols. London: Nichols and Son et al., 1805.

Magens, Nicolas. *An Essay on Insurances*. 2 vols. London: J. Haberkorn, 1755.

Mahan, Alfred T. *The Influence of Sea Power upon History 1660–1783*. London: Sampson Low, 1890; repr. New York: Dover, 1987.

———. *The Major Operations of the Navies in the War of American Independence*. London: Sampson Low, Marston, 1913.

Mahon, John K. *The War of 1812*. Gainesville: Univ. of Florida Press, 1972.

Malo, Henry. *Les derniers corsaires. Dunquerque 1715–1815*. Paris: Emile-Paul Frères, 1925.

Marder, Arthur J. *From the Dreadnought to Scapa Flow*. Vol. 4, *1917: Year of Crisis*. London: Oxford Univ. Press, 1969.

Marriot, James. *The Case of the Dutch Ships Considered*. London: R. and J. Dodsley, 1759.

Marshall, P. J., ed. *The Eighteenth Century*. Vol. 2 of *The Oxford History of the British Empire*, edited by Wm. Roger Louis. Oxford, U.K.: Oxford Univ. Press, 1998.

Marvel, William. *The Alabama & the Kearsarge: The Sailor's Civil War*. Chapel Hill: Univ. of North Carolina Press, 1996.

Marzagalli, Silvia, ed. *Bordeaux et la Marine de Guerre*. Bordeaux, Fr.: Presses Universitaires de Bordeaux, 2002.

———. *Bordeaux et les États-Unis, 1776–1815: Politique et stratégies négociantes dans la genèse d'un réseau commercial*. Geneva: Droz, forthcoming.

———. *"Les boulevards de la fraude": Le négoce maritime et le Blocus continental, 1806–1813 —Bordeaux, Hambourg, Livourne*. Villeneuve d'Ascq, Fr.: Presses Universitaires du Septentrion, 1999.

May, Ernest R. *The World War and American Isolation*. Chicago: Quadrangle Books, 1966.

McCusker, John J. *Money and Exchange in Europe and America, 1600–1775*. Chapel Hill: Univ. of North Carolina Press, 1978.

McGiffin, Lee. *Yankee of the Yalu: Philo Norton McGiffin, American Captain in the Chinese Navy (1885–1895)*. New York: E. P. Dutton, 1968.

McKee, Christopher. *A Gentlemanly and Honorable Profession: The Creation of the U.S. Naval Officer Corps, 1794–1815*. Annapolis, Md.: Naval Institute Press, 1991.

Melvin, Frank Edgar. *Napoleon's Navigation System: A Study of Trade Control during the Continental Blockade*. New York: D. Appleton, 1919 [repr. 1970].

Menkhaus, Kenneth John. "Local Security Systems in Somali East Africa." In *Fragile States and Insecure People? Violence, Security and Statehood in the Twenty-First Century*, edited by Louise Andersen, Bjørn Møller, and Finn Stepputat. New York: Palgrave Macmillan, 2007.

———. *Somalia: State Collapse and the Threat of Terrorism*. Oxford, U.K.: Oxford Univ. Press, 2004.

Mevers, Frank C. "Naval Policy of the Continental Congress." In *Maritime Dimensions of the American Revolution*. Washington, D.C.: Naval History Division, U.S. Navy Dept., 1977.

Meyer, Jean, and John S. Bromley. "La seconde guerre de Cent Ans (1689–1815)." In *Dix siècles d'histoire franco-britannique. De Guillaume le conquérant au marché commun*, edited by Douglas Johnson, François Bédarida, and François Crouzet. Paris: Albin Michel, 1979.

Michie, Alexander. *The Englishman in China during the Victorian Era*. Vol. 2. Edinburgh, U.K.: William Blackwood and Sons, 1900.

Miller, Edward S. *War Plan Orange: The U.S. Strategy to Defeat Japan, 1897–1945*. Annapolis, Md.: Naval Institute Press, 1991.

Miller, Nathan. *Sea of Glory: A Naval History of the American Revolution*. Charleston, S.C.: Nautical and Aviation, 1974.

Milner, Marc. *Battle of the Atlantic*. St. Catharines, Ont.: Vanwell, 2003.

Ministerstvo Inostrannykh Del SSSR. *Dokumenty Vneshnei Politiki SSSR*. Vol. 20. 1937; repr. Moscow: Izdatel'stvo Politicheskoi Literatury, 1976.

Ministry of Defence (Navy), ed. *The U-boat War in the Atlantic, 1939–1945*. Facs. ed. with introduction by Andrew J. Withers. 3 parts. London: HMSO, 1989.

Mitchell, Donald W. *A History of Russian and Soviet Sea Power*. New York: Macmillan, 1974.

Moore, John Bassett, and Francis Wharton. *A Digest of International Law, as Embodied in Diplomatic Discussions. . . .* Washington, D.C., GPO, 1906.

Morison, Elting E. *Admiral Sims and the Modern American Navy*. Boston: Houghton Mifflin, 1942.

Morison, Samuel Eliot. *History of U.S. Naval Operations in World War II*. Vol. 1, *The Battle of the Atlantic, September 1939–May 1943*. Boston: Little, Brown, 1947 [repr. 1954].

Morris, Richard B. Introduction to *The American Navies of the Revolutionary War: Paintings*, by Nowland Van Powell. New York: G. P. Putnam, 1974.

Morse, Hosea Ballou. *The International Relations of the Chinese Empire*. Vol. 3. Shanghai: Kelly and Walsh, 1918.

Müller, Leos. *Consuls, Corsairs, and Commerce: The Swedish Consular Service and Long-Distance Shipping, 1720–1815*. Uppsala, Swed.: Uppsala universitet, 2004.

Mulligan, Timothy P. *Neither Sharks nor Wolves: The Men of Nazi Germany's U-boat Arm, 1939–1945*. Annapolis, Md.: Naval Institute Press, 1999.

Murphy, Martin N. *Contemporary Piracy and Maritime Terrorism: The Threat of International Security*. IISS Adelphi Paper 388. London: Routledge, 2007.

———. *Small Boats, Weak States, Dirty Money: Piracy and Maritime Terrorism in the Modern World*. New York: Columbia Univ. Press, 2010.

———. *Somalia: The New Barbary? Piracy and Islam in the Horn of Africa.* New York: Columbia Univ. Press, 2011.

Museum of the Chinese People's Revolutionary Military Affairs [中国人民革命军事博物馆], ed. 中国战争史地图集 [Map Collection of Chinese Military History]. Beijing: 星球地图出版社, 2007.

Mutsu Munemitsu. *Kenkenryoku: A Diplomatic Record of the Sino-Japanese War, 1894–95.* Translated by Gordon Mark Berger. Princeton, N.J.: Princeton Univ. Press, 1982.

Netherlands. Departement van Defensie. *Jaarboek van de Koninklijke Marine, 1936–37.* The Hague: Algemeene Landsdrukkerij, 1938.

Newpower, Anthony. *Iron Men and Tin Fish: The Race to Build a Better Torpedo during World War II.* Westport, Conn.: Praeger Security International, 2006.

Nish, Ian, ed. *British Documents on Foreign Affairs: Reports and Papers from the Foreign Office Confidential Print.* Part 1, series E, vol. 5. Bethesda, Md.: University Publications of America, 1989.

———. *The Origins of the Russo-Japanese War.* New York: Longman, 1985.

North, Douglass C. "The United States Balance of Payments, 1790–1860." In *Trends in American Economy in the Nineteenth Century.* Princeton, N.J.: Princeton Univ. Press, 1960.

O'Connor, Raymond G. *Origins of the American Navy: Sea Power in the Colonies and the New Nation.* Lanham, Md.: University Press of America, 1994.

Offer, Avner. *The First World War: An Agrarian Interpretation.* Oxford, U.K.: Clarendon, 1991.

Official Records of the Union and Confederate Navies in the War of the Rebellion. Series I, vols. 1–3, 5. Washington, D.C.: GPO, 1880–1901.

Oh, Bonnie Bongwan. "The Background of Chinese Policy Formation in the Sino-Japanese War of 1894–1895." PhD diss., Univ. of Chicago, 1974.

Oi Atsushi. "Why Japan's Antisubmarine Warfare Failed." In *The Japanese Navy in World War II: In the Words of Former Japanese Naval Officers,* translated and edited by David C. Evans, with an introduction and commentary by Raymond O'Connor. 2nd ed. Annapolis, Md.: Naval Institute Press, 1986.

O'Kane, Rear Admiral Richard H. *Wahoo! The Patrols of America's Most Famous World War II Submarine.* Novato, Calif.: Presidio, 1987.

Okumura Fusao [奥村房夫] and Kuwada Etsu [桑田 悦], eds. 近代日本戦争史 [Modern Japanese Military History]. Vol. 1, 日清・日露戦争 [The Japanese-Qing and Russo-Japanese Wars]. Tokyo: 同台経済懇話会, 1995.

Olivier, David H. *German Naval Strategy, 1856–1888: Forerunners of Tirpitz.* London: Frank Cass / Routledge, 2004.

Owsley, Frank L., Jr. *The C.S.S. Florida: Her Building and Operations.* Tuscaloosa: Univ. of Alabama Press, 1987.

Padelford, Norman J. *International Law and Diplomacy in the Spanish Civil Strife.* New York: Macmillan, 1939.

Padfield, Peter. *Dönitz: The Last Führer—Portrait of a Nazi War Leader.* New York: Harper and Row, 1984.

Paine, S. C. M. *The Sino-Japanese War of 1894–1895: Perceptions, Power, and Primacy.* Cambridge, U.K.: Cambridge Univ. Press, 2003.

Pares, Richard. *Colonial Blockade and Neutral Rights, 1739–1763.* Oxford, U.K.: Oxford Univ. Press, 1938.

———. *War and Trade in the West Indies, 1739–1763.* London: Frank Cass, 1963.

Parillo, Mark P. *The Japanese Merchant Marine in World War II.* Annapolis, Md.: Naval Institute Press, 1993.

Patton, Robert H. *Patriot Pirates: The Privateer War for Freedom and Fortune in the American Revolution.* New York: Pantheon, 2008.

Pérotin-Dumon, Anne. "Economie corsaire et droit de neutralité. Les ports de la Guadeloupe pendant les guerres révolutionnaires." In *L'Espace Caraïbe, théâtre et enjeu des luttes impériales, XVIe–XIXe siècle,* edited by Paul Butel and Bernard Lavallé. Bordeaux, Fr.: Maison des Pays Ibériques, 1996.

———. *La ville aux îles, la ville dans l'île, Basse-Terre et Pointe-à-Pitre, Guadeloupe, 1650–1815.* Paris: Karthala, 2000.

Peters, A. R. *Anthony Eden at the Foreign Office, 1931–1938.* New York: St. Martin's, 1986.

Porter, David D. *Naval History of the Civil War.* New York: Sherman, 1886; repr. Secaucus, N.J.: Castle, 1984.

Powell, J. W. Damer. *Privateers and Ships of War.* Bristol, U.K.: J. W. Arrowsmith, 1930.

Pritchard, James. *Louis XV's Navy: A Study of Organization and Administration.* Montreal: McGill-Queen's Univ. Press, 2009.

Pyle, Kenneth B. *The Making of Modern Japan.* 2nd ed. Lexington, Mass.: D. C. Heath, 1996.

Raeder, E., and Eberhard von Mantey. *Der Kreuzerkrieg in den ausländischen Gewässern.* 3 vols. Berlin: E. S. Mittler, 1922–37.

Rahn, Werner. "The Atlantic in the Strategic Perspective of Hitler and Roosevelt, 1940–1941." In *To Die Gallantly: The Battle of the Atlantic,* edited by Timothy J. Runyan and Jan M. Copes. San Francisco: Westview, 1994.

———. "The Campaign: The German Perspective." In *The Battle of the Atlantic 1939–1945: The 50th Anniversary International Conference,* edited by Stephen Howarth and Derek Law. London: Greenhill Books, 1994.

———. "Die deutsche Seekriegführung 1943 bis 1945 [German Naval Warfare 1943–1945]." In *Das Deutsche Reich und der Zweite Weltkrieg.* Vol. 10, *Der Zusammenbruch des Deutschen Reiches 1945.* Part 1, *Die militärische Niederwerfung der Wehrmacht.* Munich: Deutsche Verlags-Anstalt, 2008.

———. "The Development of New Types of U-boats in Germany during World War II: Construction, Trials and First Operational Experience of the Type XXI, XXIII and Walter U-boats." In *Les marines de guerre du dreadnought au nucléaire.* Paris: Service historique de la Marine, 1990.

———. "German Naval Strategy and Armament, 1919–39." In *Technology and Naval Combat in the Twentieth Century and Beyond,* edited by Phillips Payson O'Brien. London: Frank Cass, 2001.

———. "Long-Range German U-boat Operations in 1942 and Their Logistical Support by U-Tankers." In *Die operative Idee und ihre Grundlagen. Ausgewählte Operationen des Zweiten Weltkrieges,* edited by Militärgeschichtliches Forschungsamt. Bonn: Herford, 1989.

———. "The War at Sea in the Atlantic and in the Arctic Ocean." In *Germany and the Second World War.* Vol. 6, *The Global War: Widening of the Conflict into a World War and the Shift of the Initiative 1941–1943,* edited by Horst Boog, Werner Rahn, Reinhard Stumpf, and Bernd Wegner. Oxford, U.K.: Clarendon, 2001.

Rahn, Werner, and Gerhard Schreiber, eds. *Kriegstagebuch der Seekriegsleitung 1939–1945.* Teil [part] A. 68 vols. Bonn: Herford, 1988–97.

Ranft, Bryan. "Restraints on War at Sea before 1945." In *Restraints on War: Studies in the Limitation of Armed Conflict,* edited by Michael Howard. Oxford, U.K.: Oxford Univ. Press, 1979.

Rapalino, Patrizio. *La Regia Marina in Spagna, 1936–1939.* Milan: Mursia, 2007.

Rawlinson, John L. *China's Struggle for Naval Development 1839–1895.* Cambridge, Mass.: Harvard Univ. Press, 1967.

Régent, Frédéric. *Esclavage, métissage, liberté. La Révolution française en Guadeloupe, 1789–1802.* Paris: Grasset, 2004.

Rhys-Jones, Graham. "The Riddle of the Convoys: Admiral Dönitz and the U-boat Campaign 1941." Unpublished paper, April 1992, Newport, R.I.

Ribadieu, Henry. *Histoire maritime de Bordeaux. Aventures des corsaires et des grands navigateurs bordelais.* Bordeaux, Fr.: Dupuy, 1854.

Ritter, Gerhard. *The Sword and the Scepter: The Problem of Militarism in Germany.* 4 vols. [English translation.] Coral Gables, Fla.: Univ. of Miami Press, 1969–73.

Robidou, F. *Les derniers corsaires malouins. La course sous la Révolution et l'Empire, 1793–1815.* Paris: Oberthür, 1919.

Robinson, Charles M., III. *Shark of the Confederacy: The Story of the C.S.S. Alabama.* Annapolis, Md.: Naval Institute Press, 1995.

Robinson, William Morrison, Jr. *The Confederate Privateers.* 1928; repr. Columbia: Univ. of South Carolina Press, 1990.

Rodger, N. A. M. *The Command of the Ocean: A Naval History of Britain, 1649–1815.* New York: W. W. Norton, 2004; London: Penguin Books, 2005.

———. *The Wooden World: An Anatomy of the Georgian Navy.* New York: W. W. Norton, 1996.

Rodigneaux, Michel. *La guerre de course en Guadeloupe, XVIIIe-XIXe siècles, ou Alger sous les Tropiques*. Paris: L'Harmattan, 2006.

Rodman, Hugh. *Yarns of a Kentucky Admiral*. Indianapolis, Ind.: Bobbs-Merrill, 1928.

Rodríguez-Salgedo, M. J., ed. *Armada, 1588–1988*. London: Penguin Books, in association with the National Maritime Museum, 1988.

Rohwer, Jürgen. *Axis Submarine Successes 1939–1945*. Annapolis, Md.: Naval Institute Press, 1999.

———. *The Critical Convoy Battles of March 1943*. Annapolis, Md.: Naval Institute Press, 1977.

Røksund, Arne. *The Jeune École: The Strategy of the Weak*. Leiden: Brill, 2007.

Ropp, Theodore. *The Development of a Modern Navy: French Naval Policy 1871–1904*. Edited by Stephen S. Roberts. Annapolis, Md.: Naval Institute Press, 1987.

Roscoe, E. S., ed. *Reports of Prize Cases Determined in the High Court of Admiralty . . . from 1745 to 1859*. 2 vols. London: Stevens and Sons, 1905.

Roskill, S. W. *The War at Sea 1939–1945*. Vol. 1, *The Defensive*. London: HMSO, 1954.

———. *The War at Sea 1939–1945*. Vol. 2, *The Period of Balance*. London: HMSO, 1957.

———. *The War at Sea 1939–1945*. Vol. 3, *The Offensive*. London: HMSO, 1961.

Rössler, Eberhard. *The U-boat: The Evolution and Technical History of German Submarines*. London: Arms and Armour; Annapolis, Md.: Naval Institute Press, 1981.

Russell, John Robert. "The Development of a 'Modern' Army in Nineteenth Century Japan." Master's thesis, Columbia Univ., New York, 1957.

Rybalkin, Iurii. *Operatsiya "X": Sovetskaya Voennaya Pomoshch' Respublikanskoi Ispanii, 1936–1939*. Moscow: Aero-XX, 2000.

Sachs, William S. "The Business Outlook in the Northern Colonies, 1750–1775." PhD diss., Columbia Univ., New York, 1957.

Salewski, Michael. *Die deutsche Seekriegsleitung 1935–1945*. 3 vols. Frankfurt, Ger.: Bernard & Graefe, 1970–1975.

Schauff, Frank. *Der verspielte Sieg: Sowjetunion, Kommunistische Internationale und Spanischer Bürgerkrieg, 1936–1939*. Frankfurt, Ger.: Campus, 2005.

Scheck, Raffael. *Alfred von Tirpitz and German Right-Wing Politics, 1914–1930*. Boston: Humanities, 1998.

Scheer, Admiral Reinhard. *Germany's High Sea Fleet in the World War*. London: Cassell, 1919.

Schimpf, Axel. "Der Einsatz von Kriegsmarineeinheiten im Rahmen der Verwicklungen des spanischen Bürgerkrieges 1936 bis 1939." In *Der Einsatz von Seestreitkräften im Dienst der auswärtigen Politik: Vorträge auf der Historisch-Taktischen Tagung der Flotte*, edited by Deutsches Marine Institut. Herford, Ger.: E. S. Mittler, 1983.

Schmalenbach, Paul. *German Raiders: A History of Auxiliary Cruisers of the German Navy 1895–1945*. Cambridge, U.K.: Patrick Stephens, 1979.

Scott, Admiral Sir Percy. *Fifty Years in the Royal Navy*. New York: George H. Doran, 1919.

Semmes, Raphael. *Memoirs of Service Afloat, during the War between the States*. Baltimore: Kelly, Piet, 1869; repr. Secaucus, N.J.: Blue and Grey, 1987.

Shay, Shaul. *The Red Sea Terror Triangle*. New Brunswick, N.J.: Transaction, 2005; repr. 2007.

"Ships of the United States Navy, Winter 1811." In *The New American State Papers: Naval Affairs*, edited by K. Jack Bauer. Vol. 1. Wilmington, Del.: Scholarly Resources, 1981.

Silverstone, Paul H. *Civil War Navies, 1855–1883*. Annapolis, Md.: Naval Institute Press, 2001.

———. *The Sailing Navy, 1775–1854*. Annapolis, Md.: Naval Institute Press, 2001.

———. *Warships of the Civil War Navies*. Annapolis, Md.: Naval Institute Press, 1989.

Simpson, Benjamin Mitchell, III. *Admiral Harold R. Stark: Architect of Victory, 1939–1945*. Columbia: Univ. of South Carolina Press, 1989.

Sims, William S. *The Victory at Sea*. Garden City, N.Y.: Doubleday, Page, 1920.

Sinclair, Arthur. *Two Years on the Alabama*. Boston: Lee and Shepard, 1895.

Sondhaus, Lawrence. *Naval Warfare, 1815–1914*. London: Routledge, 2001.

———. *Preparing for Weltpolitik: German Sea Power before the Tirpitz Era*. Annapolis, Md.: Naval Institute Press, 1997.

Spector, Stanley. *Li Hung-chang and the Huai Army: A Study in Nineteenth-Century Chinese Regionalism*. Seattle: Univ. of Washington Press, 1964.

Spindler, Arno. *Der Handelskrieg mit U-Booten*. 5 vols. Berlin (Vol. 5, Frankfurt, Ger.): E. S. Mittler, 1932–66.

Starkey, David J. *British Privateering Enterprise in the Eighteenth Century*. Exeter, Devon, U.K.: Univ. of Exeter Press, 1990.

Stegemann, Bernd. *Die Deutsche Marinepolitik, 1916–1918*. Berlin: Duncker & Humblot, 1970.

Stenzel, Alfred. *Kriegführung zur See. Lehre vom Seekriege*. Edited by Hermann Kirchoff. Hannover and Leipzig, Ger.: Hahnsche Buchhandlung, 1913.

Stern, Philip Van Doren. *The Confederate Navy: A Pictorial History*. Garden City, N.Y.: Doubleday, 1962.

Still, William N., Jr., ed. *The Queenstown Patrol, 1917: The Diary of Commander Joseph Knefler Taussig, U.S. Navy*. Newport, R.I.: Naval War College Press, 1996.

Stout, Neil R. *The Royal Navy in America, 1760–1775*. Annapolis, Md.: Naval Institute Press, 1973.

Sumida, Jon Tetsuro. "Forging the Trident: British Naval Industrial Logistics, 1914–1918." In *Feeding Mars: Logistics in Western Warfare from the Middle Ages to the Present*, edited by John A. Lynn. Boulder, Colo.: Westview, 1993.

Sun Kefu [孙克复] and Guan Jie [关捷]. 甲午中日陆战史 [A History of Land Engagements of the Sino-Japanese War]. Harbin, PRC: 黑龙江人民出版社, 1984.

Swanson, Carl E. *Predators and Prizes: American Privateering and Imperial Warfare, 1739–1748*. Columbia: Univ. of South Carolina Press, 1991.

Symcox, Geoffrey. *The Crisis of French Sea Power, 1688–1697: From the Guerre d'Escadre to the Guerre de Course*. The Hague: Martinus Nijhoff, 1974.

Syrett, David. *The Defeat of the German U-boats: The Battle of the Atlantic*. Columbia: Univ. of South Carolina Press, 1994.

———. *The Royal Navy in American Waters, 1775–1783*. Aldershot, U.K.: Scolar, 1989.

———. *Shipping and Military Power in the Seven Years War: The Sails of Victory*. London: Univ. of Exeter Press, 2008.

———. *Shipping and the American War, 1775–1783*. London: Athlone, 1970.

Takahashi, Sakuye. *International Law Applied to the Russo-Japanese War*. London: Stevens and Sons, 1908.

Takahashi Hidenao [高橋秀直]. 日清戦争の道 [The Road to the Sino-Japanese War]. Tokyo: 東京創社, 1995.

Tanaka Ken'ichi [田中健一], comp. 図説東郷平八郎目でみる明治の海軍 [Illustrated Volume on the Meiji Navy through the Eyes of Tōgō Heihachirō]. Tokyo: 東郷神社・東郷会, 1996.

Tangredi, Sam J., ed. *Globalization and Maritime Power*. Washington, D.C.: National Defense Univ. Press, 2002.

Tarrant, V. E. *The U-boat Offensive, 1914–1945*. Annapolis, Md.: Naval Institute Press, 1989.

Terraine, John. *The U-boat Wars 1916–1945*. New York: Putnam, 1989.

Tirpitz, Admiral Alfred von. *My Memoirs*. 2 vols. New York: Dodd Mead, 1919.

Toussaint, Auguste. *Les frères Surcouf*. Paris: Flammarion, 1979.

Traina, Richard P. *American Diplomacy and the Spanish Civil War*. Bloomington: Indiana Univ. Press, 1968.

Trial of the Officers and crew of the privateer Savannah on the charge of piracy, in the United States Circuit Court for the Southern District of New York, Hon. Judges Nelson and Shipman, presiding. Reported by A. F. Warburton, stenographer and corrected by the counsel. New York: Baker & Godwin, 1862.

Truxes, Thomas M. *Defying Empire: Trading with the Enemy in Colonial New York*. New Haven, Conn.: Yale Univ. Press, 2008.

———, ed. *Letterbook of Greg & Cunningham, 1756–57: Merchants of New York and Belfast*. Oxford, U.K.: Oxford Univ. Press, 2001.

Tucker, Spencer C. *Raphael Semmes and the Alabama*. Fort Worth, Tex.: Ryan Place, 1996.

United States Strategic Bombing Survey, Summary Report (Pacific War). Washington, D.C.: GPO, 1946.

U.S. Adjutant-General's Office. Military Information Division. *Notes on the War between China and Japan*. Washington, D.C.: GPO, 1896.

U.S. Navy Dept. *Civil War Naval Chronology, 1861-1865*. 6 vols. Washington, D.C.: Naval History Division, 1972.

van Tuyll van Serooskerken, Hubert P. *The Netherlands and World War I*. Leiden: Brill, 2001.

Villiers, Patrick. *Les corsaires: des origines au traité de Paris du 16 avril 1856*. Paris: J. P. Gisserot, 2007.

———. *Les corsaires du Littoral de Philippe II à Louis XIV, Boulogne, Calais et Dunkerque 1560-1715*. Villeneuve d'Ascq, Fr.: Presses Universitaires du Septentrion, 1999.

———. "La Guerre de Course." In *Napoléon et la Mer, un rêve d'Empire*, edited by Jean-Marcel Humbert and Bruno Ponsonnet. Paris: Seuil, 2004.

———. *Marine royale, corsaires et trafic dans l'Atlantique de Louis XIV à Louis XVI*. 2 vols. Dunkirk, Fr.: Société dunkerquoise d'histoire et d'archéologie, 1991.

Vladimir [Zenone Volpicelli]. *The China-Japan War Compiled from Japanese, Chinese, and Foreign Sources*. Kansas City, Mo.: Franklin Hudson, 1905.

Wagner, Gerhard, ed. *Lagevorträge des Oberbefehlshabers der Kriegsmarine vor Hitler 1939-1945*. Munich: J. F. Lehmann, 1972.

Walter, John. *The Kaiser's Pirates: German Surface Raiders in World War One*. London: Arms and Armour, 1994.

Warner, Denis, and Peggy Warner. *The Tide at Sunrise: A History of the Russo-Japanese War, 1904-1905*. New York: Charterhouse, 1974.

Weigh, Ken Shen. *Russo-Chinese Diplomacy*. Shanghai: Commercial, 1928.

Weir, Gary E. *Building American Submarines, 1914-1940*. Contributions to Naval History, no. 3. Washington, D.C.: Naval Historical Center, 1991.

———. *Building the Kaiser's Navy: The Imperial Navy Office and German Industry in the Tirpitz Era, 1890-1919*. Annapolis, Md.: Naval Institute Press, 1992.

Whealey, Robert H. *Hitler and Spain: The Nazi Role in the Spanish Civil War, 1936-1939*. Lexington: Univ. of Kentucky Press, 1989.

Wheeler, Keith. *War under the Pacific*. Alexandria, Va.: Time-Life Books, 1980.

White, John. *The Diplomacy of the Russo-Japanese War*. Princeton, N.J.: Princeton Univ. Press, 1964.

White, Philip L., ed. *The Beekman Mercantile Papers, 1746-1799*. 3 vols. New York: New-York Historical Society, 1956.

Williams, Gomer. *History of the Liverpool Privateers and Letters of Marque*. London: William Heinemann, 1897.

Williams, Greg H. *The French Assault on American Shipping, 1793-1813: A History and Comprehensive Record of Merchant Marine Loss*. Jefferson, N.C.: McFarland, 2009.

Williams, Samuel Wells, and Frederick Wells Williams. *A History of China: Being the Historical Chapters from "The Middle Kingdom."* London: Sampson Low, Marston, 1897.

Wilson, Charles. *Mercantilism*. London: Historical Association, 1958.

Wylie, J. C. *Military Strategy: A General Theory of Power Control*. 3rd ed. Annapolis, Md.: Naval Institute Press, 1989.

Zhang Qiyun [張其昀], comp. 中國歷史地圖 [A Historical Atlas of China]. Vol. 2. Taipei: Chinese Culture Univ. Press, 1980.

About the Contributors

Ken-ichi Arakawa: Professor of modern war history / war and logistics in the Department of Leadership and Military History of the School of Defense Sciences at Japan's National Defense Academy. In 2003 he retired at the rank of colonel from the Japanese Ground Self-Defense Force and earned a PhD in economics from Hitotsubashi University. In 2011 he published in Japan *The Planning and Evolution of the Wartime Economic System: An Analysis of the Economic History of the Imperial Japanese Army and Navy*, winner of the 2011 Saeki Kiichi Prize for the best book on international security. He has published several book chapters and articles in English, including "The Japanese Naval Blockade of China in the Second Sino-Japanese War, 1937–41," in Bruce A. Elleman and S. C. M. Paine, eds., *Naval Blockades and Seapower* (Routledge, 2006); "Japanese War Leadership in the Burma Theatre: The Imphal Operation," in B. Bond and K. Tachikawa, eds., *British and Japanese Military Leadership* (Frank Cass, 2004); and "The Maritime Transport War: Emphasizing a Strategy to Interrupt the Enemy Sea Lines of Communication (SLOCs)" and "War and Economy: Japan's Expansion of Production," both in *NIDS Security Reports* (2002–2003).

Bruce Elleman: William V. Pratt Professor of International History and research professor, Maritime History Department, U.S. Naval War College, with an MA (1984) and PhD (1993) from the History Department, Columbia University; MS (1985) in international history, London School of Economics; and MA in national security and strategic studies, with distinction (2004), U.S. Naval War College. Recent books include *Modern Chinese Warfare, 1795–1989* (Routledge, 2001); *Naval Mutinies of the Twentieth Century: An International Perspective*, edited, with Christopher Bell (Frank Cass, 2003); *Naval Blockades and Seapower: Strategies and Counter-strategies, 1805–2005*, edited, with S. C. M. Paine (Routledge, 2006); *Waves of Hope: The U.S. Navy's Response to the Tsunami in Northern Indonesia*, Newport Paper 28 (Naval War College Press, 2007); and *Naval Coalition Warfare: From the Napoleonic War to Operation Iraqi Freedom*, edited, with Sarah Paine (Routledge, 2008).

Willard C. Frank, Jr.: Emeritus professor of history, Old Dominion University, with PhD from University of Pittsburgh, AM from College of William and Mary, and AB from Brown University. Author of "The Strategic Plight of the Spanish Republican Navy in the Spanish Civil War, 1936–39," in *New Interpretations in Naval History: Selected Papers from the Fifteenth Naval History Symposium*, ed. Maochun Yu (Naval Institute Press, 2009); "The Nyon Arrangement 1937: Mediterranean Security and the Coming of the Second World War," in *Regions, Regional Organizations and Military Power: XXXIII International Congress of Military History, 2007* (Stellenbosch: South African Commission of Military History / African Sun Media, 2008); "Multinational Naval Cooperation in the Spanish Civil War, 1936," *Naval War College Review* (Spring 1994); "Politico-Military Deception at Sea in the Spanish Civil War, 1936–1939," *Intelligence and National Security* (July 1990); and "Naval Operations in the Spanish Civil War, 1936–1939," *Naval War College Review* (January–February 1984).

Kenneth Hagan: Retired professor of strategy and war at the Naval War College program in Monterey, California, and professor and museum director emeritus at the U.S. Naval Academy. A retired captain in the U.S. Naval Reserve, he wrote *American Gunboat Diplomacy and the Old Navy, 1877–1889* (Greenwood, 1973) and *This People's Navy: The Making of American Sea Power* (Free Press, 1991). He coauthored *Unintended Consequences: The United States at War* (Reaktion, 2007) with Professor Ian J. Bickerton of the University of New South Wales. (The book has been translated and published in Japan and the Republic of Korea.) With Michael McMaster he coedited and contributed to the thirtieth-anniversary edition of *In Peace and War: Interpretations of American Naval History* (Greenwood/ABC-CLIO, 2008). He jointly edited *Strategy in the American War of Independence* (Routledge, 2009), to which he contributed the chapter "The Birth of American Naval Strategy." He coauthored, with Professor McMaster, a paper on American naval strategy, which they presented at the Fifth Conference of the International Society for First World War Studies held at the Imperial War Museum in London in September 2009. In September 2011 the same coauthors delivered a paper at the Naval Academy's Naval History Symposium on the topic of Cadet-Midshipman William S. Sims and five of his Naval Academy classmates from the class of 1880. Captain Hagan continues to be engaged in research and writing about the U.S. Navy and American foreign policy in the three transformational decades before World War I.

Paul G. Halpern: Professor emeritus, Florida State University, Tallahassee, Florida, with an MA (1961) and PhD (1966) from Harvard University. Books include *The Mediterranean Fleet, 1919–1929* (Ashgate for the Navy Records Society, 2011); *The Battle of Otranto Straits: Controlling the Gateway to the Adriatic in World War I* (Indiana University Press, 2004); *Anton Haus: Österreich-Ungarns Grossadmiral* (Verlag Styria, 1998); *A Naval History of World War I* (Naval Institute Press / UCL, 1994); *The Naval War in the Mediterranean, 1914–1918* (Allen & Unwin / Naval Institute Press, 1987); *The Mediterranean Naval Situation, 1908–1914* (Cambridge, Mass.: Harvard University Press, 1971).

John B. Hattendorf: Ernest J. King Professor of Maritime History at the U.S. Naval War College since 1984. Additionally, he is chairman of the College's Maritime History Department and director of the Naval War College Museum. He is author or editor of more than forty books, including the *Oxford Encyclopedia of Maritime History* (2007),

Talking about Naval History: A Collection of Essays (2011), and *Marlborough: Soldier and Diplomat* (2012). His academic work has been widely recognized by numerous awards.

Joel Holwitt: Currently serving as Navigator/Operations Officer of the attack nuclear submarine USS *New Mexico* (SSN 779). He earned his PhD and MA from the Ohio State University (2005) and a BS from the U.S. Naval Academy (2003). He is the author of *"Execute against Japan": The U.S. Decision to Conduct Unrestricted Submarine Warfare* (Texas A&M University Press, 2009). He has been published in the U.S. Naval Institute *Proceedings, Naval History, Submarine Review,* and the *Journal of Military History.*

Christopher Magra: Associate professor of early American history at the University of Tennessee. He has published articles related to maritime history in the *International Journal of Maritime History, New England Quarterly,* and *Northern Mariner.* The Canadian Nautical Research Society honored him with the Keith Matthews Award for his scholarship. Cambridge University Press published in 2009 his first book, *The Fisherman's Cause: Atlantic Commerce and Maritime Dimensions of the American Revolution,* which won the Winslow House Book Prize. He is completing his second book, on British naval impressment and the Atlantic origins of the American Revolution.

Silvia Marzagalli: Professor of early modern history at the Université de Nice–Sophia Antipolis and senior member of the Institut Universitaire de France. She is the author, with Michel Biard and Philippe Bourdin, of *Révolution, Consulat et Empire* (Belin, 2009); *"Les boulevards de la fraude": Le négoce maritime et le Blocus continental, 1806–1813—Bordeaux, Hambourg, Livourne* (Presses Universitaires du Septentrion, 1999); with Hubert Bonin of *Négoce, Ports et Océans. Mélanges offerts à Paul Butel* (Presses Universitaires de Bordeaux, 2000) and *Bordeaux et la Marine de Guerre* (Presses Universitaires de Bordeaux, 2002).

Kevin D. McCranie: Associate professor of strategy and policy at the U.S. Naval War College, in Newport, Rhode Island. He earned a PhD in history from Florida State University. In addition, he is the author of *Admiral Lord Keith and the Naval War against Napoleon* (University Press of Florida, 2006) and *Utmost Gallantry: The U.S. and the Royal Navies at Sea in the War of 1812* (Naval Institute Press, 2011).

Michael McMaster: Professor of joint maritime operations at the Naval War College program in Monterey, California. He served twenty-two years in the U.S. Navy as a Surface Warfare Officer and is a retired commander. In 2006 and 2007, he jointly presented two papers with Professor Hagan on the history of U.S. naval strategy at Sea Power Conferences of the Royal Australian Navy, both of which have now been published. He contributed to and served as associate editor of the thirtieth-anniversary edition of *In Peace and War: Interpretations of American Naval History* (Greenwood/ABC-CLIO, 2008), an anthology spanning the entire course of U.S. naval history. He was an editor for *Strategy in the American War of Independence* (Routledge, 2009). With Professor Hagan he was coauthor of "His Remarks Reverberated from Berlin to Washington," in the U.S. Naval Institute *Proceedings* in December 2010. He and Professor Hagan presented a paper on the U.S. Navy in World War I at the Fifth Conference of the International Society for First World War Studies in London in 2009 and in 2011 a paper titled "William Sowden Sims and Five Classmates in the Old Navy's School House, 1876–1880," at the U.S. Naval Academy Naval History Symposium.

Martin N. Murphy: Political and strategic analyst and internationally recognized expert on piracy and unconventional conflict at sea. His published works include three books—*Somalia: The New Barbary? Piracy and Islam in the Horn of Africa* (2011) and *Small Boats, Weak States, Dirty Money: Piracy and Maritime Terrorism in the Modern World* (2009), both published by Columbia University Press / Hurst and both recognized by the U.S. Naval Institute as among the most important naval titles published in their respective years; and *Contemporary Piracy and Maritime Terrorism* (2007), an Adelphi Paper published by the London-based International Institute for Strategic Studies. He is a senior fellow at the Atlantic Council of the United States; a visiting fellow at the Corbett Centre for Maritime Policy Studies at King's College London; and a research fellow at the Centre for Foreign Policy Studies, Dalhousie University, Halifax. He has taught a course on piracy, trade, and war as part of the Security Studies Program at Georgetown University, in Washington, D.C.

David H. Olivier: Assistant professor, history and contemporary studies, Wilfrid Laurier University, with PhD from the University of Saskatchewan, an MA from the University of Toronto, and BA from York University. He is the author of *German Naval Strategy, 1856–1888: Forerunners of Tirpitz* (Frank Cass / Routledge, 2004) and of "German Coastal Defence, 1859–1888" in Howard Fuller, ed., *'Twixt Sea and Shore: The Reality of "Deterrence" in the Pax Britannica* (Naval Institute Press, 2013).

S. C. M. Paine: Professor in the Strategy and Policy Department, U.S. Naval War College, with a PhD (1993) in Russian and Chinese history, Columbia University, and MIA (1984) from School for International and Public Affairs, Columbia University. Author of *The Wars for Asia, 1911–1949* (Cambridge, U.K.: Cambridge University Press, 2012); *The Sino-Japanese War of 1894–1895: Perceptions, Power, and Primacy* (Cambridge University Press, 2003); and *Imperial Rivals: China, Russia, and Their Disputed Frontier, 1858–1924* (M. E. Sharpe, 1996), winner of the 1997 Barbara Jelavich Prize for diplomatic history from the American Association for the Advancement of Slavic Studies. Editor of *Nation Building, State Building, and Economic Development: Case Studies and Comparisons* (M. E. Sharpe, 2010) and coauthor, with Bruce A. Elleman, of *Modern China: Continuity and Change 1644 to the Present* (Prentice Hall, 2010).

Werner Rahn: Captain (Retired), German Navy, with a PhD in history from Hamburg University. Author of PhD dissertation *Reichsmarine und Landesverteidigung 1919–1928* [German Navy and Defense Policy, 1919–1928] and coeditor, with Gerhard Schreiber, of the facsimile edition of *Kriegstagebuch der Seekriegsleitung 1939–1945* [War Diary of the Naval Staff], Part A, 68 vols. (Herford, 1988–97).

Thomas M. Truxes: Clinical associate professor of Irish studies and history, New York University, with PhD from Trinity College, Dublin; MA from Trinity College, Hartford, Conn.; MBA from Syracuse University, Syracuse, N.Y.; and BS from Boston College, Chestnut Hill, Mass. Coeditor, with L. M. Cullen and John Shovlin, *The Bordeaux–Dublin Letters, 1757: Correspondence of an Irish Community Abroad,* volume 53 of the British Academy's Records of Social and Economic History (Oxford University Press, 2013). Author of *Defying Empire: Trading with the Enemy in Colonial New York* (Yale University Press, 2008) and *Irish-American Trade, 1660–1783* (Cambridge University Press, 1988). Editor of *Letterbook of Greg & Cunningham, 1756–57: Merchants of New York and Belfast,* volume 28 of the British Academy's Records of Social and Economic History (Oxford University Press, 2001).

Spencer Tucker: Spencer C. Tucker retired in 2003 after thirty-six years of university teaching, the last six as holder of the John Biggs Chair in Military History at the Virginia Military Institute. He is currently the senior fellow in military history for ABC-CLIO Publishing. Dr. Tucker is the author or editor of forty-three books of military and naval history, including *Raphael Semmes and the* Alabama (Ryan Place, 1996); *Blue and Gray Navies: The Civil War Afloat* (Naval Institute Press, 2006); *Naval Warfare: An International Encyclopedia* (3 vols., ABC-CLIO, 2003); and *Civil War Naval Encyclopedia* (2 vols., ABC-CLIO, 2010).

George Walker: Research Professor of Admiralty and International Law at the Law School, Wake Forest University, with an LLM from the University of Virginia, AM from Duke University, LLB from Vanderbilt University, and BA from the University of Alabama. Stockton Professor of International Law, Naval War College, 1992–93. Author of "Self-Defense, the Law of Armed Conflict and Port Security," *South Carolina Journal of International Law and Business* (Spring 2009), and *The Tanker War, 1980–88: Law and Policy* (Naval War College International Law Studies, vol. 74, 2000) and general editor of *Definitions for the Law of the Sea* (Nijhoff/Brill, 2012).

Index

A

Abdullahi, Farah 263
Abgaal subclan 263
Abu al-Bukosh 244
Achilles' heel 197
Adams (USN) 62
Adams, Charles Francis 78, 85
Adams, John 27, 28, 30, 31
Aden, Gulf of 254, 255, 256
Admiral Scheer (German) 173
Admiralty Council 96, 101
Adriatic Sea 136, 144, 145
Adventure (French CR) 50, 56n66
Aegean Sea 176, 177, 178, 179
aerial antisubmarine warfare 284, 290 *See also* antisubmarine warfare
aerial bombardment 166, 230
Africa 10, 15, 82, 128, 193, 197, 253–67
Afweyne (Mohamed Abdi Hassan) 263, 264
Ageef, Terente 131
Agrippina (CSS tender) 78, 80, 81
aircraft 2, 7, 165, 167, 168, 170, 172, 174, 175, 176, 179, 182, 183n13, 184n26, 187, 188, 189, 190, 192, 194, 200, 202–203, 209, 221, 229, 230, 233, 242, 248–49, 277–78, 279, 282, 290
airship 139
Alabama 73
Alabama (CSS) 4, 73–87, 88n21, 225
Alabama-Kearsarge naval battle 4, 83–85
alcohol 12, 33, 36, 79, 123
Alert (French CR) 50
Alert (HMS) 59

Algameca 168
Algeria 172, 179
al-Hamadi port 247
Allen, William H. 63
Allied General Headquarters 214
Allied operations 192, 193, 194, 195, 196, 197, 200, 201, 202, 203, 209, 214, 222, 223, 225, 232, 235, 279, 283, 284, 288, 289
allies 3, 5, 6, 7, 9, 99, 137, 144, 146, 151, 153, 154, 158, 162, 165, 191, 192, 193, 194, 195, 196, 197, 200, 201, 202, 203, 209, 222, 223, 225, 232, 235, 271, 274, 275, 279, 283, 284, 285, 288, 289 *See also* coalitions
Al-Muharaq (Kuwaiti merchant) 243
American Civil War *See* Civil War (American)
American Revolution 2, 3, 19, 22, 27–36, 58, 272, 273, 274, 275, 276, 277, 278, 279, 281, 283, 284, 285, 286, 287, 288
ammunition 9, 13, 33, 78, 108, 114, 122, 123, 141
Amsterdam 13, 15
Amsterdam (merchant) 26n94
Anchuthengu *See* Anjengo
Ancona (Italian ocean liner) 144
Anglo-American relations 5, 6, 86, 151–63
Anglo-Dutch Treaty (1674) 14
Anglo-French wars 9, 93
Anglo-German Naval Agreement (18 June 1935) 170, 187
Anglo-Japanese alliance (1902) 125, 136
Anglophilia 154, 160
Angra Bay 78
Anjengo 82
Antigua 13, 17

Anti-Submarine Detection Investigation Committee (ASDIC) 188, 203
See also radar

antisubmarine warfare 5, 6, 142, 148, 158, 160, 187, 189, 200, 201, 202, 203, 222, 225, 233, 235, 273, 284, 286
See also aerial antisubmarine warfare

Aqaba, Gulf of 244

Arabian Gulf 256
See also Persian Gulf

Arabian Peninsula 254

Arabian Sea 244

Arabic (ocean liner) 142

Arab-Israeli conflict 251

Arab League 7, 240, 241, 242, 249, 251

Arab League Extraordinary Summit 242–43, 246

Arab League mutual defense treaty 241

Arcas Islands 80

Archimede (Italian to Spanish) 174

Arctic 62

Ardova (merchant) 124

Argus (USN) 61, 63

Ariel (Union merchant) 80

armed schooner 32–35

Armilla Patrol 241, 242, 246

armored cruiser 122, 126, 139

arms transport ships 166, 167, 168, 170, 171, 172, 176, 179

    Cabo Santo Tomé (Soviet) 179, 185n50

    Ciudad de Cádiz (Spanish) 176, 185n39

    Kamerun (German) 174

    Magallanes (Soviet) 172

    Usaramo (German) 174

army 1, 28, 33, 36, 41, 54n20, 73, 85, 90, 93, 95, 96, 106, 107–108, 111, 113, 114–16, 123, 131, 136, 143, 145, 147, 148, 165, 199, 209, 210, 211, 213, 214–17, 218–19, 221, 223n5, 223n9, 228, 230, 276, 279

Army Chief of Staff, Japan 210

Army General Staff, Japan 213

Arnauld de la Perière, Lothar von 145

Asan 110, 112

Asia 5, 6, 65, 82–83, 104, 105, 114, 121, 123, 124, 125–26, 128, 129, 209–23, 229, 230, 232, 234, 280

Askold (Russian) 131

Aspinwall Line 80

Atlantic Ocean 4, 6, 9, 10, 11, 12, 13, 14, 15, 16, 19, 20, 22, 28, 29, 35, 36, 50, 57, 61, 75, 77, 81, 92, 98, 99, 144, 151, 154, 155, 156, 158, 159, 160, 161, 162, 173, 179, 180, 188, 190, 191–93, 194, 197, 200–203, 272, 284

atomic bombs 235, 273, 279

attrition ix, 6, 63, 64, 116, 189, 222, 228, 273, 288

Aube, Hyacinthe-Laurent-Théophile 93–95, 99

Augusta (Prussia) 92, 95

Australia 211, 241, 265, 308

Austria 9–10, 133, 136, 140, 141, 142, 144, 181

Austro-Hungary 136, 144

Axis 7, 174–76, 177, 179, 196, 200, 209, 210, 229, 275

Azores 60, 67, 78, 79

B

B-5 (Spanish) 170

Babcock, John V. 152

Bab el-Mandeb 254

Baghdad 241

Bahadur, Jay 262

Bahama (CSS) 78–79

Bahamas 3, 19

Baharihindi (merchant) 255, 263

Bahia 78, 81

Balkans 143

Baltic Fleet 122, 132

Baltic Sea 9, 20, 50, 147, 179

Baltimore 141

Bandar Abbas 240, 244
Bandar Khomeni 241, 242
Bandar Mashahr 241
Bangkok 82
Bank of Spain 168–69
banner system 107, 108
Baralong (Q-ship) 142
Barbados 21, 33
BARBAROSSA, Operation 210
Barbuda 17
Barcelona 171, 179
barkentine-rigged ship 79
BARNEY, Operation 234
Barre, Mohammed Siad 254, 257, 260, 262, 268n24, 269n44
Barron, Samuel 84
Bastarrèche brothers 46–47, 55n34, 55n35
Batavia 82
Batsch, Carl Ferdinand 100
Battle of Britain 188
battle of Jutland 146, 156
battle of P'yŏngyang 114
battle of Sedan 92
Battle of the Atlantic 194, 284
battle of the Yalu 112, 114
battle of Trafalgar 157–58
battle of Tsushima 118n29, 132
battle of Yorktown 225, 273, 275, 288
battleship 6, 57, 93–94, 97, 99, 100, 102, 114, 115–16, 122, 152, 156–59, 160, 162, 172, 173
Battleship Division 9 156–59
Bauer, Hermann 138
bauxite 215, 219, 220
Bayly, Lewis 154, 155–56, 159, 162, 163
Bayonne 16, 44, 46–47, 54n18, 55n34
B-Dienst 200, 201
Beach, Edward L. 232

Beatty, David 154, 156, 159–63
Beijing 105, 108, 110, 112, 121, 125, 128, 129–30
Beiyang Squadron 109, 112, 113, 114, 115
Belgian Relief Commission 141
Belgium 139, 140
Belize 257
Bell, Henry H. 80
belligerents 2, 3, 5, 20, 42, 51–52, 75, 77, 90, 96, 105, 109, 111, 122, 124, 126–27, 130, 133, 152, 158, 163, 165, 167, 172, 177, 226, 239, 240–41, 242, 243, 244, 247, 248, 250, 251, 271, 273, 275, 276, 289
Benjamin Morgan (merchant) 50
Benson, William S. 158–59, 162
Berlin 2, 92, 96, 151, 170, 173, 174–75, 180
Bermuda 192
Bethmann Hollweg, Theobald von 139–40, 141, 142–43, 146
Biliu River 115
Biscay, Bay of 167, 202
Bismarck Islands 216
Black Sea 127, 167, 171, 175, 176, 177, 178, 179, 180
Blair, Clay 235
Blake, Homer C. 80–81
Blakely gun 79, 83
Blakely shell 85
Blanquilla Island 80
blockade 4, 16, 48, 63–64, 65, 73, 74, 76, 78, 90, 91, 93, 94, 97, 98, 116, 126, 136, 139, 156, 158, 165, 226, 228–29, 230, 235, 240, 241, 272, 273, 274, 279, 280, 283, 288, 289
blockade-runner 73, 74, 76, 77, 81
Blomberg, Werner von 171
blood payment *(diya)* 259
Bogatyr (Russian) 123
Bo Hai, Gulf of 105

Bolshevik Russia 181
See also Russia; Soviet Union

Bolsheviks 181, 186n58

Bonaparte, Napoleon 2, 42, 45, 48, 51, 56n50, 56n57, 56n64, 67–69, 143, 282, 287
See also Napoleonic Wars (1793–1815)

Bonsella (merchant ship used as Somali mother ship) 255, 263, 267n7

Boosaaso 255

Bordeaux 4, 12, 14, 20, 44, 46–47, 48, 49, 50, 54n18, 54n24, 56n57, 56n66, 97

Boston 11, 18, 31, 33, 35–36, 62

Bourgois, Siméon 99–100

Boyah 263, 267n9

Brazil 77, 78, 81

Brest 12, 15, 16, 20, 50, 162

Bridgeton (U.S. flag oil tanker) 245–46

brigantine 36

Bristol 12

Bristol (HMS) 12

Britain *See* Great Britain

British Borneo 211, 216, 217

British Canada *See* Canada

British Empire 1, 10, 23n5, 27, 135, 153, 187, 196

British Isles *See* Great Britain

British Malaya 212, 216, 217
See also Malaya

British West Indian islands 13, 17, 19, 60, 66, 67

Brooklyn (Union) 75, 77, 80

Broughton, Nicholson 33

Brunswick 75

Buchanan, Franklin 85

Bulloch, James D. 74, 76, 77–79, 84, 86

Bureau of Ordnance (USN) 230–31

Burma 211, 216

Bushire 241, 242

businessmen 14, 58, 73

Butcher, Mathew J. 78

C

C-3 (Spanish) 170–71, 174, 176, 183n19

Cabo Santo Tomé (Soviet arms transport) 179, 185n50

Cádiz 15, 77, 169

Calais 83, 138

Calhoun (CSS CR) 75

California 80

Campeador (Spanish tanker) 176, 185n38

Canada 58–59, 61, 74, 87, 182n6

Canarias (Spanish) 171–72, 177

Canaries 67

cannon 14, 17, 46, 97

Canton (Guangzhou) 82

Cape François 18

Capelle, Eduard von 146

Cape Town 81–82, 193

Cape Verde Islands 191

Caprivi, Leo von 96–97, 100

CAPTOR *See* Mark 60 Encapsulated Torpedo

cargo vessels 5, 155, 161
See also merchant shipping: merchant ships

Caribbean Sea 10, 13, 14, 17, 19, 20, 21, 22, 77, 96, 192

Carlton Hotel 154

Carrickfergus 25n71

Cartagena 167, 168, 170, 172, 173

cartel trade 18

Catalonia 165, 179

Cattaro, Gulf of 144

Celebes 216

Central Powers 143

Ceuta 171

Chamberlain, Neville 177

Champlain, Lake 58

Channel Islands 12

Charleston 44, 54n19, 73, 75

Charmes, Gabriel 94–95, 99, 102

chartering ships 86, 90, 175, 239–51, 252n7

Chase, Samuel 30

Chateleau, Soubier du 16

Chefoo *See* Qifu

Cherbourg 4, 48, 82, 83, 85

Chesapeake (USN) 61, 63, 70n43

Chickamauga (CSS) 74

China 5, 105–17, 121–33, 153, 180, 181, 182, 210, 218, 236n2, 247, 272, 276, 280, 283, 284

Chinese navy 109, 111, 116, 281, 284
See also Beiyang Squadron

Choiseul, duc de (Étienne-François de Choiseul) 22

choke point 123, 144, 254, 289

Christmas Day 152, 171

Churchill, Winston (American novelist) 158

Churchill, Winston S. (British politician) 163n6

Ciano, Galeazzo 177, 179, 180–81

Ciudad de Cádiz (Spanish arms transport) 176, 185n39

Civil War (American) 1, 2, 4, 73–87, 89, 91, 92, 96, 98, 272, 273, 275, 276, 278, 279, 281, 283, 284, 285, 286, 287, 288

Civil War (Spanish) 6, 165–82, 272, 274, 275, 276, 277, 278, 279, 280, 281, 285, 286, 287, 288

clan structure 7, 254, 255, 259–61, 262, 263, 265, 266, 267, 269n41

coal 2, 78, 79, 80, 81, 92, 96, 97–98, 103, 114, 123, 124, 127, 137, 157, 158, 159, 226

coalitions 152–55, 174, 177, 250, 252n2, 256, 275, 276, 278, 280, 285
See also allies

coastal defense 30, 90, 95, 97, 105, 113, 157

Cobh *See* Queenstown

colliers 137

collusive capture 19

colonies 3, 10, 14, 15, 22, 27, 28, 29, 32, 33, 41–42, 48, 55n39, 93, 95, 96, 286, 288

colonist ships
Franklin 32, 38n31
Hancock 32, 38n31
Hannah 32–33, 38n33
Lee 32, 34
Warren 32

Combined Joint Task Force–Horn of Africa (CJTF-HOA) 256

combined operations 6, 140, 151, 153, 161, 255–56, 264, 271, 277–80, 289

Combined Task Force 150 255–56, 264

Comedian (merchant) 128

Comintern *See* Communist International

Commander in Chief, Asiatic Fleet (CINCAF) 229–30

commerce protection 90, 233

commerce raiding ix, 1–8, 9–22, 27, 28, 29, 32, 33, 36, 39n55, 44, 57–69, 73, 76–77, 89, 90, 91, 92, 93, 94, 95–99, 101, 102, 103, 105, 109, 111, 113, 114, 116, 117, 121–33, 135, 136, 137, 138, 139, 140, 142, 143, 146, 148, 151, 157, 162, 225, 234, 239, 240, 253, 254, 258–59, 264, 266, 271–90

commerce raiders 2, 3, 5, 7, 65, 67, 73–87, 90, 91, 94, 95, 96, 98, 101, 121, 124, 126, 129, 133, 161, 257, 267, 276, 279, 280, 284, 285, 286
See also surface raiders
Adventure (French) 50, 56n66
Alert (French) 50
Calhoun (Confederate) 75
Confiance (French) 48–49, 55n50
Gascon (French) 48
Herliquin (American) 12
Psyché (French) 46
Rôdeur (French) 48
Savannah (Confederate) 75
Terrible (British) 17
Triton (Confederate) 75

commerce raiding *(continued)*
   commerce raiders *(continued)*
      Vengeance (French) 17
      Venus (French) 46
      Washington (American) 35

Committee of Public Safety 54n26

Committee of Safety 31, 32

Communist International 166, 175, 179

Comoro Islands 82

Confederacy (Confederate States of America) 4, 74, 75, 76, 95, 103, 276, 281, 285, 286, 288
   Confederate ships
      Agrippina 78, 80, 81
      Alabama 4, 73–87, 88n21, 225
      Bahama 78–79
      Chickamauga 74
      Enrica 78
      Florida 74, 78, 79, 81
      Georgia 74, 81, 83
      Habana 76
      Rappahannock 74, 83
      Shenandoah 74, 86
      Sumter 76–77, 79, 83, 86
      Tallahassee 74
      Tuscaloosa 81–82

Confiance (French CR) 48–49, 55n50

Congress (USN) 60, 62, 65

Congress, Confederate 74

Congress, U.S. 58, 152, 164n26, 226–27

Conrad (Union merchant) 81

Conseil des Prises 45, 54n24, 54n27
See also Prize Council

Constantinople 147

Constitution (USN) 57, 65, 69n2, 70n43

containerization 250

Conte, Jacques 47, 49

Continental Congress 3, 27, 28, 29–30, 31, 32–33, 34, 35, 36

Continental Navy 35

continental power 1, 2, 41, 87, 274, 285, 288
See also land power

contraband 13, 36, 43, 99, 123, 124, 240–41, 245, 248

contre-torpilleurs (four-stack destroyer class) 178

control of sea-lanes 89, 105, 112, 123, 127, 161, 165, 226, 228–29, 260
See also sea control

convoys 5, 6, 12–13, 16, 17, 22, 50, 51, 57, 60, 66, 67, 68, 116, 148, 154, 155, 156, 157, 160–62, 172, 174, 176, 188, 189–91, 192–93, 194, 200–201, 202–203, 222, 227, 232, 233, 240, 241, 242, 244, 245, 246, 249, 250, 256, 268n18, 273, 282, 283, 284–85

cooperative adversary 280, 284

Copenhagen 15

copper 215, 220

copper-plated hulls 2, 82

Corbett, Julian S. ix, 2, 126, 155, 290n2

Corfu 162

Cork 12, 16, 26n94

Corregidor 218

corvette 62, 95

council *(shir)* 259

Council of Islamic Courts (CIC) 264

countermeasures 3, 5, 17, 49–51, 141, 145, 148, 200, 203, 271, 274, 275, 279, 282–85, 287, 290

crew 15, 17, 21, 28, 31, 33, 34, 35, 42, 45, 46, 47, 48–49, 52, 53n13, 54n24, 55n36, 64, 75, 78, 79, 81, 82, 83, 84, 85, 94, 98, 99, 101, 110, 129–30, 131–32, 133, 138, 143, 158, 170, 172, 174, 176, 178, 194, 197, 206n45, 226, 233–34, 239, 247, 249, 250, 253, 254, 255, 256, 258, 261, 262, 264

Crewe Hall (merchant) 127

cruisers 5, 11, 12, 16, 18, 61, 63, 74, 77–78, 86, 94, 96, 97, 100, 101, 103, 110, 113, 114, 115, 121,

122, 123, 124, 126, 127, 136–37, 139, 157, 160, 162, 179

Cuba 77

cumulative strategy ix, 8, 29, 274, 288
See also strategy

Curaçao 13, 15, 18

customary law *(xeer)* 259

Customs Enforcement Act of 1763 22

cut military supply line 32, 34, 112, 123, 188, 222, 226, 288

cutter 31, 32

Cyane (HMS) 69n1

Cyprus 247

Czechoslovakia 181

D

Dagu 110

Dahlgren gun 76, 83, 84

Dalian 113

Daniels, Josephus 152, 153, 154, 156, 158, 159

Danish War 95

Danish West Indies 18

Dardanelles 127, 144, 179

Darood clan 260

Davis, Jefferson 73, 74, 75

Dayton, William 83

Deane, Silas 33

Death, William 17

Decatur, Stephen 60

de Chair, Dudley R. S. 153

declaration of neutrality 75

declaration of neutrality, Chinese 5, 121, 124–25, 130, 133

Declaration of Paris (1856) 74, 90–91, 92, 98, 99, 278

Declaration of the Rights of Man and of the Citizen (1789) 41

declaration of war 12, 13, 14, 15, 43, 101, 110, 111, 114, 147, 172, 222, 252n7

Deerhound (British yacht) 85

Delaware (USN) 157, 159

Delaware River 65, 81

Delbos, Yvon 177–78, 180, 185n42

Denmark 3, 14, 15, 18, 20, 60, 95, 241, 244

Department of the Navy (U.S.) 59, 152, 154, 156, 159, 161

destroyer 6, 113, 114, 115, 122, 130, 136, 138, 154, 155, 156, 158, 159, 160, 161, 162, 167, 171, 174, 176, 177, 178, 232, 244, 257

Deutschland (German) 172, 173, 184n26

Diamond (HMS) 50

Diet (Japan) 113, 235

Die U-Bootswaffe (Dönitz) 187–88

Digil clan 260

diplomacy 15, 18, 44, 47, 51, 69, 95, 103, 109, 121, 125, 126–29, 133, 136–37, 139, 144, 145–46, 151, 152, 154, 161, 178, 181, 210, 212, 246, 252n7, 279, 287

Dir clan 260

Directory (1795–99) 44

Distinguished Service Order (DSO) 155

Djibouti 256, 260

Dogger Bank 139

Don (Russian) 127

Dongwon-ho (South Korean fishing) 253

Dönitz, Karl 170, 187–88, 189, 190–93, 194, 195–96, 197, 200–201, 202, 204n3, 206n45

Dover 138

Dover Strait 135, 138

Dreadnought (HMS) 138, 152–53, 156–57, 158

dream scenario 283–84

dual-propulsion ship 76, 79, 86, 96, 103

Dubai 248, 265

Duke of Bourbon (French merchant) 12

Dulhabante clan 260, 269n44

Dunkirk 15, 16, 44, 48, 49, 53n16, 145

Dutch Borneo 216

Dutch East Indies 175, 210, 211, 212, 213, 217, 219

Dyson, John 35

E

East Asia 5, 105, 121, 123, 125–28, 129, 209, 210, 212, 217
See also Far East

Eastern Front 6

Eastern Seaboard (U.S.) 191–92, 257, 275

East Indies 49, 60, 61, 68, 82, 175, 210, 211, 212, 213, 217, 219

East Kent 16

Eastleigh 265

East London 128

economic factors 1, 4, 10, 18, 20, 29, 33, 41, 42, 47, 52, 53n7, 73, 89, 92, 94, 95, 99, 101, 103, 146, 151, 156, 165, 175, 180, 184n36, 189, 192, 196, 209, 210, 211, 215, 216, 217, 220, 225, 226, 228–29, 235, 244, 246, 249, 258, 265, 266, 279, 280, 282, 285, 288, 290

Eden, Anthony 177–78, 180, 185n42

Edwardian era (1901–10) 152

Egypt 98, 124, 126–28, 193, 241, 257

El Alamein 193

elders *(oday, duk)* 259

Elsa Cat (Danish merchant) 241

Emden (German) 137

Emergency Plan for Materials Mobilization 210

enemy commerce 90, 91, 94, 96, 227, 234

enemy countermeasures 3, 5, 17, 49–51, 141, 145, 148, 200, 203, 271, 274, 275, 279, 282–85, 287, 290

English Channel 16, 45, 67, 94, 98, 138, 139

Enigma machine 284

Enrica (CSS) 78

Entente 141, 148, 151, 156, 279, 286, 288, 289

Enterprize (USN) 64

Epervier (HMS) 62

Era of Humiliations 112

Erie (USN) 167

Essex (USN) 70n43

Etappe system 136–37

Ethiopia 178, 254

Ethiopian National Defense Force 264

Europe 3, 5, 6, 10, 11, 15, 27, 41, 44, 45, 46, 47, 74, 75, 82, 90, 96, 100, 103, 108, 109, 110, 124, 125, 147, 148, 151, 162, 166, 167, 168, 172, 174–75, 176, 181, 188, 195, 196, 202, 210, 211, 228, 240, 246, 249, 251, 252n6, 262, 277, 285

European Community (EC) 240, 249, 251

European Union (EU) 251, 252n6, 257, 262

Ever-Victorious Army 107

exclusive economic zone (EEZ) 257, 266–67

Exmouth (U.S. steamer) 167

extraterritoriality 121, 129–31, 133, 219

F

failed state 7, 253, 261, 276, 290
See also pariah state

Falconer, William 34

Falkenhayn, Erich von 143, 145

Falkland Islands 137

Far East 82, 95
See also East Asia

Far East Squadron 137

Fascist Italy 175, 177, 181
See also Italy

Favourite (merchant) 54n30

Feisty Gas (merchant) 263

Feng Island 110, 111, 112

Fernando de Noronha 81

First Army, Japan 111–12, 114–15

First Lord of the Admiralty 59, 66

First Sea Lord 152, 153, 155, 158, 160

First Sino-Japanese War (1894–95) *See* Sino-Japanese War (1894–95), First

Fischel, Hermann von 172, 173, 184n26

Fisher, John 152

fisheries 46, 53n13, 258, 262

fishermen 16, 42, 115, 257–58, 262, 264, 268n28

fishing vessel 32, 33, 35, 38n27, 39n50, 139, 253, 254, 258, 266, 268n29

Flanders Submarine Flotilla 140, 145

fleet-on-fleet engagement 57, 93, 116

fleet submarines (USN) 227–28, 231

Florida 3, 28

Florida (CSS) 74, 78, 79, 81

Florida (USN) 157, 159

Flour Act of 1757 19

foodstuffs 18, 19, 35, 44, 99, 102, 122, 123, 151, 153, 166, 197, 212, 226, 234, 235, 261

force 3, 6, 7, 8, 20, 27, 28, 29, 30, 31–32, 33, 35, 41, 58, 61, 62, 63–64, 66, 67, 68, 80, 85, 94, 105, 107, 108, 111, 112, 113, 114, 115, 116, 122, 123, 126, 131, 132, 138, 140, 144, 147, 153, 154, 159, 161, 162, 165, 170, 171, 172, 173, 176, 177, 178, 180, 188, 189, 190, 191, 193, 194, 196, 200, 202, 203, 209, 210, 211, 217, 218, 222, 226–27, 228–29, 230, 231, 232, 233, 234, 235, 236, 238n29, 238n41, 242, 243, 244, 245, 246, 248, 254, 255–56, 257, 258, 261–62, 263, 264, 266, 271, 272, 273, 277–80, 282, 284, 288, 289

Foreign Enlistment Act of 1819 77

Fort-Dauphin 18

Fort-de-France 80

Forte (HMS) 128

Fort Sumter 73, 74

France ix, 1, 3, 4, 9–22, 28, 41–52, 53n12, 67, 69, 75, 87, 89–95, 97–100, 101, 102–103, 124, 125, 135, 136, 151, 156, 160, 162, 166–67, 175, 177, 178–79, 184n24, 187, 188, 193, 194, 195, 243, 245, 247, 275, 276, 279, 281, 282, 284, 286, 287, 288

    French navy ix, 10, 12, 15–16, 41, 42, 43, 44, 50, 52n2, 54n20, 67, 82, 92, 93, 99, 100, 178

Franco, Francisco 174, 180

Franco-Prussian War (1870–71) 89, 91, 92, 95, 96, 97, 98, 101

Franco-Russian declaration (1902) 125

Frankby (merchant) 123–24

Franklin (colonist) 32, 38n31

freedom of navigation 1, 240, 242, 243, 244, 246

freedom of the sea 240, 243

Freetown 193

free trade 1, 13

French and Indian War *See* Seven Years' War (1756–63)

French Convention *See* National Convention

French Indochina 210, 213, 217, 218

French Wars (1793–1815) 41–52, 53n12, 271, 272, 273, 274, 275, 276, 278, 279, 281, 283, 284, 286, 287, 288

See also Napoleonic Wars (1793–1815)

frigate 16, 28–29, 50, 57, 59, 60, 61, 62, 63, 64, 65, 66, 67, 68, 69n2, 80, 97, 245, 248

Frolic (HMS) 59

Frolic (USN) 63, 70n41

Führer der Unterseeboote *See* U-boat flotilla in the North Sea

Fujian Squadron 113, 114

Full City (merchant) 255

funnel 76, 79, 171

Fuss, Richard 146–47

G

Gadsden, Christopher 30–31

Galguduud region 261

Galsworthy, Thomas Ryder 110

Galveston 80–81

Gara'ad 255

Garad, Mohamed 263, 264

Gardiner, Robert 34

Gascon (French CR) 48

Gato-class submarine 227

Gulf Cooperation Council (GCC) 7, 241, 242, 243, 247, 249, 250, 251

Gedo region 260

General Assembly, Rhode Island 29

Genêt, Edmond-Charles 44, 54n19

Geneva 87, 177–78, 179

Geneva Conventions (1949) 242

Genova, Arturo 170

George II 11, 13, 15

Georgetown 56n66

Georgia 75

Georgia (CSS) 74, 81, 83

Germany 2, 4, 5–6, 87, 89–103, 104n28, 108, 115, 122, 125, 127, 132, 133, 135–49, 151–63, 165–82, 184n24, 184n26, 184n33, 184n36, 187–203, 210, 225, 228, 229, 232, 255, 272, 273, 274, 275, 276, 277, 278, 279, 280, 281, 282, 283, 284, 286, 287, 288, 289

    German ships

        Admiral Scheer 173

        Deutschland 172, 173, 184n26

        Emden 137

        Leipzig 173

        Moltke 161

        U.27 142

        U-33 170

        U-34 170, 176, 183n19

        U-35 (Spanish Civil War) 174

        U.35 (World War I) 145

        U-156 206n45

        U-459 192–93

        Wolf 137, 149n9

Gerry, Elbridge 31

Ghent 171–72, 183n23

Giangreco, D. M. 235

Gibraltar 77, 128, 144, 162, 165, 175, 184n34, 186n60, 191, 193

Gibraltar (merchant) 77

Gibraltar, Strait of 144, 165, 184n34, 186n60

Giovanni da Verrazzano (Italian) 171

Glasgow 12

Glete, Jan 66

Glitra (merchant) 138

global commerce 8, 51, 95, 96, 98, 282, 287

Global Positioning System (GPS) 2, 261, 272, 290

Glover, John 32–34, 38n29

Glucksburg (merchant) 255

gold 79, 80, 168–69, 183n10, 183n11

Golden Rocket (merchant) 77

Gonzalez (USN) 257

Good Hope, Cape of 60, 82

Government Liaison Conference 211, 212, 216

Grand Alexander (merchant) 17

Grand Fleet 6, 138, 154, 156, 157, 158, 159, 160, 161, 162

Granville, 2nd Earl (John Carteret) 21

Graves, Samuel 28, 31

Great Britain ix, 1, 2, 3, 4, 5, 6, 9–22, 27, 28, 29, 30, 31, 33, 35, 36, 37n12, 41, 42, 43, 45, 48, 49, 50, 51, 53n5, 54n19, 57–69, 70n41, 74, 75, 76, 77, 78, 79, 81, 85, 86, 87, 89, 90, 91, 92, 93, 94, 95, 97, 98, 99, 101, 105, 109, 110, 111, 122, 123, 124, 125, 126, 127, 132, 135, 136, 137, 138, 139, 140, 141, 142, 143, 144, 145, 146, 147, 148, 151, 152, 153, 154, 155, 156, 157, 158, 159, 160, 162, 163, 170, 171, 172, 175, 176, 177, 178, 179, 180, 181, 184n24, 185n40, 186n55, 187, 188, 189, 190, 191, 193, 194, 195, 196, 200, 201, 202, 206n45, 209, 210, 211, 212, 218, 222, 225, 228, 241, 242, 244, 245, 246, 248, 250, 254, 258, 261, 272, 273, 275, 279, 280, 281, 284, 287, 288, 289

    British ships

        Alert 59

        Bristol 12

        Cyane 69n1

        Diamond 50

        Dreadnought 138, 152–53, 156–57, 158

        Epervier 62

        Forte 128

        Frolic 59

        Guerriere 57, 59, 65, 69n1

COMMERCE RAIDING 321

Havock 176
Java 57, 59, 69n1
Macedonian 57, 59, 69n1
Queen Elizabeth 159
Shannon 63
Venus 127
Woolwich 15
Greater East Asia Co-prosperity Sphere 210
Greater East Asia War 209, 212, 218
See also Pacific War; World War II
Great Lakes 59
Great Wall 109–10
Great War for Empire 10
See also Seven Years' War (1756–63)
Greece 141, 179
Green Standard Army 107, 108
Gris-Nez, Cape 138
Grivel, Richild 93
Gromoboi (Russian) 122, 126
Grosvenor Square 154
Guadalcanal 221
Guadeloupe 14, 17, 48, 50, 51, 55n37, 55n41
Guam 216
Guangdong Squadron 113, 114
Guardafui, Cape 254
guerre de course See commerce raiding
Guerriere (HMS) 57, 59, 65, 69n1
Guildhall speech 153, 163n6
Gulf maritime exclusion zone (GMEZ) 240, 241, 242
Gulf Wars 239, 251
gunboat 74, 94, 99, 100, 111, 167, 247, 248
guns 9, 12, 14, 15, 16, 17, 28, 30, 31, 57, 59, 62, 63, 64, 69n1, 69n2, 75, 76, 79, 81, 83, 84, 85, 86, 91, 97, 99, 100, 106, 108, 113, 114, 122–23, 127, 140, 142, 145, 146, 152, 153, 157, 160, 164n36, 172, 176, 232, 239, 255, 256, 261
See also small arms

H
Habana (CSS) 76
Habir Gedir Ayr 260
Habir Gedir subclan 263
Habsburg 144
Hague, The 145
Hai River 110
Halifax 192, 205n27
Hamburg 56n66, 184n33
Hamilton, Paul 58
Hampton Roads 155
Han Chinese 107, 108
Hancock (colonist) 32, 38n31
Hannah (colonist) 32–33, 38n33
Hansen, Stig 264, 267n9
Haradheere 253, 256, 263, 264
Hardinge, Charles 127, 128
Hardwicke, 1st Earl of (Philip Yorke) 15
Hart, Robert 107
Hart, Thomas C. 229, 230
Harti clan 260
Hartlepool 173
Hart Security 258, 261–62, 270n54
Hatteras (Union) 80–81
Havana 76, 85
Havock (HMS) 176
Hawaii 230
Hawiye clan 260, 263
Hay, John 132
Heidelberg 146
helicopters 246, 248–49, 251
Heligoland Bight 136
Herliquin (U.S. CR) 12
Hersi, Muhamed Muse 263
Hidalgo de Cisneros, Ignacio 179, 186n53
High Court of Admiralty 45

High Seas Fleet 136, 139, 145, 146, 157, 158, 160, 161, 162

Hindenburg, Paul von 147

Hispaniola 18
See also San Domingo

Hitachi Maru (Japanese merchant) 122–23

hit-and-run tactics 29, 32

Hitler, Adolf 171, 172, 173, 179, 180, 181–82, 186n59, 188, 195, 200, 202

Holland 13
See also Netherlands

Holsatia (German merchant) 127

Holtzendorff, Henning von 143, 145, 146, 147–48

Hong Kong 152, 211, 216, 244

honor 83, 100–102, 103, 104n28, 131, 156

Hopkins, Stephen 29

Hormuz, Strait of 240

Hornet (USN) 65

Hossbach conference 181, 186n59

hostages 47, 253, 256

House, Edward M. 158

howitzer 83
See also cannon, guns

Huayuankou 112, 115

Hull, F. S. 77–78

Hundred Years' War (1337–1453) 52n5

HX.229 convoy 201
See also convoys

I

Ibiza 172, 184n26

Ibrahim, Abdirahman 263

Igrek 167, 179, 185n51

Imperial German Navy (IGN) 95, 96, 97, 100, 102, 103
See also Germany: German ships

Imperial Japanese Navy 105, 109, 113, 115–16, 279

Imperial Maritime Customs 107

Imperial Navy Office, Germany 135, 147

Imperial Supreme Command 210

Inchŏn 110, 114, 115, 130

India 10, 82, 95, 98, 193, 211, 242

Indian Ocean 47, 48, 55n50, 60, 82, 137, 193, 241, 242, 254

industrial base 21, 38n29, 44, 73, 95, 106, 109, 113, 150n25, 154, 165, 182n1, 195, 202, 214, 223, 226, 233, 234, 235, 251

Industrial Revolution 89, 91

Influence of Sea Power upon History, 1660–1783, The (Mahan) 157, 287

insurance rates 4, 7, 21, 29, 46, 50, 73, 74, 86, 93, 95, 122, 124, 143, 175, 218, 242, 249, 251, 289, 290

intelligence 30, 32, 33, 34, 60, 168, 171, 173, 175, 185n40, 188, 190, 191, 194, 196, 254, 266, 277–78, 283

Intelligence Department, Politburo 167

interisland commerce 20

international banking 95, 165, 183n10, 218

International Brigades 165

International Chamber of Shipping 243

International Committee of the Red Cross 240, 251

international community 2, 7, 241, 246, 266, 273

International Expert Group 268n28

international law 1, 15, 73, 75, 78, 83, 87, 98–99, 100, 101, 110, 111, 121, 127, 129, 130–31, 133, 138, 178, 200, 245, 250–51, 256–57, 277, 278

International Maritime Bureau (IMB) 255

International Maritime Organization (IMO) 256

International Military Tribunal 204n3
See also Nuremberg trials

international relations 42, 47, 52, 121, 166

international shipping 42, 263

International Shipping Federation 243
international tension 6, 42
international trade 2, 5, 239, 288, 290
International Transport Workers Federation 243
international tribunal 87
international warrior-diplomat 152
international waters 78–79, 242, 243
Internet 239
intern ships 103, 121, 128–31, 132, 133, 137, 161, 173
Interpol 265
Iran 2, 7, 239–51, 252n8, 252n9, 273, 276, 283
Iranian Revolutionary Guards 245, 246
Iran-Iraq Tanker War 2, 7, 239–51
See also Tanker War (1980–88)
Iraq 3, 7, 239–51, 252n2, 252n4, 252n15, 272, 275, 283
Ireland 3, 10, 15, 16, 18, 19, 20, 26n94, 60, 63, 67, 78, 138, 139, 154, 155, 190
Ireland, northern 138
Irish Sea 16
ironclad 73, 74, 84, 102, 116
Isaaq clan 254, 260
Islamic Conference Organization 241
Italy 6, 79, 93, 144, 165, 166, 167, 169, 170, 171, 172, 174, 175, 176, 177, 178, 179, 180, 181, 183n13, 184n24, 184n26, 185n37, 185n40, 185n49, 186n55, 186n60, 192, 196, 246, 248, 262, 289
    Italian ships
        Archimede 174
        Giovanni da Verrazzano 171
        Saetta 176
        Torricelli 170, 174
Iwatake Study 214, 215
Iwatake Teruhiko 223n5
Iwo Jima 235

J

Jachmann, Eduard 96
Jaime I (Spanish) 173
Jamaica 3, 13, 19, 67, 81
James River Squadron 85
Japan 2, 5, 6–7, 105–17, 118n29, 121–33, 135, 136, 158, 163, 182, 196, 209–23, 225–36, 240, 249, 272, 273–74, 275, 276, 277, 278, 279–80, 281–82, 284, 286, 287, 288–89, 290
Japan, Sea of 227, 234
Japanese Foreign Trade Annual 214
Japan Mail Steamship Company 114
Jardine, Matheson and Company 111
Java 82, 216, 218
Java (HMS) 57, 59, 69n1
Jeddah 254
Jefferson, Thomas 58
Jellicoe, John 138, 153, 156, 158, 160
Jeune École ix, 1, 89–103, 110, 122, 272, 273, 274, 276, 278, 281, 283, 286, 287
Johnston, Joseph E. 85
joint operations 116, 155, 173, 228, 271, 277–80, 282
Jones, Jacob 61
Jones, William 4, 57–69, 71n65
Joseph (Union) 75
J. R. Anderson & Co. *See* Tredegar Iron Works
Jubba Valley Alliance 260

K

Kamerun (German arms transport) 174
Kamimura Hikonojō 126
Kane (USN) 167
Katwijk (Dutch merchant) 141
Kearsarge (Union) 4, 77, 83–85
Kell, John McIntosh 79, 81, 84, 85
Kent (British merchant) 49
Kenya 265

Kersaint, Armand de 15

Kert, Faye 66

Kharg Island 241, 242, 243, 244

Kleinekrieg 101

Kniphausen 44–45

Komsomol (Soviet merchant) 171–72, 183n23

Konoe Fumimaro 212

Korea 5, 7, 105, 109–10, 111–12, 114–15, 116, 122, 126, 130, 236n2
See also South Korea

Korean War 7, 251

Koreets (Russian) 130

Kowshing (British transport leased by China) 5, 105, 109–11, 117, 122, 274, 282

Kraft durch Freude (Strength through Joy cruise ships) 180

Krupp guns 122–23

Kuwait 241, 242, 243, 244, 245, 247, 249

Kuwaiti Oil Tanker Company 245

Kuznetsov, Nikolai Gerasimovich 170

L

Lacombe, Daniel 46, 48

Laconia order 200, 206n45

land lines 2, 271, 285

land power 1, 273, 274, 275, 276, 277, 286, 287
See also continental power

Larak terminal 244, 248

Largo Caballero, Francisco 168

La Rochelle 20

Latvia 124

Lavan Island 243

Lavan terminal 244

Laveille, Captain 46

Lawrence, James 63

League of Nations 167, 177

Lee (colonist) 32, 34

Leeward Islands 13, 21, 67

Le Havre 20, 83, 145

Leipzig (German) 173

Lena (Russian) 132

Leningrad 167, 179
See also St. Petersburg

Lessar, Pavel Mikhailovich 131

Lesser Antilles 50

Lesser Sunda Islands 216

letters of marque 12, 27, 29, 35, 36, 41, 42, 45, 46, 47, 52, 54n19, 74, 75, 284

Levy, Hermann 146–47

Lewis, Ioan 259, 261

Liaodong Peninsula 111, 112, 115

Libau 124, 128

Liberia 247, 248, 249

Li Hongzhang 107, 108, 109, 113, 116, 117

limited war 64, 92, 188, 227, 229, 280, 281–82, 289

Lincoln, Abraham 74–75

Little, Peter 261

little war 101

Litvinov, Maxim 177, 178

Liverpool 12, 16, 74, 77, 78, 79, 86

Lloyd's of London 50

Lockwood, Charles A. 231, 234, 238n26

locomotives 106, 122–23, 174

logistics 3, 5, 6, 7, 15–16, 17, 18, 20, 25n52, 28, 29, 32, 33, 34, 36, 65, 74, 78, 80, 92, 96, 97, 98, 99, 103, 105, 108, 112, 113, 114, 116, 117, 122, 123, 124, 128, 129, 137, 155, 156, 158, 165–82, 188, 189, 192, 193, 194, 195, 196, 197, 201, 205n26, 209–23, 226, 229, 232, 233, 254, 258, 275, 281, 283, 288

London 2, 10, 12, 13, 14, 15, 16, 17, 19, 20, 21, 22, 60, 78, 86, 130, 139, 152, 153, 154, 156, 158, 159, 160, 162, 163, 166, 169, 172, 175, 242, 267, 281

"London Flagship" 163

London Naval Treaty (1930) 178, 226, 227

London Submarine Protocol (1936) 169–70, 178, 226

L'Orient 20

Louis XIV 43

Louis XV 11, 13

Low, John 81–82

Luftwaffe 188, 189, 196

lumber 18, 33

Lüshun 111, 112, 113, 114, 115

Lusitania (British ocean liner) 141

M

Macedonian (HMS) 57, 59, 69n1

Macedonian (USN) 60, 71n43

Madeira 60, 67

Madison, James 58, 59, 64–65, 68, 71n65

Madrid 168

Magallanes (Soviet arms transport) 172

Mahan, Alfred Thayer ix, 2, 29, 36, 39n55, 157, 158, 161, 162, 287, 290n2

Maine 62, 75

Maiskii, Ivan 169

Majarteen clan 255, 260, 263

Malacca, Strait of 82

Málaga 170, 171

Malaya 211, 212, 213
See also British Malaya

Mallorca 174, 183n13

Mallory, Stephen R. 73–74, 75–78, 86

Malta 94

Manchu 107, 108, 115, 116

Manchuria 105, 108, 111, 112–13, 121, 122, 125, 129, 218

Manila 82

Marblehead 31, 32, 33, 34, 38n29

Marehan clan 260

Margate 16

Marine-Akademie 100–101

Marine Committee 33

Marinekorps Flandern *See* Flanders Submarine Flotilla

maritime commerce ix, 59, 89, 90, 91

maritime power *See* sea power

Mark VI exploder 230–31

Mark XIV torpedo 230–31, 232, 238n26

Mark XVIII torpedo 232

Mark XXVII torpedo 232

Mark 60 Encapsulated Torpedo 239

Marseilles 20, 44

Marshall, George C. 228

Martinique 14, 17, 48, 80

Maryland 30, 141

Massachusetts 27, 28, 30, 31, 32, 33, 34, 35, 79

Mauritius 48, 56n66, 60, 82

Mayo, Henry T. 158

Mediterranean Sea 6, 9, 14, 20, 46, 47, 55n38, 65, 68, 98, 127, 143–45, 147, 169, 171, 173, 174, 175, 176, 177–79, 180, 181, 182, 186n59, 190, 191, 192, 193, 194, 197, 243, 272

Meiji Restoration 107, 113

Melville (USN) 154, 155

Mémoire Instructif 13

Menelaus (merchant) 127

merchant marines 7, 90, 95, 122, 146, 174, 221–22, 225, 226, 229, 232–33, 234, 235, 274, 284

merchant shipping ix, 3, 6, 7, 28, 29, 73, 74, 127, 137, 138, 175, 178, 188, 220, 221, 227, 229, 232, 249, 256

    merchant ships 17, 41, 42, 43, 44, 47, 50, 52, 58, 63, 67, 73, 77, 78, 82, 86, 93, 97, 98, 99, 100, 103, 113, 123, 136, 138, 141, 143, 145, 154, 167, 168, 169–70, 172, 173, 175, 176, 179, 202, 209, 222, 223, 226, 227, 229, 233, 234, 239, 243, 246, 247, 274, 275, 276, 280, 288–89
    See also cargo vessels

        Al-Muharaq 243

        Amsterdam 26n94

merchant shipping *(continued)*
   merchant ships *(continued)*
      Ardova 124
      Ariel 80
      Baharihindi 255, 263
      Benjamin Morgan 50
      Bonsella 255, 263, 267n7
      Comedian 128
      Conrad 81
      Crewe Hall 127
      Duke of Bourbon 12
      Elsa Cat 241
      Exmouth 167
      Favourite 54n30
      Feisty Gas 263
      Frankby 123–24
      Full City 255
      Gibraltar 77
      Glitra 138
      Glucksburg 255
      Golden Rocket 77
      Grand Alexander 17
      Hitachi Maru 122–23
      Holsatia 127
      Katwijk 141
      Kent 49
      Komsomol 171–72, 183n23
      Kowshing 5, 105, 109–11, 117, 122, 274, 282
      Menelaus 127
      Molly 56n66
      Mongolia 123
      Naviluck 254
      Nimattulah 264
      Norasia Samantha 255, 267n7
      Noustar 255
      Ocean Eagle 75
      Ocmulgee 79
      Persia 127
      Prinz Heinrich 127
      Rozen 264
      Shen Kno II 255, 263
      Sussex 146

Mesopotamia 143
Mexico 76, 85, 90, 166
Mexico, Gulf of 192
Middle East 193, 282
Midway 220
Miguel de Cervantes (Spanish) 170, 171
military intelligence 30, 34, 191
Military Tribunal, Nuremberg 204n3 *See also* Nuremberg trials
Miller, Edward S. 235
Mina Ahmadi terminal 245
minelayer 136, 138, 139, 140, 142, 146, 187, 246
minesweeper 245
mine warfare 73, 106, 116, 136, 149n9, 166, 167, 196, 239, 245
missiles 7, 239, 241, 242, 245, 246, 251, 255, 257, 278
Mississippi River 75, 77–78
Mobile Bay 78
Mogadishu 254, 264
Mohican (Union) 81
Molly (merchant) 56n66
Moltke (German) 161
Molucca Islands 216
Mombasa 254
Mongol 107
Mongolia (British merchant) 123
Monte Cristi 18
Monte Cristi Bay 18
Montgomery 73

morale 63, 79, 86, 116, 272, 274

Morlaix 16

Morocco 165

Mortefontaine Treaty (1800) 44

Morton, Dudley W. "Mush" 232–33

mother ships 127, 243, 255, 262, 264, 277

Moylan, Stephen 33–34

M'Pherson, Captain 15

Mudug region 261, 263

Mukden 116

munitions factories 106

Munya, Hassan 262

Murata rifle 106

Murmansk 162

Mussolini, Benito 171, 174, 175, 176, 177, 179, 180, 184n36

N

Nairobi 265

Nantes 4, 20, 44, 46

Nantucket Chief (U.S. tanker) 167

Nanyang Squadron 113, 114

Naples 180

Napoleon *See* Bonaparte, Napoleon

Napoleon III 82–83, 102

Napoleonic Wars (1793–1815) 48, 55n50, 143, 287
See also French Wars (1793–1815)

Nassau 19, 77

National Convention 43

National Geospatial-Intelligence Agency 254

national honor *See* honor

Nationalists *See* Spain: Spanish Nationalists

NATO 7, 241, 247, 248, 251, 252n6

Nautilus (USN) 70n15

naval warfare 93, 94, 98, 102, 188, 196, 200, 249, 287

Naviluck (merchant) 254

Nazi Germany 163, 175, 181, 228
See also Germany

Negrín, Juan 168–69

Nelson, Horatio 157–58

Netherlands 3, 6, 13, 14, 15, 18, 20, 21, 24n52, 26n94, 43, 79, 83, 139, 141, 143, 145, 171, 175, 184n24, 184n34, 209, 210, 211, 212, 213, 216, 217, 218, 219, 244, 289
See also Dutch Borneo; Dutch East Indies

Neurath, Konstantin von 173

neutral powers 2, 3, 5, 13–16, 18, 19, 20, 21, 42–43, 44–45, 46, 47, 50, 51–52, 75, 77, 86–87, 90, 91, 92, 96, 99, 102, 103, 109, 111, 121–33, 137, 139, 140, 141–42, 145, 147, 151, 165, 166, 172, 175, 226, 227, 229, 240–41, 243, 244, 246, 247, 248, 249, 250, 271, 272, 273, 276, 279, 280, 283, 284, 285, 289

neutral shipping 13–16, 19, 21, 43, 44–45, 46, 47, 50, 51–52, 75, 77, 90, 91, 92, 96, 109, 111, 140, 141–42, 145, 151, 153, 176, 226, 227, 229, 240–41, 243, 244, 246, 247, 248, 249, 250, 279, 283, 285, 289

New England 28, 38n33, 79

Newfoundland 46, 53n13, 79, 192, 194

Newfoundland Grand Banks 192

New Guinea 216, 232

New Orleans 75, 76, 80

Newport 12, 18, 152, 154, 230

new school *See* Jeune École

New World 10, 201

New York 12, 15, 16, 18, 54n30, 65, 75, 97, 155, 267

New York (passenger liner) 152

New York (USN) 157, 159

New Zealand 211

Nian Rebellion 116

nightmare scenario 282, 283, 284

night-vision goggles 261

Nimattulah (merchant) 264

Nimitz, Chester W. 203

Nine Years' War (1688–97) 53n16

NKVD *See* People's Commissariat for Internal Affairs (NKVD)

Noble, Ronald 265

Nomura Naokuni 221–22

nonferrous metals 220–21

nongovernmental organization (NGO) 7, 243, 251

Non-Intervention Agreement (NIA) 166, 167, 171, 174, 175

Non-Intervention Committee (NIC) 6, 166, 172–74, 177, 184n24

Norasia Samantha (German merchant) 255, 267n7

North America 3, 10, 12, 16, 18–19, 20, 27, 28, 33, 46, 57, 59, 67, 151, 161, 191

North Atlantic 4, 57–58, 190, 192–93, 194, 200, 201, 202, 203

North Carolina 85

North German Confederation 92, 93, 95, 101

North Sea 98, 138, 139, 140, 143, 147, 156–59, 160, 161, 162, 250

Norway 138, 140, 141, 156, 160–61, 193, 195

Norwegian Sea 138, 161, 194

notice to airmen (NOTAM) 242, 245

notice to mariners (NOTMAR) 240, 242, 243, 244, 245, 247, 252n1

Noustar (merchant) 255

Nova Scotia 3, 28, 192

Nuremberg trials 200, 204n3, 206n45

Nyon Arrangement (14 September 1937) 6, 166, 176–80, 181, 182, 186n55, 186n60

O

Oahu 230

Oberkommando der Marine (OKM) 171

occupy territory 41, 105, 110, 112, 139, 210, 211, 216–18, 222, 239

Ocean Eagle (Union merchant) 75

ocean liners 94, 138, 141, 142, 143, 145, 152, 169, 180

   Ancona 144

   Arabic 142

   Lusitania 141

   New York 152

   Seabourn Spirit 256, 264

   Tubantia 145

Ocmulgee (Union merchant) 79

Odessa 168

oil 2, 7, 12, 157, 167, 176, 205n26, 211, 215, 220, 221, 226, 233, 234, 239–51, 263, 271, 272, 281, 286

oil platforms 246, 248, 252n8, 252n15

oil tankers 2, 7, 167, 176, 192–95, 205n26, 215, 232, 233–34, 239–51, 271, 272, 273, 274, 275, 276, 277, 278, 279, 280, 281, 282, 283, 284, 285, 286, 287, 289

   Bridgeton 245–46

   Campeador 176, 185n38

   Nantucket Chief 167

   Seawise Giant 248

   Sungari 247

   Texaco Caribbean 245

O'Kane, Dick 238n29

Okinawa 235

Oldenburg 44–45

Old Regime France 41
See also France

Oman, Gulf of 243

Oman Mohammed subclan 262

Onward (Union) 81

Operational Intelligence Centre, Admiralty 185n40, 194

Opium War, First (1839–42) 107

Opium War, Second (1856–60) 107

Orange Bay 13

ordnance 78, 83, 114, 115, 152, 153, 230, 231

Orkney Islands 159
Orlov, Alexander 168, 183n11
Orlov, Vladimir Mitrofanovich 169
Osaka 234
Ostende 140
outfitting ships 35, 42, 76, 78, 79
overseas bases 3, 13, 16, 17, 19, 21, 90, 95, 96, 100, 101, 103, 129, 135, 139, 140, 154, 157, 159, 162, 167, 169, 174, 189, 193, 261
Ōyama Iwao 111

P

P&O Steam Navigation Company 123
Pacific Fleet, Russia 122, 126
Pacific Fleet, United States 230–31
Pacific Ocean 5, 86, 122, 226, 227, 229, 234, 235
Pacific Squadron, Russia 132
Pacific Squadron, United States 80
Pacific War 6, 158, 211, 216, 222, 235–36, 279, 280, 282, 289
See also Greater East Asia War
Page, Walter Hines 152, 154
Palanqui, Monsieur 16
Palma de Mallorca 169, 184n26
Palmer, Mick 265
Panama 80, 174, 180, 248–49, 254
"Panama steamers" 174, 180, 184n33
Papenburg 44–45
pariah state 285, 290
See also failed state
Parillo, Mark 235
Paris 41, 44, 45, 46, 48, 83, 168, 180–81, 183n10
Paris Agreement (30 September 1937) 179, 180–81
Parliament 11, 19, 21, 22, 39n50
paroled seamen *See* prisoner parole
Peace of Amiens (1802) 50
Peace of Paris of 1763 22

Peacock (USN) 65
Pearl Harbor 218, 221, 222, 225, 228, 230, 236n2
People's Commissariat for Internal Affairs (NKVD) 168
Perry (Union) 75
Persia (merchant) 127
Persian Gulf 240, 278, 286
See also Arabian Gulf
Petersburg (Russian) 127–28
Philadelphia 15, 18, 29, 50, 75
Philippines 10, 158, 211, 213, 216, 217, 218, 235
pipelines 240, 243, 250
piracy ix, 2, 6, 7, 11, 21, 31, 178–79, 181, 186n58, 253–67, 274
pirates 2, 7, 74–76, 173, 179, 180–81, 185n45, 227, 253–67, 268n29, 270n54, 272, 276, 277, 278, 281, 283, 286
Plan DOG 163, 228, 229, 230, 235
plantation economy 12, 20
Plymouth 50, 162
Pohl, Hugo von 139
Pola 144, 145
Poland 228
Politburo 167
Port Arthur *See* Lüshun
Port-au-Prince 18
Porto Praia da Vitória 78
Port Royal 13, 81
Port Said 123–24, 127
Portugal 166, 174
Poti 171–72
Pratt, William Veazie 158
President (USN) 60, 62, 65, 70n43
press 11, 13, 15, 16, 21, 35, 62, 76, 80, 106, 108, 109, 111, 139, 234
Prieto, Indalecio 172
Pringle, Joel Roberts Poinsett 155–56

Prinz Heinrich (German merchant) 127

prisoner-of-war exchanges 18

prisoner parole 49–50, 80, 81, 85, 121, 129–31, 132

prisoners 18, 31, 33, 42, 49–50, 76, 80, 81, 94, 131, 255

prison ships 49–50, 279

privateer *See* commerce raiding: commerce raiders

Prize Council 44–45
See also Conseil des Prises

prize courts 11, 35, 36, 42, 54n19, 55n41, 74, 75, 98, 124, 142, 146, 241, 243, 247, 248, 278

prize crews 17, 31, 46, 81–82, 138

prize master 19

prize money 3, 4, 7, 9, 11–12, 13, 19, 21, 22, 28, 34–35, 42–43, 44, 45, 47–51, 52n2, 54n24, 55n36, 61, 65, 75–76, 77, 79, 80, 81, 82, 86, 92, 94, 96, 98, 101, 103, 122, 138, 142, 146, 250–51

prize shares 9, 28, 34, 43

propeller 76, 79, 161

Prussia 9, 87, 89, 91, 92, 95, 96, 97, 98, 101, 106
    Augusta 92, 95

Psyché (French CR) 46

Puerto Rico 80

Puntland 254, 255, 258, 260, 261, 263, 269n44

Pusan 114, 126

P'yŏngyang 110, 112, 114, 115, 116

Q

Qifu 128, 129, 131

Qingdao 132

Qing dynasty 106, 107, 108, 113

Q-ship 142–43, 246
    Baralong 142

Quasi-War (1798–1800) 44, 46, 51, 252n7

Queen Elizabeth (HMS) 159

Queenstown 78, 154, 155, 157, 162

R

radar 201, 202, 232, 239, 245, 284
See also Anti-Submarine Detection Investigation Committee (ASDIC)

radio 161, 188, 190, 191, 194, 195, 200, 201, 239
See also wireless communications

Raeder, Erich 200

Rahanweyn clan 260, 269n38

raid ashore 266

railways 106, 109–10, 122, 123

RAINBOW 3 war plan 229–30

ramming 81, 91, 258

Rappahannock (CSS) 74, 83

Ras Hafoon 255

Rastoropny (Russian) 130

Rattlesnake (USN) 71n65

razee 57, 67, 68, 69n2

reconnaissance 80, 188, 189, 190, 195, 200–201, 202, 282

Red Cross 240, 251

Red Sea 5, 121, 123–26, 127, 128, 129, 132, 255, 277, 289–90

Reed, Joseph 33–34

Reeves, Joseph M. 226

refueling 98, 127, 182n7, 192, 194

Reichsmarineamt *See* Imperial Navy Office, Germany

Reichstag 135

Republicans *See* Spain: Spanish Republic

Republic of Somaliland *See* Somaliland

Reshitelny (Russian) 128–29, 131, 132

Rhode Island 12, 29, 31, 152, 230

rice 123, 215, 219, 220, 221, 236n2

Richmond 74, 85

Riurik (Russian) 122, 126

Rochefort 16

Rochelais 20

rocket launcher 255

Rockland 75
Rôdeur (French CR) 48
Rodger, N. A. M. 51, 56n58
Rodgers, John 60, 62, 66
Rodman, Hugh 151, 156–63
Romania 176
Rome–Berlin Axis *See* Axis
Roosevelt, Franklin D. 182n6, 228, 230
Roosevelt, Theodore 129
Roshchakovskii, Mikhail Sergeevich 128–29
Rossiia (Russian) 122, 126
Rostum oil platform 246, 247
Rotterdam 13, 15, 141
Royal Holland Lloyd 145
Royal Navy *See* Great Britain: British ships
Rozen (merchant) 264
rules of engagement (ROE) 57, 170, 181, 186n60, 243, 245, 246, 248
Russell, John 74–75
Russia 5, 101, 112–13, 118n29, 121–33, 135, 136, 162, 168, 181, 259, 272, 273, 276, 277, 279, 280, 282–83, 284, 285, 286, 288, 289–90

    Russian ships

        Askold 131

        Bogatyr 123

        Don 127

        Gromoboi 122, 126

        Koreets 130

        Lena 132

        Petersburg 127–28

        Rastoropny 130

        Reshitelny 128–29, 131, 132

        Riurik 122, 126

        Rossiia 122, 126

        Smolensk 127–28

        Variag 130

        Vral 127

Russo-Japanese War (1904–1905) 5, 118n29, 121–33, 272, 273, 276, 277, 278, 279, 280, 281, 282, 283, 284, 285, 287, 288, 289–90

S

Saeki Bunrō 218
Saeki Memo 218–19, 221
Saetta (Italian) 176
sailor 10, 14, 16, 17, 18, 21, 32, 36, 42, 43, 44, 46, 49, 50, 52, 54n20, 85, 98, 121, 130, 131, 133, 142, 155, 157, 161, 168, 172, 234, 245
See also seamen
sail power 4, 30, 34, 75, 79, 80, 84, 89, 91, 96, 97, 100, 103, 142, 145, 261, 282
See also dual-propulsion ship
Saint Christopher 13
Saint Croix 18
Saint-Domingue 12, 13, 14, 17, 18
Saint Eustatius 13, 14, 15, 18, 26n94
Saint-Jean-de-Luz 16
Saint-Malo 16, 17, 44, 46, 48, 53n16, 55n50
Salem 34
Sampson, William T. 157
Samuel B. Roberts (USN) 248
San Domingo 80
See also Hispaniola
San Francisco 132
San Jacinto (Union) 80
Santiago de Cuba 157
Sassam oil platform 248
satellites 261, 278
Saudi Arabia 241, 242, 243, 244, 245, 246, 247, 248, 250
Savannah (CSS CR) 75
Sa Zhenping 129
SC.42 convoy 190
See also convoys
SC.122 convoy 201
See also convoys
Scandinavia 139, 160, 162

Scapa Flow 138, 157, 159, 160, 161, 162

Scarborough raid (1914) 173

Scheer, Reinhard 139, 145, 146, 148, 161

schooner 17, 30, 32, 33, 34, 35, 38n33, 39n38, 39n45, 65, 75

Scotland 67, 135, 138, 162

Scott, Percy 152

sea-air cooperation 189

Seabourn Spirit (cruise ship) 256, 264

sea control 89, 105, 133, 172, 226, 228–29
See also control of sea-lanes

Sea Island Terminal 247

Sea King (British ship) 86
See also *Shenandoah* (CSS)

Sealion (USN) 227

seamen 4, 20, 30, 31, 35, 36, 44, 46, 47, 49, 50, 51, 54n19, 54n20, 55n37, 64, 77, 79, 142
See also sailor

sea power 1, 2, 3, 14, 27, 28, 30, 32, 41, 68, 75, 157, 273, 274, 275, 276, 278, 283, 287, 288, 289

Seawise Giant (oil tanker) 248

Second Army (Japan) 111, 115

Second Empire (France) 93

Secretary of the Navy 4, 57, 58, 60, 64, 65, 68, 69n5, 73, 76, 152, 154, 159, 160, 161

Sedan 92

Semmes, Raphael 76–86

Seoul 110, 115

sequential strategy ix, 29
See also strategy

Service Afloat and Ashore during the Mexican War (Semmes) 76

Sevastopol 124

Seven Years' War (1756–63) 3, 9–22, 43, 53n16, 90, 272, 273, 274, 275, 276, 278, 279, 281, 283, 284, 286, 287, 288, 289

Seville 184n33

Shanghai 82, 109, 130, 131, 132, 133

Shanghai murder case 131–32

Shanhaiguan 109–10

Shannon (HMS) 63

shareholders 42, 46, 47, 48, 49, 52, 55n34

Shatt al-Arab waterway 240

Shenandoah (CSS) 74, 86

Shen Kno II (Taiwanese trawler) 255, 263

shipbuilder 4, 76, 154, 195, 197, 202, 233, 274

ship of the line 57

ship-on-ship battle 4, 57

shipowner 3–4, 29, 35, 36, 41, 42, 44, 45, 48, 51, 52, 98

Shipping Administrative Association (SAA) 214, 215

shipyards 17, 106

Sicilian Channel 176

side-wheeler 81

Sierra Leone 193

Sims, William S. 6, 152–63

Sinclair, Arthur F. 79, 84–85

Singapore 82, 211, 218, 265

Sino-French War (1884–85) 107, 109

Sino-Japanese War, First (1894–95) 5, 105–17, 122, 124, 125, 272, 273, 274, 275, 276, 277, 278, 279, 280, 281, 283, 284, 285, 286, 287

Sino-Japanese War, Second (1937–45) 210, 211, 216

Sirri oil platform 243, 244, 248

6th Battle Squadron 159, 160, 161

SJ radar 232
See also radar

Slade, Edmond J. W. 124

sloop 4, 17, 30, 59, 62, 63, 65, 67, 68, 75, 77, 78, 81, 82, 83

small arms 14, 46, 78
See also guns

Smith, John 60–61

Smolensk (Russian) 127–28

Socotra 255, 268n11

Sóller 174

Somalia 2, 7, 253–67, 271, 272, 273, 274, 275, 276, 277, 278, 279, 281, 283, 285, 286, 287

Somali Coast Guard 255

Somali High Seas Fishing Company (SHIFCO) 262

Somaliland 254, 260, 261, 269n44

Somali marines 253, 263

Somali National Movement (SNM) 254, 260, 269n44

Somali Salvation Democratic Front (SSDF) 255, 263, 266, 269n44

SomCan security company 258, 262, 263

Somerville, James F. 170

Sŏnghwan 112

sonic boom gun 256
See also guns

South Africa 82, 128, 193

South America 61, 77, 137

South American Station 67

Southampton 78, 85

South Carolina 30, 44, 60, 73

Southeast Asia 7–8, 209–23, 234

Southern Resources Area 7, 209, 210, 211, 215, 218

Southern Resources Shipment System 7, 209, 210–12, 216–18, 222

South Korea 253, 257

Soviet Union 165, 166, 168, 169, 174, 175, 177, 180, 181, 182, 183n11, 210, 240, 245, 247, 248
See also Russia

Spain 3, 6, 9, 10, 13, 14–15, 18, 20, 21, 39n50, 43, 53n16, 77, 79, 90, 92, 94, 96, 157, 165–82, 183n10, 183n11, 183n13, 184n24, 184n26, 184n36, 186n52, 186n53, 186n57, 186n60, 250, 257, 272, 273, 274, 275, 276, 277, 278, 279, 280, 281, 282, 285, 286, 287, 288

    Spanish Armada 39n50

    Spanish Civil War (1936–39) See Civil War (Spanish)

Spanish Nationalists 6, 165–82, 183n13, 184n24, 186n60, 272, 275, 276, 277, 278, 279, 280, 281, 283, 285, 286, 287

Spanish Republic 6, 165–82, 184n24, 184n29, 186n60, 272, 273, 276, 277, 278, 279, 280, 281, 282, 283, 285, 286, 287, 288

Spanish ships

    B-5 170

    C-3 170–71, 174, 176, 183n19

    Canarias 171–72, 177

    Jaime I 173

    Miguel de Cervantes 170, 171

Special Imports and Early Imports Programs 210

Spee, Maximilian von 137

Stalin, Joseph 167, 168, 169, 179, 180, 181, 182, 186n52, 186n53

Stark (USN) 245, 252n4, 252n15

Stark, Harold R. 163, 228, 229–30

Stark, Oskar Victorovich 122

steam power 74, 76, 79, 81, 82, 91, 93, 96, 97–98, 100, 103, 157

steel 91, 226

steel hulls 98

Stenzel, Alfred 100–101, 104n31

St. Lawrence, Gulf of 61

St. Lawrence River 58, 61

Stoddert, Benjamin 58

St. Petersburg 125, 127, 128, 131
See also Leningrad

strategic resources 180, 184n33, 209–23

strategist ix, 93, 108, 155

strategy ix, 1, 2, 3, 4, 5, 6, 12, 13, 27, 28, 29, 30–32, 33, 34, 36, 41, 42, 47, 49, 50, 57, 58, 59, 60, 61–65, 66, 68–69, 73, 76, 89, 90, 95, 97, 99, 100, 101, 103, 105, 109, 110, 111–13, 116–17, 122, 123, 127, 128, 129, 135–36, 146, 151, 153, 154, 157, 160, 162, 173, 175, 179, 180, 182, 188–91, 193–94, 196, 197, 200, 202, 209, 210, 211, 222–23, 226–27, 228, 229, 230, 235, 247,

strategy *(continued)*
253, 264–67, 271, 272, 273, 274, 279, 280, 281, 282, 284, 285, 286, 287, 288, 289, 290

submarine ix, 2, 5, 6–7, 73, 135–48, 149n10, 151, 152, 154, 158, 160, 161, 162, 166, 167, 168, 169–71, 173–76, 178, 181, 184n29, 185n40, 185n45, 186n60, 187–203, 209, 213, 220, 221, 222, 225–36, 238n29, 238n41, 273, 275, 277, 278, 279, 282, 283, 284, 286
See also U-boats

Suez Canal 123, 127–28, 144, 179, 254

sugar 10, 17, 18, 33, 211, 221, 236n2

Sugiyama Hajime 210, 213

Suleiman subclan 263

Sultan (Somali clan head) 259

Sumatra 82, 216

Sumter (CSS) 76–77, 79, 83, 86

Sunda Strait 82

Sungari (Liberian-flagged oil tanker) 247

Supreme High Command of the Armed Forces (OKW) 191

Surcouf, Robert 48, 55n50

surface raiders 5, 6, 137, 156, 160–61
See also commerce raiding: commerce raiders

Sussex (French steamer) 146

Syria 241

T

Taar group 262

tactical goals 1, 7, 12, 20, 27, 29, 30, 32, 43, 58, 62, 99, 132, 146, 155, 160, 187–88, 189, 200, 201, 227, 232, 250

Taft, William Howard 153

Taiping Rebellion 107–108, 116

Taiwan 112, 211–12, 219, 255, 257

Takahira Kogorō 129

Tallahassee (CSS) 74

Tanaka Documents 214–15, 220

Tanker War (1980–88) 2, 7, 239–51, 252n9, 272, 273, 274, 275, 276, 277, 278, 279–80, 281, 282–83, 284–85, 286, 287, 289–90

tanks 165, 168

Tarragona 171

Taussig, Joseph K. 155

telegraph 83, 94, 106, 107, 108, 123, 154

telephones 123, 152

Terashima, Lieutenant 128–29

Terceira Island 78

Terdman, Moshe 259–60

Terrible (British CR) 17

territorial sea 242

Territorial Sea Convention (1958) 246

Terror (France, 1795) 45

Terror (USSR, 1937) 179, 182, 186n52

Texaco Caribbean (oil tanker) 245

Texas 80

Thailand 210, 213, 217, 218

3rd Light Cruiser Squadron 161

thirty-two-pounder gun 75, 79, 81, 83

Thurot, François 25n71

Tianjin 109–10

Tirpitz, Alfred von 135, 136, 139, 140, 141–42, 143, 145, 146

Tōgō Heihachirō 110–11, 118n29, 123, 126

Tōjō Hideki 211, 212

Tokyo 124, 125, 128, 129, 131, 209, 210, 222, 226, 234, 282

Tokyo Bay 122, 126

Toronto Economic Summit 249

torpedo 7, 91, 102, 136, 138, 141, 142, 145, 146, 170–71, 173, 176, 187, 191, 192, 200, 209, 221, 222, 230–33, 238n26, 239, 271–72, 289

    torpedoes, bubble-less ejection 187

    torpedoes, noncontact pistols 187

    torpedoes, trackless 187

torpedo boat ix, 4, 94, 97, 98, 99–100, 102–103, 114, 115, 122, 123, 126, 147, 171, 203

Torpedo Flotilla (USN) 155

Torricelli (Italian to Spanish) 170, 174

Toulon 20, 99

transatlantic convoys 22, 161, 162
See also convoys

transatlantic route 5, 12–13, 22, 29, 151, 156, 161, 162

Treaty of Paris (1856) 127

Treaty of Shimonoseki (1895) 112

Tredegar Iron Works 74

Tribunals of Commerce 45

Triple Entente 151
See also Entente

Triton (CSS CR) 75

Troop, John 54n30

troopships 5, 28, 105, 109, 110–13, 114, 115, 116–17, 122, 124, 138, 141, 147, 155, 162, 169, 180, 184n33, 201, 207n45, 272, 273, 274, 279, 281, 282, 284, 286

Truman, Harry 203

Tsugaru Strait 122, 125

Tsushima Strait 118n29, 132

Tubantia (Dutch ocean liner) 145

Tunisia 94

Turkey 178, 243

Turner, Richmond Kelly 229, 230

Tuscaloosa (CSS) 81–82

Tuscarora (Union) 78

Type VII C submarine 192, 202

Type IX B submarine 192–93

Type IX C submarine 192–93

Type IX D_2 submarine 193

Type XIV U-tankers 205n26

Type XXI submarine 203, 207n56

Type XXIII submarine 203, 207n56

U

U.27 (German) 142

U-33 (German) 170

U-34 (German) 170, 176, 183n19

U-35 (German; Spanish Civil War) 174

U.35 (German; World War I) 145

U-156 (German) 206n45

U-459 (German) 192–93

UB boat 140, 144, 147

U-boat Command Staff, German 187, 190, 192, 194, 200, 202, 204n3

U-boat flotilla in the North Sea 138

U-boats 5, 6, 135–49, 151, 153, 154, 156, 162, 170, 171, 173, 187–203, 204n3, 225, 232, 281, 283, 284
See also submarine

UC boat 140, 142, 146, 147

ULTRA 190, 191, 200, 203

unemployed 42, 258

unemployment 102

Union 73–87, 98, 103, 225, 272, 274, 276, 279, 281, 284, 287
  Union ships
    Brooklyn 75, 77, 80
    Hatteras 80–81
    Joseph 75
    Kearsarge 4, 77, 83–85
    Mohican 81
    Onward 81
    Perry 75
    San Jacinto 80
    Tuscarora 78
    Vanderbilt 81, 82
    Wachusett 78
    Wyoming 82

Union of Soviet Socialist Republics *See* Soviet Union

United Arab Emirates (UAE) 241

United Colonies of America 3, 28, 32, 33
See also colonies

United Kingdom 242, 244, 248
See also Great Britain

United Nations 248, 251

United Nations Convention on the Law of the Sea (UNCLOS) 257

United Nations Security Council 240, 242, 244, 246, 249, 251, 255–56

United Press 234

United Somali Congress (USC) 260

United States ix, 5, 6, 9, 27, 44, 45, 51, 54n19, 57, 58–59, 62, 64, 67, 68, 74–75, 78, 87, 90, 91, 92, 97, 125, 140, 143, 145, 151–63, 166, 167, 168, 183n11, 188, 189, 191, 194, 195, 209, 210, 211–12, 214, 221, 222, 225–36, 240, 242, 243, 244–45, 247, 248, 250, 252n15, 257, 273, 275, 276, 279, 280, 282, 283, 284, 287, 288, 289

    U.S. Navy ships

        Adams 62

        Argus 61, 63

        Chesapeake 61, 63, 70n43

        Congress 60, 62, 65

        Constitution 57, 65, 69n2, 70n43

        Delaware 157, 159

        Enterprize 64

        Erie 167

        Essex 70n43

        Florida 157, 159

        Frolic 63, 70n41

        Gonzalez 257

        Hornet 65

        Kane 167

        Macedonian 60, 71n43

        Melville 154, 155

        Nautilus 70n15

        New York 157, 159

        Peacock 65

        President 60, 62, 65, 70n43

        Rattlesnake 71n65

        Samuel B. Roberts 248

        Sealion 227

        Stark 245, 252n4, 252n15

        United States 60, 70n43

        Vincennes 248–49

        Viper 70n15

        Vixen 70n15

        Wahoo 232, 238n29

        Wasp 70n15

        Wyoming 157, 159, 161

United States (USN) 60, 70n43

unlimited war 280–82
See also limited war

unrestricted submarine warfare 2, 5, 7, 139–40, 143–48, 151, 152, 166, 225–36, 273, 278, 279, 282, 283

URSULA, Operation 170–71, 173, 183n16

URSULA training exercise See URSULA, Operation

Usaramo (German arms transport) 174

U.S. Naval Academy 157, 160

U.S. Naval War College 152, 157, 230

USSR See Soviet Union

U.S. Strategic Bombing Survey 214, 215, 221, 223n9

U-tankers 192, 193, 194

V

Valencia 171, 173

Vanderbilt (Union) 81, 82

Variag (Russian) 130

Venezuela 80

Vengeance (French CR) 17

Venus (French CR) 46

Venus (HMS) 127

Versailles 13, 15, 16, 22

Versailles Peace Treaty (1919) 187

Vienna Economic Summit (1987) 246
Vigo 174, 180, 184n33
Villiers, Patrick 47–48, 49, 55n38
Vincennes (USN) 248–49
Viper (USN) 70n15
Virginia 74
Vixen (USN) 70n15
Vladivostok 121, 122–24, 125–26, 132–33
Voge, Dick 227
volunteer cruisers, Russia 121, 127–28, 132
Vral (Russian) 127

W

Wachusett (Union) 78
Waddell, James 86
Wahoo (USN) 232, 238n29
Ward, Samuel 29
War of 1812 2, 4, 57–69, 74, 273, 275, 277, 279, 281, 284, 286
War of the Austrian Succession (1744–48) 9–10
War of the Spanish Succession (1702–13) 53n16
War Plan ORANGE 157, 158, 226, 228–29, 235
Warren (colonist) 32
Warren, John Borlase 60, 66–67
War Risks Rating Committee 242
Warsangeli clan 260, 269n44
warship 3, 4, 10, 11, 12, 13, 16, 18, 20, 21, 27, 28, 30, 32, 34, 35, 36, 57, 58, 59, 61, 62, 63, 64, 65, 66, 67, 68, 69n1, 69n2, 70n15, 73–74, 75, 76, 77, 79, 80, 81, 83, 86, 91, 92, 94, 95, 96, 98, 100, 102, 103, 107, 110, 112, 115, 121, 123, 124, 125–26, 127, 128, 130, 132, 135, 137, 147, 152, 157, 158–59, 160, 165, 166, 167, 168, 169, 170, 172, 173, 174, 175, 176, 178, 179, 181, 184n24, 185n40, 187, 234, 239, 241, 242, 243, 244, 245, 247, 248, 251, 256, 266
See also battleship; cruisers; gunboat
War Shipping Administration 221
Wars of Unification 95

war zone 129, 139, 145, 158, 165, 230, 240, 241
Washington (U.S. CR) 35
Washington, D.C. 2, 62, 75–76, 152, 153, 154, 156, 157, 159, 230, 247
Washington, George 33–34, 54n19
Wasp (USN) 70n15
Wehrmacht 180
Weihaiwei 105, 111–12, 113, 114, 115, 116
Welles, Gideon 73–74, 81
Western Hemisphere 87
West Indian Squadron 81
West Indian Station 60, 66
West Indies 13, 15, 17, 18, 19, 22, 33, 34, 44, 47, 48, 50, 60, 65, 67, 77, 81
See also British West Indian islands
white flag 84–85
Wilhelm I 100
Wilhelm II 136, 139, 140, 141, 143, 145, 146, 151
Wilson, Woodrow 141, 142, 145–46, 151, 152, 154, 158
Winslow, John A. 83–85
wireless communications 137, 139, 146, 148
See also radio
Wolf (German) 137, 149n9
wolf pack 148, 187
Wŏnsan 114–15
wood-hulled ships 2, 74, 79, 82, 91, 98, 157
Woolwich (HMS) 15
World War I ix, 1, 2, 5, 6, 133, 135–48, 151–63, 188, 233, 246, 250, 272, 273, 274, 275, 277, 278, 279, 280, 282, 283, 284, 285, 286, 287, 288, 289
World War II ix, 1, 2, 6, 7, 159, 162, 176, 181, 182, 185n40, 187–203, 209–23, 225–36, 239, 249, 250, 251, 271, 273, 274, 275, 277, 278, 279, 280, 282, 283, 284, 285, 286, 287, 288, 289
Wyoming (Union) 82
Wyoming (USN) 157, 159, 161

Y

Yalu River 112, 114, 116

Yamagata Aritomo 111

Yemen 254, 255, 257

Yorktown 225, 273, 275, 288

Young School *See* Jeune École

Yucatan 80

Yusuf, Abdullahi 260, 262–63

Z

Z, Operation 182

Zanzibar 128

Zeebrugge 140

Zhili Province 105, 108

The Newport Papers

Influence without Boots on the Ground: Seaborne Crisis Response, by Larissa Forster (no. 39, January 2013).

High Seas Buffer: The Taiwan Patrol Force, 1950–1979, by Bruce A. Elleman (no. 38, April 2012).

Innovation in Carrier Aviation, by Thomas C. Hone, Norman Friedman, and Mark D. Mandeles (no. 37, August 2011).

Defeating the U-boat: Inventing Antisubmarine Warfare, by Jan S. Breemer (no. 36, August 2010).

Piracy and Maritime Crime: Historical and Modern Case Studies, edited by Bruce A. Elleman, Andrew Forbes, and David Rosenberg (no. 35, January 2010).

Somalia . . . From the Sea, by Gary Ohls (no. 34, July 2009).

U.S. Naval Strategy in the 1980s: Selected Documents, edited by John B. Hattendorf and Peter M. Swartz (no. 33, December 2008).

Major Naval Operations, by Milan Vego (no. 32, September 2008).

Perspectives on Maritime Strategy: Essays from the Americas, edited by Paul D. Taylor (no. 31, August 2008).

U.S. Naval Strategy in the 1970s: Selected Documents, edited by John B. Hattendorf (no. 30, September 2007).

Shaping the Security Environment, edited by Derek S. Reveron (no. 29, September 2007).

Waves of Hope: The U.S. Navy's Response to the Tsunami in Northern Indonesia, by Bruce A. Elleman (no. 28, February 2007).

U.S. Naval Strategy in the 1990s: Selected Documents, edited by John B. Hattendorf (no. 27, September 2006).

Reposturing the Force: U.S. Overseas Presence in the Twenty-first Century, edited by Carnes Lord (no. 26, February 2006).

The Regulation of International Coercion: Legal Authorities and Political Constraints, by James P. Terry (no. 25, October 2005).

Naval Power in the Twenty-first Century: A Naval War College Review *Reader,* edited by Peter Dombrowski (no. 24, July 2005).

The Atlantic Crises: Britain, Europe, and Parting from the United States, by William Hopkinson (no. 23, May 2005).

China's Nuclear Force Modernization, edited by Lyle J. Goldstein with Andrew S. Erickson (no. 22, April 2005).

Latin American Security Challenges: A Collaborative Inquiry from North and South, edited by Paul D. Taylor (no. 21, 2004).

Global War Game: Second Series, 1984–1988, by Robert Gile (no. 20, 2004).

The Evolution of the U.S. Navy's Maritime Strategy, 1977–1986, by John Hattendorf (no. 19, 2004).

Military Transformation and the Defense Industry after Next: The Defense Industrial Implications of Network-Centric Warfare, by Peter J. Dombrowski, Eugene Gholz, and Andrew L. Ross (no. 18, 2003).

The Limits of Transformation: Officer Attitudes toward the Revolution in Military Affairs, by Thomas G. Mahnken and James R. FitzSimonds (no. 17, 2003).

The Third Battle: Innovation in the U.S. Navy's Silent Cold War Struggle with Soviet Submarines, by Owen R. Cote, Jr. (no. 16, 2003).

International Law and Naval War: The Effect of Marine Safety and Pollution Conventions during International Armed Conflict, by Dr. Sonja Ann Jozef Boelaert-Suominen (no. 15, December 2000).

Theater Ballistic Missile Defense from the Sea: Issues for the Maritime Component Commander, by Commander Charles C. Swicker, U.S. Navy (no. 14, August 1998).

Sailing New Seas, by Admiral J. Paul Reason, U.S. Navy, with David G. Freymann (no. 13, March 1998).

What Color Helmet? Reforming Security Council Peacekeeping Mandates, by Myron H. Nordquist (no. 12, August 1997).

The International Legal Ramifications of United States Counter-Proliferation Strategy: Problems and Prospects, by Frank Gibson Goldman (no. 11, April 1997).

Chaos Theory: The Essentials for Military Applications, by Major Glenn E. James, U.S. Air Force (no. 10, October 1996).

A Doctrine Reader: The Navies of the United States, Great Britain, France, Italy, and Spain, by James J. Tritten and Vice Admiral Luigi Donolo, Italian Navy (Retired) (no. 9, December 1995).

Physics and Metaphysics of Deterrence: The British Approach, by Myron A. Greenberg (no. 8, December 1994).

Mission in the East: The Building of an Army in a Democracy in the New German States, by Colonel Mark E. Victorson, U.S. Army (no. 7, June 1994).

The Burden of Trafalgar: Decisive Battle and Naval Strategic Expectations on the Eve of the First World War, by Jan S. Breemer (no. 6, October 1993).

Beyond Mahan: A Proposal for a U.S. Naval Strategy in the Twenty-First Century, by Colonel Gary W. Anderson, U.S. Marine Corps (no. 5, August 1993).

Global War Game: The First Five Years, by Bud Hay and Bob Gile (no. 4, June 1993).

The "New" Law of the Sea and the Law of Armed Conflict at Sea, by Horace B. Robertson, Jr. (no. 3, October 1992).

Toward a Pax Universalis: A Historical Critique of the National Military Strategy for the 1990s, by Lieutenant Colonel Gary W. Anderson, U.S. Marine Corps (no. 2, April 1992).

"Are We Beasts?" Churchill and the Moral Question of World War II "Area Bombing," by Christopher C. Harmon (no. 1, December 1991).

www.ingramcontent.com/pod-product-compliance
Lightning Source LLC
Chambersburg PA
CBHW060230240426
43671CB00016B/2899